HALOCARBONS

Effects on Stratospheric Ozone

The National Research Council

INSTITUTE OF MEDICINE · NATIONAL ACADEMY OF ENGINEERING
NATIONAL ACADEMY OF SCIENCES

HALOCARBONS:
Effects on Stratospheric Ozone

Panel on Atmospheric Chemistry
Assembly of Mathematical and Physical Sciences
National Research Council

NATIONAL ACADEMY OF SCIENCES
Washington, D.C.   1976

Library of Congress Catalog Card No. 76-51910
International Standard Book No. 0-309-02532-X

Available from
Printing and Publishing Office, National Academy of Sciences
2101 Constitution Avenue, N.W., Washington, D.C. 20418

Printed in the United States of America

PANEL ON ATMOSPHERIC CHEMISTRY

OF THE

COMMITTEE ON IMPACTS OF

STRATOSPHERIC CHANGE

H. S. Gutowsky, University of Illinois at Urbana-
  Champaign, *Chairman*

Julius Chang, Lawrence Livermore Laboratory
Robert Dickinson, National Center for Atmospheric Research
Dieter Ehhalt, Institute for Chemistry, West Germany
James P. Friend, Drexel University
Frederick Kaufman, University of Pittsburgh
R. A. Marcus, University of Illinois at Urbana-Champaign
George Pimentel, University of California at Berkeley
H. I. Schiff, York University
John H. Seinfeld, California Institute of Technology
Brian Thrush, University of Cambridge
Cheves Walling, University of Utah
Geoffrey Watson, Princeton University

Bruce N. Gregory, *Executive Secretary*
Richard Milstein, *Staff Officer*

PREFACE

Awareness that man's activities can affect the ozone
shield stems from studies of the potential modification
of the stratosphere by high-flying aircraft, especially
the supersonic transport (SST). Concern about such possible effects led the Department of Transportation (DOT)
to institute the Climatic Impact Assessment Program (CIAP)
in 1971. Its objective was to study and report to Congress
by the end of 1974 on the possible physical, biological,
social, and economic effects that might result from future
aircraft operations in the stratosphere. In 1972, the
National Academies of Sciences and Engineering established
the Climatic Impact Committee (CIC) to advise DOT and CIAP.
Early in 1975, CIAP released a summary of findings, and
the CIC issued its report entitled *Environmental Impact
of Stratospheric Flight: Biological and Climatic Effects
of Aircraft Emissions in the Stratosphere* (National Academy of Sciences, Washington, D.C., 1975). The principal
feature of both studies is the possible reduction of
stratospheric ozone by the nitrogen oxide (NO_x) emissions
of the SST.

In 1973, research carried out at the University of
Michigan by R. J. Cicerone and R. S. Stolarski[1] led them
to suggest that the stratospheric ozone could be dimin-
ished by a catalytic chain reaction involving chlorine
atoms (Cl). F. S. Rowland and M. J. Molina of the

---

[1]R. S. Stolarski and R. J. Cicerone. 1974. *Can. J.
Chem.* 52:1582-1591.

v

University of California, Irvine, independently reached
the same conclusion[2] and proposed that stratospheric photodissociation of chlorofluoromethanes (CFMs), which accumulate in the atmosphere in large amounts as a result of
their widespread use as spray propellants and refrigerants, is a significant source of Cl.

In the summer of 1974, during the development of its
report on CIAP, the CIC considered the extent of catalytic
ozone reduction by Cl atoms produced from stratospheric
hydrogen chloride (HCl) in the exhaust gas of the Space
Shuttle. Preliminary estimates reviewed by the CIC indicated that such reduction will probably be small, but
attention was drawn to the possibly more serious effects
of Cl from the CFMs.

As a result of informal discussions with Rowland, as
well as the published predictions[2] of reduction in stratospheric ozone by the CFMs, R. B. Bernstein, chairman of
the National Research Council Division of Chemistry and
Chemical Technology (DCCT), proposed to the Executive
Committee of the newly organized Assembly of Mathematical
and Physical Sciences (AMPS) early in September 1974 that
an *ad hoc* committee be formed to examine the short- and
long-range implications of CFM injection in the atmosphere.
The *ad hoc* Panel* of the DCCT was convened on October 26,
1974, to assess the data, to define the urgency of the
problem, and to advise the National Academy of Sciences
(NAS) about further action it should take. The Panel concluded that the problem was indeed serious and recommended
the formation of a committee to see that the problem of
stratospheric chlorine be given broad, prompt attention,
and that the NAS report its findings within a year.

These recommendations were transmitted to Philip Handler,
president of the Academy, and to the Governing Board of
the National Research Council (NRC), by N. E. Bradbury,
chairman of AMPS, with a strong endorsement from Bernstein
on behalf of the DCCT. The Board, at its next meeting on
November 16, 1974, approved the establishment of a panel
under the aegis of the Climatic Impact Committee, to
examine the effects of halogens on stratospheric ozone.

---

[2]M. J. Molina and F. S. Rowland. 1974. *Nature* 249:810.
*Donald M. Hunten, Kitt Peak Observatory, chairman;
Francis S. Johnson, University of Texas, Dallas; Harold S.
Johnston, University of California, Berkeley; F. S. Rowland,
University of California, Irvine; Michael B. McElroy,
Harvard University; William Spindel, DCCT, staff officer.

Funding for the study was sought and obtained from the National Aeronautics and Space Administration (NASA), the National Science Foundation (NSF), the Environmental Protection Agency (EPA), and the National Oceanic and Atmospheric Administration (NOAA). Computer studies were made possible by support from the Federal Aviation Administration of the Department of Transportation (DOT).

The Panel on Atmospheric Chemistry was appointed on March 25, 1975, and charged with assessing the extent to which man-made halocarbons, particularly CFMs, and potential emissions from the Space Shuttle might inadvertently modify the stratosphere. The Panel was asked to examine critically the existing atmospheric and laboratory measurements as well as the mathematical models used to assess the impact of such pollutants on stratospheric ozone and to make recommendations on studies needed to improve understanding of the processes involved. The charge to the Panel did not include and our report does not deal with the question of whether the probable extent of ozone reduction warrants restrictions on the use of CFMs. This question is being addressed by the Committee on Impacts of Stratospheric Change (CISC), which was established by the National Academies in the summer of 1975 as a successor to the CIC. Its findings in the matter are being presented in a separate report.[3]

In order to fulfill its particular part of the overall task, the Panel held meetings throughout 1975 and the first half of 1976. Its initial meeting, organized by William Spindel (executive secretary, DCCT), was held in Philadelphia, Pennsylvania, on April 19, 1975, following a symposium on "Chlorine Reactions and Stratospheric Ozone," sponsored jointly by the AMPS Committee on Kinetics of Chemical Reactions and the American Chemical Society. Subsequent meetings included an intensive 6-day working session in July 1975. In addition to talking directly with a number of scientists most directly concerned with the question, the Panel solicited and received information and assistance from a broad range of interested parties in government, industry, and the universities. We are grateful to the many who aided us throughout the course of this study.

---

[3]*Halocarbons: Environmental Effects of Chlorofluoromethane Release.* National Academy of Sciences, Washington, D.C., 1976.

An early concern of the Panel was the availability of the wide range of accurate information necessary for a conclusive and convincing determination of the stratospheric ozone reduction that might occur. Therefore, the Panel prepared and, in July 1975, issued an interim report that focused attention on 19 experimental and theoretical studies required to understand the detailed atmospheric processes involved in the destruction of ozone by halogen atom catalysis and to assess their importance in quantitative terms. Much of what is needed has been provided by the large amount of effort and research over the past 3 years on the stratospheric chemistry of the NO_x compounds, from natural sources and high-flying aircraft, which is analogous to that of the ClO_x produced from chlorine compounds. CIAP has given a greatly improved understanding of the natural ozone production and destruction cycle. Furthermore, many of the additional data still needed for more accurate assessment of the CFM question are being obtained by redirection or extension of research initiated in connection with CIAP and supported by many federal agencies, including ERDA, NASA, NOAA, NSF, EPA, DOT, and the Department of Defense (DOD). Nonetheless, there were and still are some gaps in the observations available and in our understanding of them.

The present report considers the extent to which catalysis by halogen atoms reduces the amount of stratospheric ozone. Chapter 1 summarizes the various aspects of the problem and presents the Panel's findings. Chapter 2 provides a brief introduction to the scientific principles involved; it gives a qualitative description of the stratosphere and the natural ozone cycle, of how they are perturbed by pollutants, and how the magnitudes of such perturbations are determined. Chapter 3 reviews what is known about the sources and amounts of pollutants, while Chapter 4 considers the large variety of conceivable removal processes that have been suggested for the halogen compounds. Chapter 5 examines how these compounds are transported from one region of the atmosphere to another. Chapter 6 discusses the experimental measurements of these compounds and related species in the atmosphere, and Chapter 7 describes the way in which these measurements are used in developing and applying the atmospheric models. Chapter 8 evaluates the extent to which stratospheric ozone will be affected as a result of releasing the pollutants into the atmosphere, and Chapter 9 discusses several other considerations germane to the question of ozone reduction.

Many of the details and literature references for the subjects of Chapters 4-7 are given in Appendixes A-D, respectively. Appendix E compares the amounts of CFM observed in the atmosphere with the amounts released as a means (materials balance) for determining whether there are any significant tropospheric sinks. Appendix F is a revision of our July 1975 interim report recommending further atmospheric and laboratory studies; it stresses the difficulties that we have encountered since then in seeking a quantitative understanding of the processes affecting stratospheric ozone.

The major emphasis of this report is upon the CFMs, especially $CFCl_3$ (F-11) and CF_2Cl_2 (F-12), which have been released into the atmosphere by man in the largest amounts. Considerable attention is given to several other halocarbons that might present similar problems or be less troublesome alternatives. The emissions of hydrogen chloride and particulates by the Space Shuttle are also considered, as is the nitrous oxide (N_2O) from nitrogen fertilizers. There is a diverse and growing list of stratospheric pollutants. Their perturbations of stratospheric chemistry and composition are complex, they may interact with one another, and the consequences of their presence may not be simply additive. Because of this, the Panel emphasizes the need to consider the collective effects as well as the individual effects of all such pollutants in any regulatory actions.

CONTENTS

                      xiii

1 FINDINGS

I. INTRODUCTION

The stratosphere is a region that extends from about 16 to
50 km (52,000 to 160,000 ft) above the surface of the earth
at low latitudes and from 8 to 50 km at high latitudes.
In contrast to the lower atmosphere, where there is turbu-
lence and vertical mixing, the stratosphere is relatively
quiescent. As a consequence, it is particularly susceptible to contamination, because pollutants introduced there
tend to remain for long periods of time--several years or
more.

One of the trace constituents of the stratosphere is
ozone (O_3). Although ozone represents only a few parts
per million of the gases in the stratosphere, potential
threats to this ozone have become a focus of scientific
interest and public concern during the past few years.
This is because, even in its small amount, stratospheric
ozone absorbs virtually all of the solar ultraviolet (uv)
radiation with wavelengths less than 290 nanometers (nm)
and most of that in the biologically harmful 290 to 320 nm
(uv-B) wavelength region, thus preventing the radiation
from reaching the surface of the earth, where it could
adversely affect human, plant, and animal life. As a
consequence of this absorption, and that in the visible
and infrared, ozone also helps to maintain the heat
balance of the globe and is directly responsible for the
temperature inversion (temperature increasing with
increasing altitude) that characterizes the upper stratosphere. Thus, any decrease in ozone would increase the

amount of harmful uv-B radiation reaching the earth's surface; it would also perturb the atmospheric heat balance and thereby might trigger a change in the world's climate.

Concern over human effects on stratospheric ozone was first raised as a possible consequence of emissions from high-flying aircraft. This particular problem was intensively studied by scientists during the Department of Transportation's Climatic Impact Assessment Program (CIAP). The effects of aircrafts and other activities of man upon stratospheric ozone depend on the natural pro- cesses that determine the distribution of ozone in the stratosphere, unperturbed by man. Understanding of those natural processes is extensive and on a demonstrably firm foundation. The ozone distribution is maintained as the result of a dynamic balance between creation and destruc- tion mechanisms. Ozone is produced in the upper strato- sphere by the action of solar uv radiation upon molecular oxygen and is destroyed by several processes. The most important of these, which accounts for about two thirds of the total destruction rate, is a catalytic chain reaction involving various oxides of nitrogen (NO_x). Other relevant destruction mechanisms include direct reaction of oxygen atoms with ozone (Chapman reaction) and catalytic chain reactions involving several species containing hydrogen or chlorine (HO_x and ClO_x).

The stratospheric production of ozone is relatively insensitive to man's activities. The rate is determined by the intensity of solar radiation of wavelengths shorter than 242 nm, as well as by the distribution in altitude of molecular oxygen and of the ozone itself. The absorption of solar radiation by pollutants can affect the amount and distribution of the uv light that is available to dissoci- ate oxygen. In this indirect way, pollutants can affect the ozone production, but such secondary effects are small.

The time required to destroy an ozone molecule can, however, be influenced appreciably by man's activities. As mentioned above, several naturally occurring catalytic chemical reactions have been identified as ozone-destruc- tion mechanisms. The chemical species involved in these reactions (NO_x, HO_x, and ClO_x) are referred to as catalysts because they are not used up by the reactions. The indi- vidual reactants are regenerated and thereby are capable of reacting with ozone over and over again. Each of them can remove thousands of ozone molecules, before being destroyed itself by some other process. Consequently, even though the concentration of these catalytic molecules in the stratosphere is quite low (1 to 10 parts in 10^9), they have important effects.

Unfortunately, artificial introduction of these catalysts into the atmosphere in the large amounts now associated with man's activities can lead to a significant increase in their stratospheric concentrations. As a consequence, the average lifetime of an ozone molecule is decreased relative to that in the unperturbed stratosphere. Since the overall production of ozone is not increased, while the individual molecules are *destroyed* more rapidly, the result is a net reduction in the amount of ozone present. One such example of human ability to modify stratospheric ozone is the direct emission of NO_x into the stratosphere from the exhausts of SST's and other high-flying aircraft, referred to above. Another is the release of chlorofluoromethanes (CFMs) in the use of spray cans, air conditioners, and refrigerators. To give an idea of the magnitudes involved, 1 percent of the global ozone is about 33 million metric tons; the total world production of the CFMs $CFCl_3$ (F-11) and CF_2Cl_2 (F-12) in 1974 was nearly a million metric tons.

The main purpose of the present report is to evaluate the extent to which stratospheric ozone will be affected by the CFMs and other chlorine compounds introduced by man. We also consider the hydrogen chloride and particulates emitted by the Space Shuttle that is now being developed. Moreover, in the course of our work there have been suggestions that additional stratospheric pollutants, such as nitrous oxide (N_2O), derived from nitrogen fertilizers, or methyl bromide (CH_3Br), used as a fumigant, could produce appreciable reductions in stratospheric ozone. Our treatment of them has been limited to assuring that their effects are less immediate than those of the CFMs. Nonetheless, the importance of stratospheric ozone to life on earth requires that all such suggestions be thoroughly investigated and that attention be focused upon the aggregate effects.

II. THE CFMs $CFCl_3$ (F-11) AND CF_2Cl_2 (F-12)

The two CFMs most widely used and about which there has been the greatest concern are $CFCl_3$ (F-11) and CF_2Cl_2 (F-12). *All the evidence that we examined indicates that the long-term release of F-11 and F-12 at present rates will cause an appreciable reduction in the amount of stratospheric ozone. In more specific terms, it appears*

*that their continued release at the 1973 production rates
would cause the ozone to decrease steadily until a probable
reduction of about 6 to 7.5 percent is reached, with an
uncertainty range of at least 2 to 20 percent, using what
are believed to be roughly 95 percent confidence limits.
The time required for the reduction to attain half of this
steady-state value (3 to 3.75 percent) would be 40 to 50
years.**

There is little question about the fundamental aspects
of the problem. F-11 and F-12 have been produced and used
in large quantities that are a matter of quite accurate
record. The large fraction of the production that enters
the atmosphere can be readily inferred from the types of
use. The compounds do accumulate in the troposphere; they
have been measured there at steadily increasing concentrations consistent with the estimated release rates. Moreover, recent measurements confirm that F-11 and F-12 are
transported into the stratosphere. Laboratory experiments
show that once in the stratosphere F-11 and F-12 must
undergo photolytic dissociation to produce Cl atoms. The
CFM concentrations have been observed to decrease in the
middle and upper regions of the stratosphere at a rate
corresponding to the combined effects of photolysis and
the "transport lag." Finally, as soon as the Cl atoms are
generated in the stratosphere, they will react with O_3 in
the catalytic cycle by which Cl and ClO destroy ozone.
These reactions have been measured individually in the
laboratory, and they must occur in the stratosphere. It
is inevitable that CFMs released to the atmosphere do
destroy stratospheric ozone. The more difficult problem
is evaluating such effects quantitatively.

The numerical values for ozone reduction by F-11, F-12,
and other pollutants are determined in general by the
aspects just described--the amounts released, transport in
the atmosphere, and the particular photochemical and
chemical reactions involved. There is, however, the important qualification that alternative removal mechanisms
for these pollutants, if any exist, could reduce the
results accordingly. Moreover, there might be other
processes tending to diminish or to amplify the effects
upon stratospheric ozone of the ClO_x generated from the
CFMs. As in the case of all physical and chemical phenomena,
none of these factors can be measured exactly. There are

---

*For a discussion of uncertainties in the time dependence
see Chapters 5 and 8.

uncertainties in each, and much of our effort has been
spent on identifying the possibilities and uncertainties
and reducing them to the extent feasible within the time
available. Each of these aspects is reviewed separately
below.

Many of the uncertainties in predicting the ozone
reduction are difficult to evaluate. Whenever possible
we have given numerical estimates, as ranges or percentages
(±) about the value that seems most probable at this time.*
Frequently, for the normal, symmetrical distribution of
measured values, the experimental errors are expressed in
terms of the standard deviation about the average value
(±σ). We have elected to use uncertainty limits *equivalent* to two standard deviations (±2σ). This, of course,
does not affect the uncertainties, but it may influence
the way in which they are perceived. If one takes the
smaller ±σ range, the probability that the actual value
falls between +σ and -σ is 68 percent, i.e., these
are 68 percent confidence limits. There is one chance in
three (32 percent) that the actual value lies outside the
±σ range. Because of the importance of the ozone reduction, it seems better to use wider limits that are more
likely to include the actual value. With uncertainty
ranges corresponding to ±2σ, the confidence limits are 95
percent; there is only one chance in twenty that the
actual value falls outside that (larger) range.

A. Release Rates

The recent surveys of F-11 and F-12 production, use, and
release rates cited in Chapter 3 have improved significantly the completeness and reliability of the data
available. The totals for the amounts produced and
released so far are more accurate than the annual figures,
which require year-end inventory estimates. The data from
Eastern Bloc countries are still approximate, but these
involve a minor part (5 percent) of the total. *The stated
uncertainty in the total amounts of F-11 and F-12 that have
been released so far (through 1975) has now been reduced
to ±5 percent.*
The calculations of ozone reduction are for specified
release rates, so errors in the actual release rates do

---

*The shapes of the probability distributions are also un-
certain in most of the cases considered; the contributing
factors are quite different, of doubtful symmetry and conceivably "long-tailed."

not affect directly the numerical predictions. However, the uncertainties in release rates do enter when one applies the predictions to the actual releases. Also, the uncertainties in release rates are highly important in determining whether there are unknown processes removing CFMs from the atmosphere.

B. Transport

The history of a particular pollutant molecule from point of release to the time of its degradation and/or removal from the atmosphere as long as 50 or 100 yr later on the average, is complicated by the atmospheric motion. Furthermore, once ClO_x is generated from a CFM, one needs to know what happens to it. The reduction in stratospheric ozone is the net effect of enormous numbers of such histories. The calculation and adding up of the histories to obtain the net effect is a time-consuming mathematical problem made feasible only by replacing particular histories with averages. Some averages involve time period (diurnal and seasonal); others are for location (latitude and longitude). The type of averaging employed defines a model for the atmospheric motions and chemistry and introduces characteristic approximations (see Chapter 7). These approximations produce uncertainties in the results in addition to those associated with the various constants that describe the rates of reaction.

So far, the calculations available to us of the effects of the CFMs on stratospheric ozone have been made by the one-dimensional (1-D) model. The approach employed is equivalent to averaging the concentrations, motions, and reactions over latitude and longitude, leaving only their dependences on altitude and time. There are physical reasons why such simplification is reasonable; for example, any longitudinal (east-west) differences are expected to be small. In any case, the equivalents of averaged transport rates appear in the model as the vertical eddy mixing or transport coefficient K, which depends only on the altitude. One approximation made in the 1-D model is the choice of K, which is adjusted empirically to fit the experimentally observed distribution in altitude of trace substances in the atmosphere.

A closely related approximation, less apparent but implicit in the 1-D model, is the use of space- and time-averaged concentrations to calculate the reaction rates that determine the ozone reduction. The nature of the approximation is described in Chapter 7, but its accuracy is difficult to establish. These approximations can be avoided, in principle, with two- or three-dimensional (2- or 3-D) calculations. However, this was not feasible within the time period established for our study. In practice, as described in Chapters 5 and 7 and Appendix B, K was chosen to fit the altitude profile (concentration as a function of altitude) observed for a tracer gas, such as N_2O or CH_4, that is itself released at ground level and undergoes destruction in the stratosphere. The effects of the chemistry are included in the 1-D calculations made to develop the fit; hence the choice of K depends on the correctness of the tracer-gas chemistry involved as well as on the concentration profiles used.

We have explored the approximations of the 1-D model in three ways. First, a thorough investigation was made of the procedures used to choose K (Chapter 5 and Appendix B) and how these affect the uncertainties of CFM histories in the atmosphere. Second, concentration profiles were calculated with various choices of K for a number of important, reactive atmospheric species and compared with the all-too-limited measurements that have been made so far (see Chapter 7). Also, some comparisons are made with the results of 2-D calculations. Finally, as described in Chapter 8, ozone reductions were calculated using different choices of K but keeping other model parameters the same. Much of the interpretation of these diverse studies is subjective, and combination of their results in an overall numerical uncertainty range is very difficult. However, the various comparisons generally agree well within a twofold or at most a threefold range.

At present, we estimate that use of the 1-D model to approximate the distribution and transport of the chemical species involved in the reduction of stratospheric ozone by the CFMs causes uncertainty by a factor of 1.7 in either direction (+70 to -40 percent) in the predicted amount of the globally averaged reduction (a threefold range).

C. *Stratospheric Chemistry*

In addition to the transport approximations just described, each calculation is subject to uncertainties in the quantitative details of stratospheric chemistry. Some workers may treat the chemistry more completely than others or use more efficient computer programs, but, in principle, any 1-D calculations can include all the known photochemical and chemical processes affecting ozone. The difficulty of concern here is that the factors governing the catalytic destruction of ozone by CFMs for any *specific* reaction scheme are subject to experimental error in their determination. These factors include the chemical reaction rate constants, the solar flux, the photolysis rates, the temperature distribution in the stratosphere, and the concentrations (or source and sink strengths) of trace species in the unperturbed atmosphere (see Chapter 7 and Appendix D). Of these, the largest source of uncertainty that we have identified is a relatively small number of reactions that dominate the ClO_x chemistry in the stratosphere. These reactions involve unstable, highly reactive species, and the determinations of some of their rate constants are extremely difficult. Lesser uncertainties are identified with the photochemical processes and with the concentrations of natural species (Chapter 8).

The sensitivity of the calculated ozone reduction to the rate constants, or to any other input parameter, can be investivated simply by calculating the reduction for different values of the particular input parameter, keeping all other aspects of the 1-D model the same. We have done this with the rate constants for seven reactions for which the uncertainties presently have a large effect on the outcome, obtaining the results given in Table 8.3. *For the particular reaction scheme employed, uncertainties in seven of the rate constants cause a fivefold uncertainty range in predictions of ozone reduction by the CFMs. The largest contributions are from the $HO + HO_2$ and $HO + HCl$ reactions; the other reactions included in the analysis are $ClO + NO_2$, $Cl + CH_4$, $ClO + O$, $ClO + NO$, and $Cl + O_3$. Additional uncertainties in the photochemical processes and the concentrations of natural species are estimated to increase the overall uncertainty range associated with the stratospheric chemistry to a factor of 2.5 in either direction (a sixfold range).*

The initial calculations of ozone reduction by the CFMs were more sensitive to experimental errors in the rate constants than was *generally* appreciated at the time, because of the lack of systematic studies such as those described in Chapter 8. Moreover, the recent inclusion of $ClONO_2$ in the reaction scheme has increased the contribution of these rate constants to the uncertainty of the predictions from a fourfold to a fivefold range. The sensitivity studies show that the dependences of the ozone reduction on changes in individual rate constants are largely independent of one another, within the ranges given for each. Therefore, the consequences of future improvement in the seven rate constants can be estimated by a simple scaling of the results given in Table 8.3.

D. Other Factors

So far, our discussion assumes that all the CFM released will contribute to the ozone reduction according to the reaction system employed in the calculations (Appendix D). However, if this is not the case, and some additional mechanism modifies the effects of the CFMs, the ozone reductions otherwise calculated would have to be scaled up or down accordingly. Several types of possibilities have been considered: inactive removal (that does not lead to ozone destruction), competing reactions, feedback mechanisms, and large natural sources of stratospheric chlorine.

1. *Inactive Removal* The importance of inactive removal may be seen by considering a particular pollutant such as F-12 (CF_2Cl_2) under steady-state conditions, for a given release rate. Photolysis and reaction with $O(^1D)$ in the stratosphere remove about 1 percent per year of the total amount of F-12 in the atmosphere, giving products that destroy ozone. If they are the only removal processes, the atmospheric residence (or removal) time of F-12 is nearly 100 yr. If, however, there were also a process (sink) that removed an additional 2 percent per year, but did not destroy ozone, the fraction of F-12 that destroyed ozone would be one third instead of unity, and the ozone reduction would also be one third of what it would other-wise be. Similarly, the total residence time and the amount of F-12 in the atmosphere would be multiplied by 1/3 in the steady state for this case, which would be attained more rapidly.

We see that inactive removal processes become signi-ficant when their rate approaches or exceeds the overall rate of removal via stratospheric photolysis. That removal, however, is a slow process, and therefore other processes need not be very fast to compete with photolysis and be important. In principle, since any natural sources of F-11 and F-12 are negligible, the best way of searching for such processes would be to carry out an *accurate* materials balance, i.e., a comparison of what has been released into the atmosphere with what is measured to be actually present plus what is calculated to have been used up in the stratospheric processes that destroy ozone.

Any missing CFM would be the amount removed without affecting stratospheric ozone. However, such comparisons are of limited value unless the total amount in the atmosphere (the global burden) and the amount released are *both* known to high accuracy (≤ 5 percent).

Attempts have been made at a materials balance, as presented in Chapters 3 and 6 and Appendix E. More extensive atmospheric measurements are available for F-11, which is easier to detect, than for F-12, which is more abundant. Unfortunately, even for F-11 the observations are limited. The inadequacy of available observations combines with the difficulty of the measurements to give an uncertainty in the global atmospheric burden that is too high (± 40 percent) for the comparison with the release rates that are known with greater accuracy (± 5 percent) to have much significance. Although the materials balance can be interpreted as consistent with little or no inactive removal of the CFMs, the uncertainty limits range from zero inactive removal to a rate sixfold faster than that of the stratospheric photolysis (see Appendix E).

We have also taken the other approach to this question--looking at each of the individual inactive removal mechanisms. An intrinsic difficulty with this approach is that an important possibility might be overlooked. The large number of suggested mechanisms may be classified according to the site of removal (surface, troposphere, or stratosphere) and according to the nature of the process, e.g., a nondestructive reservoir or chemical degradation. Examples of the latter two categories that have been frequently cited are incorporation in the polar ice caps and decomposition via reactions with neutral species (HO, O, etc.) in the troposphere. Neither is significant; the rate of the first is demonstrated to be no more than 0.001 percent per year, while for the latter the concentrations involved and/or the reaction rates are small. In Chapter 4 and Appendix A these and the other possibilities suggested have been analyzed carefully on the basis of *known* chemical reactions and *known* physical processes. The results are summarized in Table 4.2.

Three processes have estimated inactive removal times for F-11 and F-12 that are short enough to warrant further, more detailed study. Lower limits of $\sim 10^2$ (70 and 200), 10^3, and 5×10^3 yr have been placed, respectively, on the removal times for solution in the surface waters of the oceans (followed by some unknown degradation process) and by ion-molecule reactions and photodissociation in the troposphere. If each of these processes actually

removed F-11 and F-12 in the time corresponding to the
lower limit set for it, the maximum combined effect would
be a decrease in the predicted ozone reductions by at most
2/5 of what they would be in the absence of such inactive
removal. However, we expect the effect to be no more
than 20 percent (a decrease by 1/5), based on the limited
data available for the oceanic sink.

2. *Competing Reactions* Stratospheric processes that remove Cl or ClO from the ClO_x catalytic chain limit the
amount of ozone that they destroy. Thus, the formation of
HCl and $ClONO_2$ (by the reactions $Cl + CH_4$ and $ClO + NO_2 +$
M) provides temporary "reservoirs" that store the chlorine
from decomposed CFMs in inactive forms pending downward
transport from the stratosphere followed by rain out from
the top of the troposphere. There are also reactions that
convert HCl and $ClONO_2$ back into active species, so that
the importance of the reservoirs depends on the balance
struck between formation and reconversion.

At first, HCl was considered to be the only reservoir
of any consequence for the ClO_x catalysts. However, a
re-examination in early 1976 of the possible role of $ClONO_2$
indicated that $ClONO_2$ might be a significant reservoir,
doubly important because it removes not only ClO from the
ClO_x cycle but also NO_2 from the NO_x cycle. Since then,
the intensive laboratory studies of the formation and
destruction processes for $ClONO_2$ (see Appendix A) have
confirmed its probable importance in stratospheric
chemistry. Therefore, its reactions have been incorporated in our calculations and in the results we report
(Appendix D and Chapter 8). Its inclusion reduces the
predicted ozone reductions by a factor of about 1.85,
modifies to some degree the distribution with altitude
of the ozone, and increases the kinetics-related range
of uncertainty in the ozone reduction from fourfold to
fivefold.

The important role of HCl is supported by stratospheric measurements of its occurrence, as compared with
a calculated distribution in Chapter 7. In the case of
$ClONO_2$, infrared measurements have placed an *upper bound*
of ~1 ppb on its current stratospheric concentration at
25 km (Chapter 6). The corresponding concentrations
calculated for that region are on the order of 0.5 ppb
(Figure 7.16). Therefore, more sensitive measurements are
needed to provide a definitive check on the inclusion of
$ClONO_2$ in the reaction scheme; such observations should
be available in the latter part of 1976.

Another possible competitor with the catalytic effectiveness of ClO is its photolysis, but this proves too slow to be important (Appendix A, Section III). A wide variety of other processes that might affect Cl and ClO have been considered. One example is the coupling between the ClO_x and NO_x cycles (see Chapter 9), which occurs in addition to that provided by $ClONO_2$. It is included in the reaction scheme employed (Appendix D), along with a number of less important processes. The probable importance of $ClONO_2$ was a surprise. Whether or not there is another such surprise in store remains to be seen. However, a modest number of reactants is involved, the number of their possible reactions is finite, and these possibilities have for the most part already been examined.

3. *Feedback Mechanisms* Several types of feedback mechanisms have been proposed that might alleviate or amplify the effect of CFMs on stratospheric ozone. One is the partial "self-healing" of the stratosphere. This argument states that if ozone is destroyed in the upper stratosphere by CFMs (or other pollutants), the uv solar radiation penetrates deeper into the stratosphere, photolyzing more O_2 and generating more O_3 at lower levels. This does occur to some extent, but its importance is limited, and the effect is included in the 1-D calculations that we have employed (see Chapter 7).

Another, more speculative suggestion is that redistribution of the ozone to lower altitudes by the effects of the CFMs will increase the temperature at the tropical tropopause and in the lower stratosphere and enable more H_2O to enter the stratosphere. This would, in turn, generate more HO radicals by reaction of H_2O with $O(^1D)$, to convert HCl back to catalytically active Cl by the important HO + HCl reaction, and increase the rate of ozone destruction. A better understanding of water vapor transport between troposphere and stratosphere and of the temperature changes is needed before this mechanism can be regarded as established. But it might have a significant effect, and its further study is desirable. Details are given in Chapter 9, along with comments about other less likely possibilities.

4. *Natural Sources of Stratospheric Chlorine* It has been suggested that the injection into the atmosphere from natural sources of large amounts of chlorine compounds, such as HCl, CH_3Cl, and perhaps CCl_4, casts doubt upon

or reduces the significance of the man-made sources.
Such suggestions have not borne up under close scrutiny.
Although the "natural" chlorine was not included in the
early calculations, this has since been done, and, as
described in Chapter 7, the effects are modest. The
importance of HCl is reduced because of its rapid "washout"
from the troposphere by rain, and that of CH_3Cl by destruction processes in the troposphere. The present total
reduction in stratospheric ozone by HCl, CH_3Cl, and CCl_4
(from whatever sources) is calculated to be less than 1
percent. These three compounds now contribute *roughly* the
same amount of chlorine to the stratosphere as do the
CFMs (Chapters 3 and 6), and the ozone reductions they
produce are also comparable.

The most important fact, however, is that the
reduction in stratospheric ozone by chlorine from man-
made sources is increasing and will be in addition to
whatever is caused by chlorine from natural sources
(Chapter 3). The latter are already at their steady-
state amounts, while continued release of CFMs at
recent rates probably cause their atmospheric concentrations, and ozone reduction, to increase tenfold or
more. The significance of the man-made sources would be
reduced if the effects of ClO_x were nonlinear in the
correct direction, i.e., if the stratospheric addition
of chlorine from CFMs to that from natural sources
produced a less than proportionate increase in the ozone
reduction. Such effects are to be expected only at
catalyst concentrations that give reductions of ozone
greater than 15 to 20 percent, compared with an ozone reduction of ≤ 1 percent by ClO_x from natural sources.

*We have examined a variety of proposals that might
alleviate or amplify the reduction of stratospheric ozone
by CFMs, such as competing reactions, feedback mechanisms,
and natural sources of chlorine. Several of them have
been incorporated in our calculations; others have been
eliminated as inconsequential; a few are considered unlikely to have major effects but warrant further attention.
The role of $ClONO_2$ as an inert reservoir for stratospheric chlorine seems particularly important, the available data indicating that it reduces the effects of the
CFMs by a factor of nearly 2 (1.85).*

*5. Predicted Ozone Reduction and Its Overall Uncertainty
With inclusion of $ClONO_2$ in the reaction scheme, the reduction in stratospheric ozone by the CFMs is predicted
to be 7.5 percent at steady state for constant 1973*

*release rates (Chapter 8). The analyses summarized above
indicate that this value might be decreased by about 20
percent if inactive removal by an oceanic sink does
indeed occur. If this is confirmed by future measurements, it could reduce the predicted value of the ozone
reduction to 6 percent.*

The approximations and uncertainties involved in
predicting this value have been reviewed above. Each
source of uncertainty produces a distribution of less
likely values for the ozone reduction about the central
value of 6 to 7.5 percent. Thus, insofar as the atmospheric
chemistry taken alone is concerned, its sixfold range
of uncertainty means that the real value for the reduction
might be as little as 2.4 percent or as large as 15 percent, i.e., 6 to 7.5 percent times (1/2.5) or 2.5, with
roughly a 95 percent chance of finding the real value
between these limits. Equivalent statements apply to the
other two sources that we have estimated in numerical
terms. The three distributions are combined to give a
new distribution of uncertainty. Their uncertainties are
independent so we express them as multiplicative factors
(times f) and combine their effects.* The possibility that
some as yet unidentified processes might affect the predictions is *not* included in the analysis.

*The combination of the multiplicative uncertainty
factors for the release rates (1.05), transport (1.7),
and stratospheric chemistry (2.5) leads to an overall
eightfold uncertainty range. Application of these limits
to the central values (6 to 7.5 percent) of the reduction
in ozone expected after many decades of releasing F-11 and
F-12 at the 1973 rates gives a range of about 2 to 20
percent for the uncertainties from these three sources
and the oceanic sink.*

Insofar as we know, this is the first detailed attempt
to assess the overall uncertainty in the calculation of
the reduction in stratospheric ozone by the CFMs. The
tenfold range reflects the limitations of our knowledge
as well as our use of what we believe to be demanding
(95 percent) confidence limits. The fivefold range for
rate constants reduces the relative effects of any other,
smaller uncertainties, provided they are not systematic

---

*The square root of the sum of $(\ln f)^2$ was calculated for
the three sources to obtain r. The values of $\exp(-r)$
and $\exp(+r)$ are the factors that give the lower and upper limits for the ozone reduction, when multiplied by
the most probable value (6 to 7.5 percent).

in one direction or the other. In considering the implications of these results, it is important to remember
that the most probable value is about 6 to 7.5 percent,
that there is roughly a 95 percent chance of the real
value being between the 2 to 20 percent limits given, and
that while one might prefer a particular limit, both
limits must be considered.

E. *Verification of Predicted Ozone Reductions*

Ideally, one would like to have a direct check on the ozone
reduction predicted for the CFMs or at least some more
direct means than are now available for narrowing the
uncertainty of the predictions. However, either objective will take time to accomplish, as well as effort.
If the release of CFM continues during that time, it will
increase the eventual ozone reduction beyond what it would
have been if CFM release had been curtailed. The amount
of such increase would depend on the differences between
the two CFM release schedules compared, the period of time
involved, and the actual extent of ozone reduction per
unit of CFM (Chapter 8).

Direct observation of a decrease in stratospheric
ozone atttributable to the CFMs is obscured by the
natural, long-term irregularities of about ±5 percent
that are still incompletely understood (Chapters 9 and 6).
Furthermore, the long-term trends must be determined in
the presence of much larger daily, seasonal, and latitudinal changes, with possibly a small component associated with the sun-spot cycle thrown in for good measure.
Even if a change occurs in the long-term trend, there is
the problem of deciding whether it is natural or due to
the CFMs. This requires a large enough reduction by the
CFMs over a long enough period of time to identify its
growth characteristics (as inferred from model calculations) on top of the natural background. Sophisticated
statistical analyses are being applied to the problem,
and better observations for the purpose are being
gathered.

So far, the ozone data have provided no case for the
global stratospheric ozone having been decreased (or increased) by the CFM releases. Nor would one expect a
~0.5 percent reduction, that estimated to have already
been produced by past CFM releases, to be detectable compared with the natural fluctuations of ±5 percent. Moreover, the detection and identification of an ozone reduction as small as 2 percent by the CFMs would require

carefully calibrated ozone observations extending over
several years, either by an improved network of surface
stations or balloons or by satellite, together with adequate
statistical analysis of the data. Some scientists are more
optimistic than others about the length of time and magnitude of reduction required for success.

A more immediate approach is to measure the concentrations of the other key reactants that establish the
reduction in ozone. In fact, most of the reactants have
been demonstrated to be in the stratosphere at the levels
corresponding to those assumed in, or predicted by, the
calculations of the reduction in stratospheric ozone.
For example, the stratospheric profiles observed for
total NO and NO_2 agree with the calculated profiles to
within 50 percent (Appendix C). However, observations
in the stratosphere of several of the most important
but highly reactive species of low concentration are
either very few (O, HO, ClO) or still to be accomplished
(HO_2, Cl). Although the catalytic cycle employed in the
analysis undoubtedly exists, detailed and careful
measurements of these species would help to reduce the
possibilities of unknown factors that might affect the
extent of ozone reduction actually produced by the cycle.
In particular, stratospheric measurements of Cl and ClO,
because of their direct removal of O_3 and O, should be
especially valuable in attributing an actual decrease in
ozone to the CFMs, after allowance for Cl and ClO from
other sources.

Certainly, much has been learned about the CFM
problem during the past 2 yr. In fact our evaluation
of it has been a case of "shooting at a rapidly moving
target." Further improvements will occur during the next
year or two. These will include a more complete understanding of atmospheric chemistry, better determinations
of the rate constants and absorption coefficients, and
improved atmospheric measurements, most desirably on a
global scale. The limits on inactive removal processes
should be more closely defined; more direct evidence of
the amount of reduction in ozone by the CFMs should be
provided by observations of Cl and ClO in the stratosphere;
and the approximations made in the predictions should be
improved by the application of 2- and 3-D models to the
problem.

*Continuation of CFM releases at their present (static)
levels, while waiting for improvements in our ability to
determine the ozone reduction caused by the CFMs, will
increase the eventual peak ozone reduction and its total*

*(integrated) amount that actually occurs in comparison
with what they would be if release were curtailed at
once. Each year of release will increase the peak
reduction by about 0.07 percent (central value of a 0.02
to 0.2 percent range) and the total amount of reduction
(integrated over time) by 1/10. A resumption of exponential
growth would of course give annual increments of increasing
size.*

III. EFFECTS OF OTHER POLLUTANTS

Besides the CFMs, there are several other pollutants that
require mention. These are reviewed below to give a
catalog of the growing number of ways in which man may
reduce stratospheric ozone and to consider some of the
implications.

A. *The Space Shuttle*

As now planned, combustion of the solid propellant in the
Space Shuttle will inject HCl gas and aluminum oxide
particulates directly into the stratosphere. However,
the amount of chlorine introduced into the stratosphere
per year by 50 flights per year (the number now projected
for 1986) will be only about 1 percent of that from the CFMs
for continued release at the 1973 rates; and the effects
will be relatively small (Chapter 9). Similarly, the
amount of particulates is modest compared with that
naturally present from volcanic action; and, as described
in Chapter 9, there is no reason to believe that such
materials have significant effects on stratospheric
chemistry. *We conclude that the combustion products
from the Space Shuttle at the presently planned launch
schedule of 50 per year will make a small contribution
(~0.15 percent with a range of 0.05 to 0.45 percent)
to the total reduction of stratospheric ozone by human
activities.* Furthermore, since these products are in-
jected directly into the stratosphere, their atmospheric
residence time is relatively short, so there would not
be long-lasting aftereffects should the program be
terminated.

B. *N_2O from Fertilizer*

The natural abundance of stratospheric ozone is determined
to a large degree by the NO_x produced from N_2O. In turn,
nitrogen fertilizers contribute to the amount of N_2O released. These facts, in combination with the increasingly

widespread use of fertilizers, have led to a number of studies, now in progress, of the possible future impact of nitrogen fertilizers upon stratospheric ozone. The data presently available are inadequate to judge the issue (Chapter 9). *More detailed studies of the reduction in stratospheric ozone associated with the use of nitrogen fertilizers are essential, especially of the biological production of N_2O and the mechanisms for its removal from the troposphere.*

C. Others

The report of the Climatic Impact Committee dealt with the NO_x emissions from large fleets of several hundred SST and high-altitude subsonic planes, projected for 1990. These were estimated to reduce the stratospheric ozone by significant amounts in the absence of adequate emission controls. (Climatic Impact Committee, 1975, p. 29, Table 4; Grobecker *et al.*, 1974). Other possible ways in which man may contribute to the reduction in stratospheric ozone include F-22 (CHF_2Cl) used largely for refrigeration, CH_3Br employed in agricultural fumigation, and a number of hydrogen-containing and unsaturated chlorocarbons listed in Table 3.2. At present release rates, the ozone reduction that will be caused by each is relatively minor (≤ 0.05 percent); however, the use and release of some of these substances seem likely to increase. Furthermore, although it is not yet clear whether the CCl_4 now found in the atmosphere is largely man-made or natural in origin, it has an appreciable effect--about 0.5 percent.

D. The Total Burden of Pollutants

The reductions in stratospheric ozone by ClO_x from different sources are additive to a useful approximation for total reductions of no more than 15 to 20 percent. Beyond this point, nonlinear responses will begin to be important. Similar considerations apply to NO_x catalysis. However, there are interactions among different reactive species (e.g., ClO_x and NO_x) from different pollutants that cause their effects on stratospheric ozone to depend on the amount of other species present. The formation of $ClONO_2$ from ClO and NO_2 is one example, and the importance of the amount of water present is another (Chapter 9). Also, it is likely that nonlinear responses will occur, at some value(s) of the ozone reduction for the biological, climatic, and any other consequences. Hence, we must not only be concerned about the sum of the individual

effects, i.e., the total burden of pollutants placed
upon the stratosphere, but also be alert for interactions
among them. *We find that procedures must be established
to follow the production and release of pollutants that
affect stratospheric ozone, to monitor their concentrations in the atmosphere, and to analyze their combined
effects.*

IV. SUMMING UP

In judging the consequences of the findings it seems
important to bear in mind several features of the analysis
that have not yet been emphasized. For purposes of simplicity, the disucssion so far has been mainly in terms
of the ozone reduction calculated for steady-state conditions (at the 1973 release rates for F-11 and F-12). This
is the most widely used of the various possibilities for
presenting the predictions. However, the time scale of
events is highly significant. This may be seen in the
ozone reduction calculated for a constant CFM release
rate (1973) until 1978, when all release is halted
(Figure 8.5). The ozone reduction continues to grow
for a decade beyond cutoff (or cutback) and then requires an *additional* 65 yr to recover one half of its
maximum loss. A competitive sink of 1 percent per
year would change but little the maximum ozone reduction of this scenario, but it would accelerate the
subsequent recovery.
    Another important aspect of the findings is that
the ozone reduction calculated for the CFMs lies in a
critical range of values. If it were an order of magnitude smaller, it might be viewed as relatively minor.
If it were an order of magnitude larger, a reduction
probably would have been detected by now and action taken
to curtail release of the CFMs. Therefore, one must
consider the likelihood and consequences of future increases in release rates of the CFMs.
    When the CFM problem was first recognized 2 yr
ago, CFM use had had two decades of exponential growth at
a rate of 10 percent per year. If such growth had continued, the uncertainties we have discussed translate into
uncertainties in the time required to achieve the release
rate corresponding to a particular ozone reduction at
steady state. However, the actual releases in 1975 and
1976 experienced a 15 percent drop from the exponential
growth curve and are comparable with the 1973 release
rate used by us for the steady-state predictions.

But there is no assurance that future releases will remain constant. *Resumption of exponential growth in the production and use of CFMs could well occur and lead to a doubling of their release rate within 10 yr. Even if the release rates became constant at that point, they would cause a doubling in the expected ozone reduction, to a value of about 12 to 15 percent, with a range of 4 to above 25 percent, once a steady state was reached.*

Furthermore, CFM production and use are worldwide. If U.S. release is curtailed but other use continues the more rapid exponential growth evident in Figures 3.1 and 3.2, the magnitude of the overall reduction in stratospheric ozone could still reach much higher levels even though a longer time would be required. *Clearly, although any action taken by the United States to regulate the production and use of CFMs would have a proportionate effect on the reduction in stratospheric ozone, such action must become worldwide to be effective in the long run.*

Finally, while our knowledge of stratospheric ozone has become extensive during the past few years, it should be apparent from the disucssion given above that significant uncertainties remain. An interim report of the Panel was issued in July 1975, identifying a wide range of observations, experiments, and actions needed to deal more adequately with the CFM problem in particular and with threats in general to stratospheric ozone. Many of these needs are brought into sharper focus by the problems faced in preparing the present report, as summarized in Appendix F. *Additional improvements in our knowledge of the atmosphere and of stratospheric chemistry are essential to permit more accurate assessments to be made of the extent of potential reductions in the stratospheric ozone.*

REFERENCES

Climatic Impact Committee. 1975. *Environmental Impact of Stratospheric Flight: Biological and Climatic Effects of Aircraft Emissions in the Stratosphere.* National Academy of Sciences, Washington, D.C.

Grobecker, A. J., S. C. Coroniti, and R. H. Cannon, Jr. 1974. *Report of Findings: The Effects of Stratospheric Pollution by Aircraft.* U.S. Dept. of Transportation, Washington, D.C.

2 A DESCRIPTION OF THE PROBLEM

I. THE ATMOSPHERE

The air temperature falls with increasing altitude above
the earth's surface to a minimum value of about 210 K
(-80°F) and then rises. This temperature minimum is known
as the tropopause, and its height varies from about 8 km
(25,000 ft) near the poles to about 16 km (50,000 ft) in
tropical latitudes (Figure 2.1). The region between ground
level and the tropopause is known as the troposphere. It
is a region of relatively rapid circulation and vertical
mixing of constituents because warmer air from near ground
level tends to rise and be replaced by cooler air from
above.

The stratosphere lies above the troposphere; here the
temperature rises from its minimum at the tropopause to a
maximum value of about 280 K (50°F) at the stratopause
(~50 km, 160,000 ft). The stratosphere is a stable, virtually cloudless region with slow vertical circulation because the denser, cooler air is at lower altitudes and does
not readily rise. This temperature inversion is analogous
to those that occur at low altitudes over Los Angeles and
other natural basins.

Atmospheric pressure decreases with altitude, falling to
about one tenth of its ground-level value at the base of
the stratosphere and by a further factor of 100 through
the stratosphere.* Although vertical circulation in the

---

*The "scale height" is the vertical distance over which
the atmospheric pressure falls by a factor of e ($e = 2.72$).

FIGURE 2.1 The variation of atmospheric temperature with altitude.

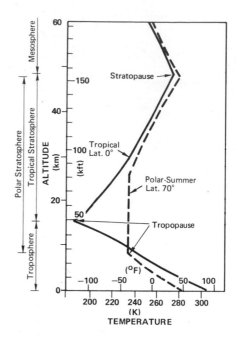

stratosphere is slow for the reason just given, gases are transported and mixed by the large-scale, nearly horizontal movement of air masses. With this form of transport, there is negligible settling out of the heavier molecules because of gravity, and the rate of transport does not depend on the nature of the molecule concerned.

II. THE OZONE SHIELD

The most important trace constituent of the stratosphere is ozone (O_3). Although present in small amounts (i.e., a few parts per million), ozone, nevertheless, is responsible for shielding the earth from ultraviolet (uv) radiation that is harmful to life.

   The amount of ozone in the stratosphere is maintained as the result of a dynamic balance between formation and destruction processes. Formation occurs predominantly at altitudes above 30 km (Figure 2.2), where solar uv radiation ($h\nu$) with wavelengths less than 242 nanometers (nm) slowly dissociates molecular oxygen (O_2) [Reaction (2.1)] into oxygen atoms (O).

$$O_2 + h\nu \rightarrow O + O \qquad (2.1)$$

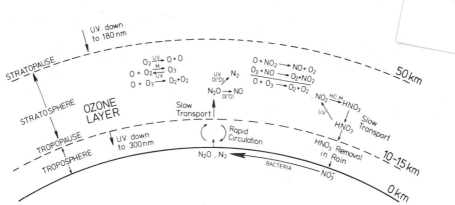

FIGURE 2.2 Ozone formation and removal processes in unperturbed atmosphere.

$$O + O_2 + M \rightarrow O_3 + M \text{ (twice)}$$

$$\text{NET:} \quad 3\,O_2 \rightarrow 2\,O_3 \tag{2.2}$$

These oxygen atoms rapidly combine with molecular oxygen to form ozone [Reaction (2.2)] in the presence of another molecule (M), which stabilizes the ozone by removing its excess energy.

The actual distribution of ozone in the stratosphere is determined to a large extent by transport processes. Ozone is produced mainly in the tropics between 25 and 35 km (80,000 to 115,000 ft), but as a result of the motions (and compression) of air masses in the stratosphere, its highest concentrations (in molecules per cm^3) are found near the poles at altitudes of about 15 km (50,000 ft).

Ozone itself absorbs solar radiation strongly [Reaction (2.3)] in the longer wavelength region 240-320 nm.

$$O_3 + h\nu \rightarrow O_2 + O \tag{2.3}$$

It is this absorption that shields the earth from harmful uv radiation. The photolysis of ozone (2.3) is not, how-ever, a true destruction mechanism because almost all of the oxygen atoms produced by this process will rapidly com-bine once again with molecular oxygen to reform ozone [Reaction (2.2)]. Nevertheless, these two processes [(2.3) and (2.2)] do have an important net effect; they convert solar energy into heat, particularly in the upper strato-sphere. Thus, besides providing a "shield" against the biologically harmful uv, the presence of ozone in the stratosphere produces the temperature inversion character-istic of that region.

Balancing the formation process [Reactions (2.1) and (2.2)] are several processes that *destroy* ozone. One example is the reaction of ozone and oxygen atoms to produce molecular oxygen:

$$O_3 + O \rightarrow O_2 + O_2 \tag{2.4}$$

If we group ozone and oxygen atoms together as "odd oxygen," it can be seen that processes (2.1) and (2.4), which are much slower than processes (2.2) and (2.3), control the amount of ozone plus atomic oxygen in the atmosphere; but the rapid processes (2.2) and (2.3) determine how this odd oxygen is distributed between ozone and oxygen atoms. Ozone is the dominant form of odd oxygen, and it is the slowness of processes (2.1) and (2.4) that makes the stratospheric ozone vulnerable to other removal processes.

The above scheme, involving only species derived from oxygen, was suggested by Chapman (1930) and has provided the basis for discussions of stratospheric ozone ever since. However, it has been learned over the last 25 yr that chemical processes other than (2.4) destroy large amounts of ozone. Partly as a result of the interest in stratospheric chemistry engendered by the development of supersonic aircraft, it is now possible to prepare an approximate budget for the stratospheric ozone balance (Johnston, 1975).

The original Chapman scheme given above accounts for about 20 percent of the total natural destruction rate for stratospheric ozone, while transport of ozone to the earth's surface contributes an additional 1/2 percent.

About 10 percent of the destruction is caused by catalytic cycles involving hydrogen-containing species: free hydrogen atoms (H), hydroxyl (HO), and hydroperoxyl (HO_2), which can achieve the same effect as Reaction (2.4) without being themselves removed. For example, Reaction (2.5) followed by Reaction (2.6) is identical in result to Reaction (2.4).

$$O + HO_2 \rightarrow HO + O_2 \tag{2.5}$$

$$HO + O_3 \rightarrow HO_2 + O_2 \tag{2.6}$$

NET: $\quad O + O_3 \rightarrow O_2 + O_2$

These hydrogen-containing species are produced by the reaction of naturally occurring water vapor (H_2O) and methane (CH_4) with excited oxygen atoms, $O(^1D)$, which are formed when ozone is decomposed by uv light with wavelengths shorter than 310 nm.

$$O_3 + h\nu \to O_2 + O(^1D) \tag{2.7}$$

$$O(^1D) + H_2O \to HO + HO \tag{2.8}$$

$$O(^1D) + CH_4 \to CH_4 + HO \tag{2.9}$$

A catalytic cycle involving nitric oxide (NO) and nitrogen dioxide (NO_2), collectively called NO_x, provides the most important destruction process for ozone. This process accounts for most of the remaining 70 percent of the natural ozone destruction rate, but there is also a small contribution from chlorine compounds (natural and man-made), the details of which are given in the next section. For the nitrogen oxides the dominant processes are

$$O + NO_2 \to NO + O_2 \tag{2.10}$$

$$NO + O_3 \to NO_2 + O_2 \tag{2.11}$$

$$\text{NET:} \quad O + O_3 \to O_2 + O_2$$

These processes again produce the same effect as Reaction (2.4) without the nitrogen oxides being consumed. In addition, the following cycle affects the partitioning of NO_x between NO and NO_2:

$$NO_2 + h\nu \to NO + O \tag{2.12}$$

$$NO + O_3 \to NO_2 + O_2 \tag{2.11}$$

$$O + O_2 + M \to O_3 + M \tag{2.2}$$

This cycle is important at lower altitudes during daylight because it competes with the catalytic cycle (2.10)-(2.11) but does not remove either ozone or nitrogen oxides.

It is now known that the major natural source of NO_x in the stratosphere is provided by the oxidation of nitrous oxide (N_2O), which is produced by bacteria in soil and water. Although almost all of this nitrous oxide is converted by uv light into N_2 and O, about 1 percent reacts with the excited oxygen atoms, $O(^1D)$, formed by the action of uv radiation on ozone, to yield nitric oxide and thereby start the NO_x cycle.

$$O(^1D) + N_2O \to NO + NO \tag{2.13}$$

The NO_x molecules, in turn, are removed mainly by the formation of nitric acid (HNO_3) in the following process, which occurs mainly in the lower stratosphere:

$$HO + NO_2 + M \rightarrow HNO_3 + M \tag{2.14}$$

Just as the atmospheric motions slowly carry nitrous oxide up from the troposphere to the altitudes where it is decomposed, these motions also carry nitric acid downward from where it is formed to the troposphere, where it is rapidly removed by rain.

In addition to the natural source of NO_x discussed above, large fleets of supersonic aircraft are capable of releasing significant amounts of NO_x directly into the stratosphere (Climatic Impact Committee, 1975). This would increase the overall destruction rate of ozone (above its natural value), shifting the dynamic balance between formation and destruction processes, to give a net decrease in the total amount of ozone present.

III. THE EFFECT OF HALOGEN COMPOUNDS

Other substances can catalyze the destruction of ozone, notably atomic chlorine (Cl) and bromine (Br) and their oxides (ClO and BrO), as indicated by the following cycle:

$$O + ClO \rightarrow Cl + O_2 \tag{2.15}$$

$$Cl + O_3 \rightarrow ClO + O_2 \tag{2.16}$$

$$\text{NET:} \quad O + O_3 \rightarrow O_2 + O_2$$

In fact, Reaction (2.15) has a rate coefficient about five times greater than the corresponding Reaction (2.10) in the cycle by which NO_2 and NO destroy ozone. However, the relative contributions of the two processes also depend on the amounts of NO_x and ClO_x present.*

The possible importance of ClO_x for catalytic destruction of ozone in the stratosphere has only recently been recognized (Stolarski and Cicerone, 1974; Molina and Rowland, 1974; Wofsy and McElroy, 1974). There are a number of chlorine compounds, from natural as well as human

---

*The term ClO_x is often used to describe both Cl and ClO as in the usage of NO_x for NO and NO_2.

sources, that can serve as sources for the ClO_x. The
nant compounds of chlorine in the troposphere are now
known (cf. Chapter 3) to be methyl chloride (CH_3Cl),
little of which is of industrial origin; the man-made
chlorofluoromethanes $CFCl_3$ (F-11) and CF_2Cl_2 (F-12); and
carbon tetrachloride (CCl_4), natural as well as man-made.
Human sources of lesser importance include trichloroethylene ($CCl_2{=}CHCl$) and the substances that are replacing it
as cleaning agents: methyl chloroform (CH_3CCl_3) and F-113
($CF_2ClCFCl_2$). The Space Shuttle, as presently planned,
will use an ammonium perchlorate/aluminum powder propellant, which would release relatively small amounts of
chlorine compounds directly into the stratosphere, mostly
as hydrogen chloride (HCl). Volcanic emissions and sea-
salt spray transported upward by air movements are other
minor sources (Crutzen, 1974; Cicerone, 1975).

Within a few years after release, these chlorine compounds, like the nitrous oxide generated at the earth's
surface by bacteria, are distributed throughout the troposphere where their concentrations tend to a uniform fraction by volume (mixing ratio). They rise more slowly
into the stratosphere (Figure 2.3). What happens to them
in this journey depends on their chemical reactivity and
their sensitivity to solar radiation. Substances that
contain hydrogen atoms, such as methyl chloride (CH_3Cl) or
double bonds such as trichloroethylene ($CCl_2{=}CHCl$) are
quite readily attacked by hydroxyl (HO). Reactive compounds of this sort are largely decomposed in the troposphere, where the HCl produced is quickly removed in rain.
However, the chlorofluorocarbons (F-11, -12, -113, etc.)
are highly inert in this region, and they will be

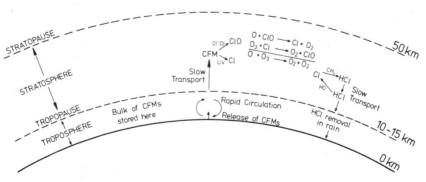

FIGURE 2.3 Simplified atmospheric behavior of
chlorofluoromethanes (CFMs).

transported upward in the stratosphere to altitudes of
25-50 km (80,000-160,000 ft). There they are decomposed
by uv light at wavelengths around 200 nm, with the production of Cl atoms.

The chlorine atoms then participate in the catalytic
cycle described by Reactions (2.15) and (2.16). This
cycle may be interrupted by conversion of the highly reactive Cl and ClO into inactive forms that do not destroy
ozone. Two such processes have been identified. The
chlorine atoms are rendered inactive mainly by reaction
with methane to form hydrogen chloride (HCl)

$$Cl + CH_4 \rightarrow HCl + CH_3 \qquad\qquad (2.17)$$

which acts as a temporary reservoir for active chlorine
species in the stratosphere. Chlorine atoms are regenerated from hydrogen chloride by reaction with hydroxyl
radicals:

$$HO + HCl \rightarrow H_2O + Cl \qquad\qquad (2.18)$$

Recently chlorine nitrate ($ClONO_2$) has been proposed as
another such temporary reservoir (Rowland *et al.*, 1976).
It is formed from ClO in the recombination reaction

$$ClO + NO_2 + M \rightarrow ClONO_2 + M \qquad\qquad (2.19)$$

The potential effect of this reaction is enhanced because
it removes the catalytically active NO_2 in addition to the
ClO. Chlorine nitrate is destroyed by photolysis

$$ClONO_2 + h\nu \rightarrow products \qquad\qquad (2.20)$$

and by other processes that probably have much smaller
rates than the photolysis (see Appendix A, Section III.D).
The importance of these reservoirs depends on the proportion of potentially active chlorine that they constitute,
and this, in turn, is governed by the rates of formation
and destruction of the reservoir species.

The destruction and regeneration of the active chlorine
species, ClO_x, can occur many times before ultimate removal of the chlorine from the stratosphere. Removal is
accomplished largely by net transport of the HCl from the
stratosphere into the upper troposphere, where it is
washed out by rainfall as in the case of HNO_3 from the NO_x
cycle. The time scale from release of CFM at ground level
to the removal from the atmosphere of its chlorine as HCl

in rain is several decades. This means that the effect on stratospheric ozone of the tremendous growth in the use of CFMs over the past 20 yr has yet to be felt fully.

IV. AN APPROACH TO THE PROBLEM

Our primary task is to convert our current knowledge of the rates of the various processes described in the preceding section into a quantitative prediction of the extent to which CFMs and other halocarbons of human origin will affect the stratospheric ozone in the future. In order to accomplish this, several key features of the problem must be considered in numerical terms. These include (1) release of the compounds into the atmosphere, (2) transport of them and other important chemical species by atmospheric motions, (3) absorption of solar radiation and photolytic production of chlorine atoms in the stratosphere, (4) the catalytic destruction of ozone by ClO_x and its dependence on related chemical reactions, and (5) removal from the atmosphere of the compounds and their end products. The dependence of most of these features on one another prevents their discussion in a simple logical progression.

The production and release data are an exception in being separable from the other features. The chemical industry has been able to provide us with increasingly complete tabulations of F-11 and F-12 production and to make more sophisticated estimates of the percentages released in the various types of usage. The most recent data for these two CFMs are presented in Chapter 3, along with information about a variety of other compounds that we have considered.

Another important, separable topic is the removal of halogen compounds from the atmosphere by processes that do not contribute chlorine atoms to the active ClO_x species catalyzing the destruction of stratospheric ozone. This leads to a smaller reduction in the ozone, which is proportional to the fraction of the compound that contributes its chlorine atoms to the catalytic cycle. A large variety of such possible sinks has been considered. The results are summarized in Chapter 4, and details are given in Appendix A. This question can also be approached by comparing the actual inventory of a compound in the atmosphere, as determined by direct measurements, with the amount known to be released. Such a comparison (materials balance) is given in Appendix E for F-11 and F-12.

Although these two features are separable from the others, all five listed above are included, at least implicitly, in calculating the reduction of stratospheric ozone expected for a given set of conditions. The mathematical description incorporates these items into the *equation of mass conservation* for each significant chemical species involved.* Briefly, this equation relates the variation in concentration of a particular species at a particular altitude to its rate of formation, destruction, and transport. Since the composition and temperature of the stratosphere vary with altitude, latitude, and, to a lesser extent, longitude, a complete solution to the mass conservation equations would involve all three dimensions as variables. Furthermore, at least 20 to 30 chemical species have to be considered. The resulting set of equations is time-consuming to solve simultaneously in three dimensions, even with powerful computers. Also, preparation and testing of the computer programs is a lengthy and demanding procedure.

Such considerations have led to the widespread use of approximations to reduce the complexity of the calculations and to make them more tractable. Also, the scarcity of atmospheric concentration measurements at different altitudes (concentration profiles) on a global basis has inhibited the development of multidimensional models. Although more sophisticated methods are now being developed, the only ozone reduction calculations available to us within our time schedule are those based on the one-dimensional (1-D) approximation. In this approach, the longitudinal and latitudinal dependences are averaged out, leaving altitude as the sole spatial variable.

Much of our effort has been devoted to a study of the uncertainties associated with the 1-D calculations. Chapter 5 describes atmospheric transport and its mathematical representation in general. It also discusses the particular approximation of transport in 1-D models using an eddy-mixing coefficient obtained empirically from the observed dependence on altitude of the concentrations of trace species such as methane and nitrous oxide. Appendix B analyzes in detail the dependence on transport of the ozone reduction by CFMs. The various chemical and

---

*Some species react so rapidly that is is necessary only to consider their chemical formation and loss processes, the transport processes being negligible by comparison. Such species are said to be in photochemical steady state.

photochemical reactions involved in such calculations and the rate constants and absorption coefficients for them are described in Chapter 4 and Appendix A.

Chapter 6 and Appendix C summarize the current state of atmospheric measurements of the various trace species that bear upon the ozone reduction problem. They include the CH_4 and N_2O concentration profiles used to obtain the eddy-mixing coefficient, as well as the profiles of the species directly involved in the NO_x, HO_x, and ClO_x catalytic destruction cycles.

Besides their limitations in representing atmospheric transport, 1-D calculations incorporate an approximate treatment of the chemical and photochemical processes. The nature of the approximation is discussed in Chapter 7, and Appendix D includes a detailed listing of the reactions and parameters employed in our 1-D calculations. An impression of the possible importance of the approximation is gained by comparison, in Chapter 6, of the concentration profiles observed for various species with those calculated.

The final results of our 1-D calculations of the reduction in stratospheric ozone by F-11 and F-12 are presented in Chapter 8. The major points treated are (1) the sources of uncertainty in the predictions, (2) the dependence over periods of several decades of the ozone reduction produced by different schedules of release, (3) the sensitivity of the ozone reduction to the choice of eddy-mixing coefficients, and (4) the dependence of the results on uncertainties in several of the key rate constants. Chapter 9 deals with a variety of other topics, and Appendix F is an updated version of the interim report that we issued a year ago identifying areas requiring further study.

REFERENCES

Chapman, S. 1930. A theory of upper-atmosphere ozone, *Mem. R. Meteorol. Soc.* 3:103-125.
Cicerone, R. J. 1975. Comment on "Volcanic emissions of halides and sulfur compounds to the troposphere and stratosphere," by R. D. Cadle, *J. Geophys. Res.* 80(27): 3911-3912.
Climatic Impact Committee. 1975. *Environmental Impact of Stratospheric Flight: Biological and Climatic Effects of Aircraft Emissions in the Stratosphere.* National Academy of Sciences, Washington, D.C.

Crutzen, P. J. 1974. A review of upper atmospheric photochemistry, *Can. J. Chem.* 528:1569-1581.

Johnston, H. S. 1975. Pollution of the stratosphere, *Ann. Rev. Phys. Chem.* 26:315-338.

Molina, M. J., and F. S. Rowland. 1974. Stratospheric sink for chlorofluoromethanes: chlorine atom catalysed destruction of ozone, *Nature* 249:810-812.

Rowland, F. S., J. E. Spencer, and M. J. Molina. 1976. Stratospheric formation and photolysis of chlorine nitrate, $ClONO_2$. Submitted to *J. Phys. Chem.*

Stolarski, R. S., and R. J. Cicerone. 1974. Stratospheric chlorine: a possible sink for ozone, *Can. J. Chem.* 52:1610-1615.

Wofsy, S. C., and M. B. McElroy. 1974. HO_x, NO_x, and ClO_x: their role in atmospheric photochemistry, *Can. J. Chem.* 52:1582-1591.

3 SOURCES OF

HALOGEN-CONTAINING COMPOUNDS

I. INTRODUCTION

A variety of halogen-containing molecules, both natural
and man-made are released into the lower atmosphere. Unless some rapid process exists for their removal, these
halogen-containing materials are soon distributed throughout the troposphere. The time scale for local vertical
mixing is a few weeks; that for east-west mixing about the
globe, a few months; and that for exchange between northern
and southern hemispheres, a year or two.

For water-soluble compounds, the major method of removal is rain-out after absorption in water droplets.
Large quantities of hydrogen chloride and inorganic halides
are introduced into the lower atmosphere by natural processes (primarily volcanic emissions and evaporation of sea
spray) and by man (industrial releases and the combustion
products of halogen-containing materials). In addition,
large quantities of molecular chlorine (approximately 10^7
metric tons/yr in the United States) are produced. Although this chlorine is almost entirely consumed in further chemical processes, and emissions are vigorously
controlled because of its toxicity, some undoubtedly escapes (National Research Council, 1976). However, removal
of water-soluble compounds by rain-out (and in the case of
chlorine, by chemical reaction) is so fast that their concentrations drop off rapidly with altitude in the troposphere, cf. Chapter 6. Because of this, there is general
agreement that no significant amount of halogen is introduced into the stratosphere from these sources (Rowland

33

and Molina, 1975; Stedman *et al.*, 1975), and we will not consider them any further in our report.

On the other hand, *direct* introduction of HCl or inorganic salts into the stratosphere, in large enough quantity, could become a significant source of stratospheric halogens. One such possibility is via major volcanic eruptions, which produce thermal updrafts large enough to penetrate the local tropopause. There seems to be no reliable estimate of the magnitude of this highly erratic phenomenon, which, in any case, is out of human control and would simply contribute to the natural stratospheric halogen burden. More pertinent to this study is the potential injection of halogen from Space Shuttle exhausts, and this is discussed further in Section II.C.

Organic halocarbons have low water solubility, and their removal by rain-out to any significant extent has neither been detected nor seriously proposed. Some may, however, be destroyed by processes that convert them into water-soluble forms, which are then removed by rain. Halocarbons containing C—H bonds of C≡C bonds react with hydroxyl radicals, with rates of destruction (sinks) such that their average tropospheric lifetimes range from a few weeks to a few years. Completely halogenated hydrocarbons such as $CFCl_3$ react much more slowly, if at all, with hydroxyl radicals, and destruction by this path is unimportant. In addition to reaction with HO, a variety of other processes for halocarbon removal have been suggested. However, for the otherwise stable, completely halogenated halocarbons, such processes are much slower than the tropospheric mixing. They are chiefly of interest as possible sinks that would reduce the amount of chlorine available for catalytic removal of stratospheric ozone. Details of such aspects are discussed in Chapter 4 and Appendix A.

For halocarbons with tropospheric residence times of more than a few weeks, the distribution in parts per trillion (ppt = 10^{-12}) by volume of air is approximately uniform, i.e., there is a constant mixing ratio in the troposphere independent of altitude. Accordingly, their rate of introduction into the stratosphere is directly proportional to the average tropospheric concentration. This, in turn, is one component in the net flux across the tropopause; the other, transport downward, depends on the stratospheric distribution of the halocarbon. (Since transport within and between troposphere and stratosphere involves the vertical motion of large air masses, it is independent of the molecular weight or structure of

the species involved.) The average concentration in the
troposphere is determined by the steady-state balance between the source strength of the halocarbon, its rate of
tropospheric destruction, and the net flux through the
tropopause. This average concentration is a directly
measurable quantity. Once it is known, the current rate
at which the halocarbon enters the stratosphere (the upward component of the net flux) can be calculated independently of uncertainties in actual source and sink
strengths in the troposphere and at the earth's surface.

Reliable values of sink strengths are required for two
purposes. First, in combination with average concentrations, they are useful in determining total source
strengths, particularly for halocarbons of natural origin
for which the actual mode and amount of production may be
in doubt. Second, for halocarbons from human sources,
they are essential for predicting the fraction of halocarbon released into the troposphere that will eventually
find its way into the stratosphere and the subsequent time-
history of net transport into the stratosphere if release
at ground level were to be terminated or reduced.

In the following sections we discuss the data available
on industrial production and atmospheric release of the
major halocarbons of interest to this study and the evidence for additional formation from natural sources. We
conclude the chapter with a brief account of N_2O sources,
because there is a possibility that the increasing use of
nitrogen fertilizers will increase atmospheric N_2O concentrations and thereby decrease stratospheric ozone.

II. PRODUCTION AND RELEASE

A. *Halocarbons*

Information on industrial production of halocarbons in the
United States is available from United States Tariff Com-
mission reports, the figures of which are believed to be
reliable to ±5 percent. World production figures are
harder to obtain, but interest in the chlorofluoromethane
(CFM) problem has led to several attempts to assemble accurate data. The first, concerned primarily with CFMs,
was carried out by the DuPont Company (McCarthy, 1974,
1975) and has been the basis of most previous discussions.
In December 1975, results of three further studies (U.S.
Department of State, 1975; A. D. Little & Co., 1975;
Manufacturing Chemists Association, 1976) became available

to the Panel and provide a more accurate and detailed
picture. Production estimates for the principal halocarbons in 1973 are shown in Table 3.1.

Atmospheric release is difficult to ascertain and must
be estimated for each material on the basis of use. This
has been done in some detail in the A. D. Little report,
and its results for the principal chlorocarbons are summarized in Table 3.2 (CFMs are considered in greater detail below). The first numerical column in the table
lists the amount of each chlorocarbon estimated to have
been released in the United States during 1973, the next
column gives the percentage of the U.S. annual production
that was released, and the last column is the amount estimated to have been released worldwide. In most cases,
the latter has been calculated by assuming that the percentage released worldwide is the same as it is in the
United States. For most compounds, U.S. release is about
half of the total release.

As might be expected, volatile hydrocarbons used as
solvents are almost entirely released into the atmosphere,
e.g., trichloroethylene (84.7 percent). However, release

TABLE 3.1 Production of Principal Halocarbons in 1973[a]
(10^3 metric tons/year)

Compound	U.S. Production	World Production	United States/ World[b]
$CFCl_3$[c]	148	368	45
CF_2Cl_2[c]	221	441	55
CCl_4	476	950	50
$CHCl_3$	115	225	50
C_2H_5Cl	300	550	55
CH_2ClCH_2Cl	4224	12,000	35
CH_3Cl	247	400	60
CH_3CCl_3	249	420	60
CH_2Cl_2	236	425	55
$CCl_2{=}CCl_2$	321	750	45
$CCl_2{=}CHCl$	205	700	30
$CH_2{=}CHCl$	2432	7100	35

[a] A. D. Little & Co. (1975).
[b] To nearest 5 percent.
[c] From Table 3.3 [Manufacturing Chemists Association (1976)].

TABLE 3.2 Release of Chlorocarbons Estimated for 1973[a]
(10^3 metric tons/year)

Compound	Amount Released in the United States (10^3 tons)	U.S. Production Released (%)	Amount Released in World (10^3 tons)
CCl_4	21.0	4.4	41.7
$CHCl_3$	6.2	5.4	12.4
C_2H_5Cl	8.0	2.7	14.6
CH_2ClCH_2Cl	199.0	4.7	565.2
CH_3Cl	4.8	2.0	7.9
CH_3CCl_3	195.0	7.8	324.2
CH_2Cl_2	197.0	83.5	346.4
$CCl_2{=}CCl_2$	272.0	84.7	609.0
$CCl_2{=}CHCl$	194.0	94.6	648.3
$CH_2{=}CHCl$	89.0	3.7	351.6

[a] A. D. Little & Co. (1975).

of compounds used as chemical intermediates involves only inadvertent escape during manufacture, transportation, and subsequent processing, and it may be very low. An example is the last entry in the table, vinyl chloride, which is almost entirely converted to polyvinyl chloride, a nonvolatile solid.

Rates of halocarbon release to the atmosphere from natural sources have not as yet been determined directly with precision. Such sources must be inferred from the detection in the atmosphere of concentrations higher than would be predicted from estimates of industrial release. Indeed, methyl chloride has been detected recently in the troposphere at levels far above those expected from human activities. Less conclusive evidence also exists for natural sources of several other halocarbons (see below). In any event, it now seems likely that natural halocarbons, transported upward into the stratosphere, make only a small contribution (≤ 1 percent) to the natural ozone cycle.

Because of their tropospheric stability, completely halogenated halocarbons, including the CFMs and carbon tetrachloride, are expected to have the greatest effect on stratospheric ozone. These compounds are considered in more detail in the next two subsections.

1. CFMs and Other Fluorocarbons Fluorocarbons include
$CFCl_3$ (F-11) and CF_2Cl_2 (F-12), which are produced in the
largest amounts, as well as other compounds such as
CHF_2Cl (F-22), $CFCl_2CF_2Cl$ (F-113), and CF_2ClCF_2Cl (F-114),
which are produced in considerably smaller quantities.
The term "CFMs" refers to those fluorine- and chlorine-
containing compounds that have only one carbon atom. For
example, of the compounds just listed, F-11, F-12, and F-22
are considered CFMs. Figures for U.S. and world production from the most recent Manufacturing Chemists Association (MCA) survey (1976) for F-11 and F-12 are listed in
Table 3.3. The cumulative totals of the latter are 15
to 20 percent larger than some earlier estimates (McCarthy,
1974, 1975).

The data in Table 3.3 were compiled by Alexander Grant
and Co., an independent accounting firm, from returns sent
directly to them by the companies participating in the MCA-
sponsored survey. The 20 companies stated by MCA to represent over 95 percent of the overall world production
submitted detailed annual schedules of production and sales
data, and four others (India and Argentina, producing
about 0.1 percent of the total) provided production data.
Respondents were asked to estimate the probable error in
their total production figures. Statistical combination
of their estimates gave a probable error of ±1.5 percent
for the two production totals in the table, excluding the
production of the Eastern Bloc countries.

In order to obtain overall world production, the 20
major manufacturers were requested to estimate the production in Eastern Bloc countries so far (1949-1975) and
3 such estimates were received. Their average is given
in footnote *b* of the table. This appears to be the most
uncertain component of the production figures; however,
it comprises only about 4.5 percent of the totals, so its
effect on the overall uncertainty is believed to be small.

The production data of Table 3.3 are shown in graphi-
cal form in Figures 3.1 and 3.2, with U.S. and non-U.S.
data plotted separately. For F-11, the data correspond
to a worldwide growth rate of 13.9 percent/yr, for the
period 1965-1974, with a doubling time of 5.3 yr. This
has two components: 9.0 percent/yr growth in the United
States (doubling time 9.3 yr) and 19.8 percent/yr growth
abroad (doubling time 3.8 yr). The corresponding data
for F-12 are a worldwide growth rate of 9.9 percent/yr,
with a doubling time of 7.4 yr; this includes a 7.3
percent/yr growth rate in the United States (doubling time
9.9 yr) and 13.2 percent/yr growth abroad (doubling time

TABLE 3.3 Production and Release of F-11 ($CFCl_3$) and F-12 (CF_2Cl_2) to Date[a] (10^3 metric tons/year)

	F-11		F-12	
Year	United States	World[b]	United States	World[b]
To 1958		172.5		570.0
1958	23	29.7	60	74.0
1959	27	35.8	71	88.5
1960	33	50.0	75	100.6
1961	41	60.8	79	110.2
1962	56	78.7	94	130.6
1963	64	94.2	99	149.9
1964	67	112.0	104	175.0
1965	77	124.5	123	196.4
1966	77	141.9	130	227.1
1967	83	163.6	141	257.8
1968	93	187.1	148	277.1
1969	109	223.2	167	311.5
1970	111	245.9	170	335.4
1971	117	274.6	177	355.9
1972	136	317.5	200	398.6
1973	148	367.8[c]	221	441.0[c]
1974	158	400.2	231	473.6
1975	121	357.3	178	416.3
TOTAL		3437.3[b]		5089.5[b]
Northern hemisphere		3326.8		4891.1
Southern hemisphere		110.5		198.4
Total released		2934.1		4414.1
Percentage released		85.36		86.73

[a]Manufacturing Chemists Association (1976).
[b]This includes Eastern Bloc production, which is estimated to be 155,000 and 209,000 metric tons for F-11 and F-12, respectively, through 1975.
[c]The calculations reported in Chapter 8 used an earlier estimate of 1973 production (3.14×10^5 and 4.70×10^5 metric tons for F-11 and F-12), with a 90 percent release rate.

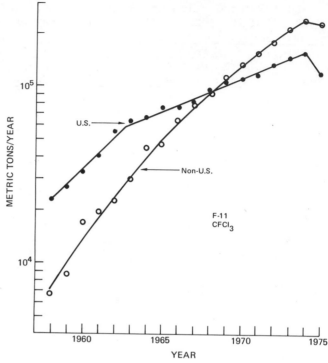

FIGURE 3.1 Annual production of F-11 (CFCl$_3$); semilog plot of metric tons per year for 1958 to 1975 (from Table 3.3).

5.6 yr). These results emphasize the worldwide nature of the CFM problem.

It is seen that the production abroad of both CFMs has grown for 15 yr at a faster (but slowing) rate than that in the United States. As a consequence, the production abroad of F-11 is now nearly double that in the United States, while for F-12 the amounts are nearly equal. Looking to the immediate future, the A. D. Little & Co. report (1975) estimates a possible 6-7 percent/yr growth in total worldwide CFM production for 1975-1980. However, Figures 3.1 and 3.2 indicate that production of both F-11 and F-12 peaked in 1973-1974 and is reported to have declined by some 15 percent in 1975. It should be noted that the decline has been primarily in U.S. production.

Although the CFMs were originally developed as refrigerants, their major current use is as aerosol propellants. Table 3.4 shows one estimate of the distribution in use

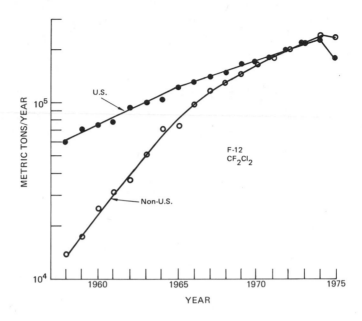

FIGURE 3.2   Annual production of F-12 (CF_2Cl_2); semilog plot of metric tons per year for 1958 to 1975 (from Table 3.3).

for 1973, which includes a small contribution from other fluorocarbons besides F-11 and F-12.   The MCA survey (Manufacturing Chemists Association, 1976) indicates that usage of F-11 for refrigeration is only 5.5 percent and of F-12, 25 percent.   The CFMs in aerosols are released on use, and those used as refrigerants escape by leakage, during recharging, or after the equipment is junked.   It seems likely that almost the entire production of CFMs eventually reaches the atmosphere.   The most important possible exception is the F-11 used as a blowing agent in closed-cell polyurethane foams, from which escape by diffusion has an estimated half-life of 14 yr (Manufacturing Chemists Association, 1976).

   The schedule of release for different uses has been discussed by Rowland and Molina (1975) and Howard and Hanchett (1975).   The problem has also been analyzed in the MCA survey (Manufacturing Chemists Association, 1976), which concludes that some 86 percent of the total production of F-11 and F-12 to date has been released (cf. Table 3.3).   The MCA estimates an uncertainty of about ±5 percent at the 2σ level in the amounts listed there.   If

TABLE 3.4 CFM Consumption by End Use (1973, percentage of total)[a]

Use	United States (%)	World (%)
Aerosol propellants	49	55
Refrigerants	28	29
Plastics and resins	4	7
Solvents	5	3
Foam-blowing agents	7	4
Other	7	2

[a] A. D. Little & Co. (1975).

all of this material were present in the troposphere, including the Eastern Bloc releases, it would correspond to an average concentration of 129 ppt (10^{-12} by volume) of F-11 and 220 ppt of F-12. Measured concentrations are 60-110 percent of these values and the question of the uncertainties in such measurements and their significance are discussed in detail in Chapter 6 and Appendix E.

The next most important CFM is F-22 (CHF_2Cl). World production through 1975 for uses other than as a chemical intermediate has been estimated as 752 metric kilotons by the MCA survey (Manufacturing Chemists Association, 1976). A comparable additional amount is made and used in the manufacture of fluorinated plastics, e.g., polytetrafluoro-ethylene, but in this use a negligible amount escapes to the atmosphere. Since the major nonintermediate use is as a refrigerant, and half of that has been in hermetically sealed units, only 40 percent (305 metric kilotons) is estimated to have escaped or been released so far. This would correspond to 12 ppt in the troposphere. However, since F-22 contains a C—H bond, it would be expected (see Section I) to have a relatively short tropospheric life, and accordingly an even lower concentration. To date it has not been detected in the atmosphere.

No significant natural source of CFMs has been detected or proposed, although minor production from volcanoes has been suggested by Stoiber *et al.* (1971). The relationship between the amounts released and the atmospheric concentrations observed is discussed in Appendix E.

2. *Carbon Tetrachloride* Carbon tetrachloride is the only other completely halogenated hydrocarbon manufactured on

a large scale and found in appreciable concentrations, approximately 120 ppt (Rasmussen, 1975b), in the troposphere. In the United States, the growth in its production has paralleled the growth in production of F-11 and F-12, for which it is an intermediate, while its use for other purposes (chiefly as a solvent and fumigating agent) has remained roughly constant, as shown in Table 3.5. Since over 90 percent of CCl_4 goes into CFM production for which world production is double that of the United States, world production of CCl_4 is also probably close to double that of the United States as well.

If it is assumed that 2.5 percent of the CCl_4 used in CFM manufacture escapes to the atmosphere during manufacture, storage, and processing; that all that is not used for this purpose is released; that figures for 1974-1975 are the same as 1973; and that world production in all categories has been double that of the United States, then 4100 metric kilotons have been released to date,

TABLE 3.5 United States Production and Use of CCl_4 to Date[a] (10^3 metric tons/year)

Year	Production	Used for Fluorocarbon	Other	Use for Fluorocarbon (%)
To 1958			1150	
1958	142	101	41	71
1959	167	122	45	73
1960	170	133	37	78
1961	174	146	38	84
1962	220	183	37	84
1963	236	197	39	84
1964	244	207	37	85
1965	270	243	27	90
1966	295	252	43	85
1967	325	272	47	84
1968	347	292	55	84
1969	401	334	67	83
1970	460	341	119	74
1971	459	356	103	78
1972	453	406	47	90
1973	477	461	16	97

[a]McCarthy (1975).

corresponding to an atmospheric burden of 160 ppt. The atmospheric residence time of CCl_4 has been estimated as 60 yr as a result of stratospheric photolysis (Table 4.1).

Integration of the data of Table 3.5 indicates that with this residence time, approximately 75 percent of the CCl_4 released (120 ppt) would still be present in the atmosphere. The agreement of this figure with the reported measurements (Rasmussen, 1975b; Lovelock, 1974) suggests that there are no major natural sources or sinks other than photolysis, although the uncertainties are large enough to accommodate some. Lovelock (1975a) has proposed on the basis of marked variations of atmospheric concentration with latitude that there are both natural sources and other sinks. However, such variability might result from the severalfold fluctuations in the recent release rates (Table 3.5).

3. Reactive Cholorocarbons As noted in Section I, halocarbons containing C=C or C—H bonds are expected to have relatively short atmospheric lifetimes because of their fast reaction with hydroxyl radicals. On the basis of measured rate constants and of estimated average hydroxyl radical concentration in the troposphere, these lifetimes can be calculated. In turn, total source strengths can be estimated from measured concentrations divided by these lifetimes, and such estimates for reactive halocarbons of major interest are listed in Table 3.6. Such calculations and sources of data are discussed in detail in Chapters 4 and 6, but it should be noted that while rate constants for hydroxyl radical reactions are quite reliable and are currently being checked in several laboratories, the average concentration of hydroxyl radicals, here taken as 2×10^6 molecules/cm^3, is uncertain and estimates run from 10^6 to 10^7 molecules/cm^3. In Table 3.6, the resulting values for total source strengths are compared with estimated release from human sources. In spite of the large uncertainties involved, the results suggest that significant natural sources exist for some of the compounds listed.

Methyl chloride, which is the most abundant halocarbon in the atmosphere (600-1200 ppt), evidently has an enormous natural source. In part it is of marine origin. Methyl iodide, its main marine precursor, is widely distributed in the oceans and found in high concentrations over kelp beds, where it is evidently produced by metabolic processes (Lovelock, 1975b). It reacts rapidly with the chloride ion in seawater to yield methyl chloride,

TABLE 3.6 Estimated Emission Rates of Halocarbons (metric kilotons/year)

Halocarbon	Average Tropospheric Mixing Ratio (ppt)	Estimated Total Global Emission Rate[a]	Estimated Global Anthropogenic Emission Rate	Natural Sources	Anthropogenic Sources
CH_3Cl	$750^{b,c}$	6690	12.4^d	Marine biological activity; smoldering vegetation	Manufacturing losses
CH_2Cl_2	$30^{b,c}$	3450	346^d	Unknown	Solvent, manufacturing loss
$CHCl_3$	20^b	2450	12.4^d	Unknown	Solvent
CH_3CCl_3	50^b	925	324^d	Unknown	Solvent, dry cleaning
$CCl_2{=}CCl_2$	5^b	1720	609^d	Unknown	Solvent, dry cleaning
$CHCl{=}CCl_2$	5^b	2030	648^d	Unknown	Solvent, dry cleaning
CH_3Br	$<5^c$	<200	4.3^e	Marine	Agricultural fumigation

[a]Computed from the average tropospheric mixing ratio divided by the estimated atmospheric residence time, cf. Chapter 4. [d]From Table 3.2 [e]Plonka (1975). [b]Lovelock (1975a). [c]Rasmussen (1975b).

45

so that its measured average life is only a few weeks (Zafiriou, 1975). It has also been suggested that methyl chloride is formed in the partial combustion of vegetable matter and is thus an inadvertent product of slash-burn agriculture, particularly in the tropics (Lovelock, 1975b). Most of the industrially produced methyl chloride is consumed as a chemical intermediate; its release rate has been estimated as 5.4 percent (Table 3.2).

Significant natural sources are also likely for methylene chloride, CH_2Cl_2, and chloroform, $CHCl_3$. It has been proposed (Yung *et al.*, 1975), but so far without quantitative confirmation, that chloroform may be produced inadvertently by man in the treatment of organic materials with chlorine, particularly in the bleaching of paper pulp.

Data on trichloroethylene and perchloroethylene, which also suggest natural sources, are equivocal, since they are based on a limited number of concentration determinations on samples gathered near ground level. Lifetimes of these materials are so short that strong vertical and horizontal concentration gradients may exist.

Table 3.6 includes only those chlorinated hydrocarbons that have been measured in appreciable quantities in the atmosphere. Others that are made in large quantities, such as ethylene dichloride, vinyl chloride, and ethyl chloride, are used primarily as chemical intermediates, so that little atmospheric release would be expected. Nevertheless, in view of the fact that the whole history of detection of halocarbons in the atmosphere only extends back a few years, and increasingly sensitive analytical techniques are being developed, the possibility remains that others will be found. An important check would be to compare total halogen in representative air samples with the sum of identified components, an experiment that has not yet been carried out.

4. Bromine-Containing Hydrocarbons Global bromine utilization has been surveyed by Plonka (1975) and is summarized in Table 3.7. Ethylene dibromide production is the major use, consuming 1.87×10^5 metric tons of bromine in 1974. Almost all of the dibromide produced is used as a gasoline additive in conjunction with tetraalkyl leads and is presumably converted to HBr and inorganic bromides during combustion. Although some may escape to the atmosphere by evaporation or spillage, it has not been detected.

The other major volatile organic bromide is methyl bromide, CH_3Br, for which Plonka estimated a global use of 17,000 metric tons/yr, approximately 80 percent for soil

TABLE 3.7 Global Bromine Utilization[a]

End Use	Bromine Produced (%)
Ethylene dibromide (gasoline additive)	63
Fire retardants	12
Agricultural	10
Inorganic bromides	7
Miscellaneous organic bromides	4
Dye stuffs	2
Disinfectants	2

[a] Plonka (1975).

fumigation. Costs and toxicity require application below ground level or under plastic sheeting to prevent loss of material. Under these conditions, Plonka estimates that 10-20 percent of the CH_3Br escapes to the atmosphere. The remaining CH_3Br is used for grain, space, and food fumigation, from which losses to the atmosphere are larger, and as a chemical intermediate.

Combining these uses, Plonka estimates current loss to the atmosphere of not over 4300 metric tons/yr, which, assuming a residence time of 0.4 yr (see Appendix A, Table A.1), would give a concentration of about 0.1 ppt. Some methyl bromide is evidently produced from natural sources, as Lovelock (1975b) has detected low concentrations in seawater. Here it must be rapidly converted to methyl chloride, since its rate of reaction with chloride ion is similar to that of CH_3I. As far as we know, it has not been detected in the atmosphere, although Wofsy et al. (1975) have estimated that marine production of CH_3Br is a source of 7.7×10^4 metric tons/yr.

B. Sulfur Hexafluoride

Sulfur hexafluoride, although not a halocarbon, is a volatile halogen-containing compound of human manufacture, which has been detected in both the troposphere and stratosphere (IMOS, 1975). Concentrations are very low, under 1 ppt, and can only be observed because of the high sensitivity of the electron capture chromatographic technique

to the material. Its chief use is as an electric insulat-
ing material, but source strengths have not been estimated. From its structure it is presumably very stable
to both hydroxyl radical attack and photolysis and therefore will continue indefinitely to accumulate in the atmosphere. For this reason, it should not be overlooked,
and we list it here to call attention to the need for assessing its possible long-range effects.

C. The Space Shuttle as a Chlorine Source

The Space Shuttle as now planned will use a solid ammonium perchlorate/aluminum powder propellant, and each vehi-
cle will introduce approximately 100 metric tons of HCl
directly into the stratosphere at between 10 and 45 km
(Whitten et al., 1975). At the level of 50 flights per
year now projected for 1986, this amounts to 5000 metric
tons/yr. For comparison, the amount of chlorine introduced into the atmosphere as F-11 and F-12 in 1973 was
approximately 5×10^5 metric tons/yr, an amount 100 times
as large. In short, Space Shuttle operations and other
uses of solid propellant rockets that release halogens
in the stratosphere, while adding to the total burden
of man-made halogens in the atmosphere, are projected
to make only a minor contribution to the total problem
(see Chapter 9).

D. Nitrous Oxide

Nitrous oxide, N_2O, is predominantly of natural origin,
produced by biochemical processes in soil and ocean. It
rises into the stratosphere, where a small portion is converted into NO by reaction with excited oxygen atoms
$O(^1D)$, and, for that reason N_2O plays an important role
in the ozone balance.

The average tropospheric concentration of N_2O is 250
ppb (10^{-9}), large compared with any of the halocarbons.
The natural net production rate has been estimated as
1.55×10^8 metric tons/yr by Hahn (1974), who assumed
that most of the production occurs in the oceans; McElroy
(1975) has questioned this assumption. There is apparently a large unidentified sink for N_2O in the troposphere
or at the earth's surface, and the atmospheric residence
time of N_2O is estimated as 10-12 yr.

N_2O is believed to be formed predominantly by biologi-

cal reduction of nitrite and nitrate, and concern has been
expressed that increased use of nitrogen fertilizers will
increase N_2O levels and in turn decrease stratospheric
ozone. Estimates of the magnitude of possible reductions
have been made by Crutzen (1974) and McElroy (1975). The
latter projects a much larger effect, but the conclusion
has been questioned by Johnston (1975). While details of
the matter are beyond the scope of this report, it may
possibly be of long-term importance and requires further
study (cf. Chapter 9).

REFERENCES

Cicerone, R. J. 1975. Comment on "Volcanic emissions
of halides and sulfur compounds to the troposphere and
stratosphere," by R. D. Cadle, *J. Geophys. Res.* 80(27):
3911-3912.
Crutzen, P. J. 1974. Estimates of possible variations
in total ozone due to natural causes and human activities, *Ambio* 3:201-210.
Hahn, J. 1974. The North Atlantic Ocean as source of
atmospheric N_2O, *Tellus* 26:160-168.
Howard, P. H., and A. Hanchett. 1975. Chlorofluoro-
carbon sources of environmental contamination, *Science*
189:217-219.
IMOS. 1975. Report of Federal Task Force on Inadvertent
Modification of the Stratosphere. *Fluorocarbons and the
Environment*. Council on Environmental Quality, Federal
Council for Science and Technology.
Johnston, H. S. 1975. Reduction of stratospheric ozone
by nitrogen fertilizers. University of California,
Berkeley. Manuscript.
A. D. Little & Co. 1975. Preliminary Economic Impact
Assessment of Possible Regulatory Action to Control
Atmospheric Emissions of Selected Halocarbons. Prepared
for the Environmental Protection Agency, publication
no. EPA-450/3-75-073.
Lovelock, J. E. 1974. Atmospheric halocarbons and strato-
spheric ozone, *Nature* 252:292-294.
Lovelock, J. E. 1975a. Bowerchalke, United Kingdom.
Private communication.
Lovelock, J. E. 1975b. Bowerchalke, United Kingdom.
Natural halocarbons; methyl chloride, bromide and
iodide in the air and in the sea. Manuscript.
Manufacturing Chemists Association, Technical Panel on
Fluorocarbon Research. March 1, 1976. The effect of

fluorocarbons on the concentration of atmospheric ozone. Preliminary data from this report supplied to the Panel on Atmospheric Chemistry by R. L. McCarthy, December 15, 1975, and revised data February 9, 1976.

McCarthy, R. L. 1974. E. I. du Pont de Nemours and Co., Fluorocarbons in the environment. Pager given at the National Geophysical Union Meeting, San Francisco, Calif., December 13.

McCarthy, R. L. 1975. E. I. du Pont de Nemours and Co., letter to C. T. Walling, November 6.

McElroy, M. B. 1975. Chemical processes in the solar system: a kinetic perspective. To be published in MTP International Review of Science.

National Research Council. 1976. *Chlorine and Hydrogen Chloride*. Report of Subcommittee on Chlorine and Hydrogen Chloride, Committee on Medical and Biologic Effects of Environmental Pollutants, National Academy of Sciences, Washington, D.C.

Plonka, J. 1975. Dow Chemical Co. Private communication.

Rasmussen, R. A. 1975a. Data presented to Panel on Atmospheric Chemistry, Snowmass, Colo., July.

Rasmussen, R. A. 1975b. Washington State University. Private communication.

Rowland, F. S., and M. J. Molina. 1975. Chlorofluoro-methanes in the environment, *Rev. Geophys. Space Phys.* 13:1-35.

Stedman, D. H., W. L. Chameides, and R. J. Cicerone. 1975. The vertical distribution of soluble gases in the troposphere, *Geophys. Res. Lett.* 2(8):333-336.

Stoiber, R. E., D. C. Leggett, T. F. Jenkins, R. P. Murrmann, and W. I. Rose, Jr. 1971. Organic compounds in volcanic gases from Santiaquitu Volcano, Guatemala, *Geol. Soc. Am. Bull.* 82:2299-2302.

U.S. Department of State. 1975. Fluorocarbons: an assessment of worldwide production, use and environmental issues. U.S. and Canada report to the OECD Environment Committee.

Whitten, R. C., W. J. Borucki, I. G. Poppoff, and R. P. Turco. 1975. Preliminary assessment of the potential impact of solid-fueled rocket engines in the stratosphere, *J. Atmos. Sci.* 32:613-619.

Wofsy, S. C., M. B. McElroy, and Y. L. Yung. 1975. Harvard University. The chemistry of atmospheric bromine. Manuscript.

Yung, Y. L., M. B. McElroy, and S. C. Wofsy. 1975. Harvard University. Atmospheric halocarbons: a discussion with emphasis on chloroform. Manuscript.

Zafiriou, O. C. 1975. Reaction of methyl halides with seawater and marine aerosols, *J. Marine Res.* 33:75-81.

4 REMOVAL PROCESSES

I. INTRODUCTION

Halocarbons, both man-made and natural, are released into
the atmosphere at the earth's surface and are transported
upward. Halocarbons can be removed from the atmosphere by
a number of processes that occur either at the surface or
in the atmosphere. It is the purpose of this chapter to
examine these removal processes.

   We first distinguish between two types of removal,
which we will call "active" and "inactive," respectively,
depending on whether they lead to ozone destruction.
Physical processes, such as trapping in polar ice, that
remove halocarbons from the atmosphere and return them to
the surface before they can affect the stratospheric ozone,
constitute inactive removal. Chemical processes, such as
photolysis and reactions with neutral and ionic species or
with aerosols, lead to breakdown of the halocarbons. It
is shown in Appendix A that such processes in the troposphere probably lead to the formation of products that are
removed by rain-out. In this case, the halocarbon can also
be said to have undergone inactive removal. However, in
the *stratosphere* the products that are formed by known
chemical processes (mainly photolysis) lead to the destruction of ozone. These processes therefore constitute
active removal.*

   In particular, active processes form Cl and ClO (jointly called ClO_x), which catalytically destroy O_3. Chlorine

---

*The only inactive removal process that might occur in the
stratosphere would be one that produces unreactive, water-
soluble chlorine products that survive for the time required
to be transported into the troposphere and washed out.

atoms, Cl, can also react with other minor constituents (mainly CH_4) to form HCl, and ClO with NO_2 to form $ClONO_2$, which do not attack O_3. However, chlorine atoms can be released again from HCl by other chemical reactions and ClO from $ClONO_2$ by photolysis or other reactions. The efficiency of the ozone destruction cycle is proportional to the fraction of chlorine products ClX(Cl, ClO, HCl, $ClONO_2$, etc.) that is in the form of ClO_x. Therefore, the processes that affect this partitioning will be discussed in considerable detail. The only known removal process for ClX from the stratosphere is transport into the troposphere and subsequent removal by heterogeneous processes (principally rain-out).

In order to assess the effect of halocarbon release on O_3 destruction, it is necessary to consider the ratio of inactive to active halocarbon removal. Let us consider the case of *continuous* release of a halocarbon at a *constant* rate. This constant release rate will, given enough time, be balanced by an equal removal rate--a condition referred to as a steady state. In this case, the maximum O_3 destruction will occur when a steady state has been achieved. At steady state, the input flux F of the halocarbon (molecules cm^{-2} sec^{-1}) must be balanced by all the removal processes for the halocarbons, i.e., $F = \Sigma\phi_i$, where ϕ_i is the removal flux for a particular process. We will assume, for simplicity, that halocarbons are mixed uniformly in latitude and longitude and consider globally averaged quantities.

The total removal flux Φ can then be expressed as an integral with respect to altitude:

$$\Phi = \int_0^\infty Lfn\,dz \qquad (4.1)$$

where n is the total number of molecules cm^{-3} at altitude z, and f, the mixing ratio, is the fraction (by volume) of the molecules that are the halocarbon. L is the (first-order) local loss rate constant (sec^{-1}) for the halocarbon by all processes. The local removal time can then be defined as $t_C = 1/L$. (For example, a bimolecular chemical reaction between the halocarbon and species X would have $L = k_X[X]$, where k is the rate constant and $[X]$ is the concentration of X).* The tropospheric removal time for

---

*It is assumed that $[X]$ and k do not change with time. This assumption is equivalent to using globally averaged values of concentrations and temperatures at a given altitude z in the 1-D models (Chapter 7).

halocarbons is generally equal to or greater than the vertical tropospheric mixing time (which is the order of a week) except for some unsaturated halocarbons; therefore, to a good approximation, the halocarbons have a constant mixing ratio, f, in the troposphere. In some parts of the stratosphere the reverse is true. In this case, the mixing ratio decreases with altitude as the photolytic loss rate increases with altitude. The total number density of the atmosphere, n, decreases approximately exponentially with altitude.

The residence time for a halocarbon in the atmosphere can be defined as the average time its molecules reside there. If at any time, the release of a particular CFM is suddenly terminated, the residence time, τ^0, is the time required for the concentration to drop to $1/e$ of its value at the time of termination.* Under steady-state conditions, the residence time, τ^0, equals the total column density divided by the input flux or by the total removal flux, i.e.,

$$\tau_i = \frac{\text{total column density (molecules cm}^{-2})}{\text{input flux (molecules cm}^{-2}\text{ sec}^{-1})}$$

$$= \frac{\text{total column density}}{\text{removal flux}} \qquad (4.2)$$

The partial residence time for a *particular* removal process, τ_i, can then be defined as the time that would be required to remove the total column by that process alone:

$$\tau_i = \frac{\text{total column density}}{\phi_i} \qquad (4.3)$$

where ϕ_i is the removal flux for that *particular* process. If the total column density is N halocarbon molecules cm^{-2}, then

$$\tau_i = \frac{N}{\displaystyle\int_0^\infty L_i f n\, dz} \qquad (4.4)$$

---

*If a particular CFM is released into the atmosphere at a constant rate (starting at $t = 0$, the residence time, τ^0, is also the time required for the fractional difference between the steady-state amount and the amount actually present to reach $1/e$.

Since most of the halocarbon will be located in the troposphere, where it has an approximately constant mixing ratio, the column density is, to a close approximation, given by $c_0 H_0$, where c_0 is the number density of the halocarbon near the surface and H_0 is the scale height* near the surface = 8.4×10^5 cm.

By definition, at steady state, the input flux, F, must be balanced by the sum of all removal fluxes ($F = \Sigma \phi_i$). It follows that the total residence time, τ^0, is related to the partial residence times for individual processes τ_i by the relation

$$1/\tau^0 = \sum_i 1/\tau_i \qquad (4.5)$$

The relative importance of any removal process is inversely proportional to its partial residence time, the fraction it removes being τ^0/τ_i. For example, an inactive removal process with a partial residence time that is 10 times τ^0 will remove 1/10 of the halocarbon, leaving 9/10 to affect the stratospheric ozone.

Because tropospheric removal is generally inactive, while stratospheric removal is generally active toward ozone removal, it is useful to compare loss processes in the troposphere with loss processes in the stratosphere. We may define the quantities

$$\tau_T = \frac{N}{\int_0^Z L_T fn dz} \qquad (4.6)$$

$$\tau_S = \frac{N}{\int_Z^\infty L_S fn dz} \qquad (4.7)$$

where L_T and L_S are the effective loss rate constants in the troposphere and stratosphere, respectively, and Z is the altitude of the tropopause. Comparison of Eqs. (4.5)-(4.7) shows that

$$\frac{1}{\tau^0} = \frac{1}{\tau_T} + \frac{1}{\tau_S} \qquad (4.8)$$

---

*The scale height, H, is the vertical distance over which the atmospheric pressure drops by a factor of e.

In terms of removal rate constants R, this becomes

$$R^0 = R_T + R_S \qquad (4.9)$$

and the ratio $R_S/R^0 = \tau^0/\tau_S$ is the fraction of a halocarbon that reaches the stratosphere. Table 4.1 lists this ratio together with other quantities for some CFMs and for CH_3Cl and CCl_4, assuming photolysis and reaction with HO are the only removal processes.

Attention should be drawn to another feature implicit in the expressions for residence times. The halocarbon number density, $c_0 = fn$, in these expressions decreases rapidly with altitude; therefore, the relative importance of loss processes is heavily weighted toward lower altitudes, where most of the halocarbons reside. Because of this, a loss process with a rate constant L_i in the troposphere will be much more effective in removing a halocarbon than a loss process with the same value in the stratosphere. Or, expressed differently, a process with a local time constant t_c of 1 week at 25 km removes as much as a process having a value of t_c of 1 yr near the ground. For example, although the photolytic local removal time, t_c, for F-12 destruction at 25-35 km, is several weeks, the total residence time $\tau^0 \sim 100$ yr. If a removal process over the *entire* troposphere had a partial residence time $\tau_T = 100$ yr, it would remove as much as does photolysis. Thus, inactive removal does not need to be fast in order to compete with stratospheric photolysis and alleviate the ozone reduction. For this reason a careful analysis has been made of all tropospheric and surface removal processes that have been suggested; this analysis is presented in detail in Appendix A. The principal results of this extensive analysis are summarized briefly below, followed by a discussion of the stratospheric processes.

Table 4.1 shows that the partial atmospheric residence times for photolysis of F-11, F-12, and CCl_4 are of the order of 50-100 yr. Any process that removes the halocarbon at the surface or in the troposphere will reduce the total atmospheric residence time τ^0 and consequently decrease the effect of the halocarbon on stratospheric ozone. In the following discussion the partial residence time for a particular process, τ_i, will be called the removal time for that process. It represents the time for the atmospheric content of the CFM to drop to $1/e$ of its value if that removal process were the only one operating and if the input flux were terminated. The reciprocal of

TABLE 4.1 Steady-State Abundances of Some Halocarbons for Constant Input Flux F at the 1973 Level, and the Corresponding Residence Time Parameters, Assuming Photolysis and HO Reactions as the Only Removal Processes[a]

	$CFCl_3$ (F-11)	CF_2Cl_2 (F-12)	CHF_2Cl (F-22)	CCl_4	CH_3Cl
Input flux F molecules/cm^2/sec	1.1×10^7	1.6×10^7	5.2×10^6	1.1×10^6	5.4×10^8
Mixing ratio f at the ground (ppb)	0.8	2.3	0.07	0.12	0.75
Column density $N = c_0 H_0$ 10^{16} molecules/cm^2	1.7	4.9	0.15	0.21	1.6
Tropospheric residence time τ_T yr	>>100	>>100	10	>>100	1
Stratospheric residence time τ_S yr	50	100	100	60	17
Total residence time $\tau^0 = \dfrac{1}{1/\tau_T + 1/\tau_S}$ (yr)	50	100	9.1	60	0.94
Fraction reaching the stratosphere τ^0/τ_S	1	1	0.09	1	0.06

[a]Since the CFMs have not been released for a sufficiently long time to achieve steady state, average values of f have been calculated for steady-state conditions using 1973 global production figures and an early version of the 1-D model described in Chapter 7. Measured values of the mixing ratio f were used for CH_3Cl and CCl_4. The values of τ_S for F-11, F-12, and CCl_4 have been calculated assuming photolysis to be the only stratospheric loss process; but for the others, reactions with HO have also been included.

the removal time $R_i = 1/\tau_i$ will be called the (atmospheric) removal rate constant for that process. Although F-11 and F-12 differ somewhat in their properties, these differences are usually small in comparison with the other factors governing their removal. Therefore, except for removal by the oceans, the two CFMs are treated jointly in our discussion of them, the results of which are summarized in Table 4.2.

II. SURFACE AND TROPOSPHERIC REMOVAL*

A. *Physical Removal (Appendix A, Sections I and II)*

Because of their low solubilities, a negligible fraction of the CFMs are removed by rain-out. If all the water droplets in the troposphere were saturated with CFMs, rain-out would have an associated removal time greater than 6×10^4 yr. Hydrolysis in water and in acid solution (characteristic of rain and liquid aerosols) is also very slow and therefore will not decrease the removal time significantly.

The oceans are potentially large reservoirs for the storage of gases. Exchange between surface air and the oceans is attained rapidly (~ 2 yr) down to thermocline depths ($\lesssim 50$ m), and the exchange mixing time of the surface layer with the deep oceans is somewhat longer (~ 15 yr). If the oceans were saturated with respect to CFMs down to the thermocline, they would contain only 0.06 percent of the total atmospheric burden.

Hydrolysis in the oceans or mixing of the surface water with the depths would make a negligible contribution to the removal rate. If other more rapid oceanic removal processes (e.g., microbial action) were operative, the overall removal rate would then be limited by diffusion of CFM through the surface boundary. This gives lower limits of 70 and 200 yr for the removal times of F-11 and F-12, respectively. These values are about twice the active (ozone-destroying) removal times for F-11 and F-12 in the stratosphere. However, estimates based on the measured concentrations of F-11 in Atlantic Ocean surface waters indicate an oceanic removal time of 270 yr for it—considerably longer than the lower limit. Additional studies of the matter are needed.

---

*Details and references may be found in the corresponding sections of Appendix A.

TABLE 4.2 Removal Times and Removal Rates for F-11 and F-12 (from Appendix A)

Section	Process	Removal Time τ (years)	Removal Rate $1/\tau$ (year^{-1})
Active Removal in Stratosphere			
III.A, III.C	Photolysis and $O(^1D)$ reaction	50; 90[a]	2×10^{-2} 1.1×10^{-2}
Surface Processes			
I.A	Removal by oceans	$>(70; 200)$	$<(14; 5) \times 10^{-3}$
I.B	Removal by soil and microbes	$>10^4$	$<10^{-4}$
I.C	Entrapment in polar ice	$>10^5$	$<10^{-5}$
Tropospheric Processes			
II.A	Photodissociation	$>5 \times 10^3$	$<2 \times 10^{-4}$
II.B	Reactions with neutral molecules	$>>100$[b]	$<<10^{-2}$
II.C.1	Direct ionization	$>10^6$	$<10^{-6}$
II.C.2	Ion-molecule reactions	$>10^3$	$<10^{-3}$
II.D	Heterogeneous processes	$>6 \times 10^4$	$<2 \times 10^{-5}$
II.E	Lightning	$>10^6$	$<10^{-6}$
II.F	Thermal decomposition	$>10^4$	$<10^{-4}$
Inactive Removal in Stratosphere			
III.H	Ionic processes	$>10^5$	$<10^{-5}$
III.I	Heterogeneous processes	$>10^8$	$<10^{-8}$

[a]Slightly different values of 54 and 80 yr were obtained as most likely values in Table 5.1.

[b]This limit is based on the detection limits of the laboratory studies of the chemical reactions. No reaction was observed, and the actual removal time is probably at least two orders of magnitude larger than this value.

The removal time by any microbial destruction of CFMs in soils will be limited by the deposition rate at the surface and is estimated to be greater than 10^4 yr. This estimate is, however, based on a limited amount of evidence, and additional work in this area would be desirable.

The possibility of entrapment of CFMs in polar or glacial snow and ice has been suggested. In a recent observation, the gas released by melting polar snow showed an enrichment in CFMs by a factor of 25 over the value in the surface air. However, even if such enrichment existed in all the permanent ice of the world, it would provide a sink with a removal time longer than 10^5 yr.

Most of the tropospheric aerosols are liquid and acidic. Solubility and hydrolysis in such aerosols will be negligible. No chemical reactions are known to occur with these acidic solutions. Solid particulates constitute a minor fraction of tropospheric aerosols. Adsorption on their surface at the low partial pressures of CFMs will be too small to provide a removal time by dry deposition less than 10^5 yr, even if smoggy urban atmospheres are included in the considerations.

B. Chemical Processes (Appendix A, Section II)

On the basis of bond energies, halocarbons can be photo-dissociated by radiation of wavelengths shorter than 400 nm. Stratospheric ozone and oxygen effectively prevent radiation at wavelengths less than 280 nm from penetrating into the troposphere. If the observed cross sections for absorption by halocarbons are extrapolated into the spectral region between 280 and 400 nm, a removal time for tropospheric photodissociation of 5×10^3 yr is obtained. There is a remote possibility that some weak forbidden transition exists in this wavelength region. Although no evidence for such a transition exists, it would be worthwhile to undertake an undoubtedly difficult laboratory search for such transitions. Photosensitized reactions, in which another molecule absorbs light and then transfers its energy by collision to a CFM molecule, have also been considered and dismissed as an insignificant removal process.

Removal rates of halocarbons by reaction with neutral molecules in the troposphere are determined by the product of the rate constant for the reaction and the concentrations of the reactant species. The magnitude of this product in the troposphere is such that only reactions of HO and O_3 with the halocarbons need be considered. Such reactions are very effective in removing unsaturated halocarbons from the troposphere. Reactions of HO with saturated halocarbon molecules containing one or more hydrogen atoms are also fast enough that only a small fraction of these halocarbon molecules will survive long enough to be transported into the stratosphere. This also applies to naturally produced CH_3Cl, which reacts sufficiently rapidly with HO that only about 6 percent reaches the stratosphere (cf. Table 4.1). In contrast, there is no evidence that HO reacts with CFMs.

At the high temperatures existing near lightning flashes, thermal decomposition of CFMs can occur, but the frequency of such flashes is so small that it leads to removal times of about 10^8 yr for this process. Photolysis by uv radiation from lightning would give removal times of about 10^6 yr. Decomposition of CFMs in combustion processes is much greater than that in lightning, but it is still negligible with a removal time of 10^4 to 10^5 yr.

Since CFMs have large cross sections for electron capture, bombardment by electrons from cosmic rays or natural radioactivity has been suggested as a removal process. However, oxygen, which also has an appreciable electron capture cross section, is nine orders of magnitude more abundant than CFMs, and the latter therefore cannot compete for the electrons. The removal time for this process is greater than 10^6 yr.

The primary positive ions that are formed in the troposphere rapidly react with atmospheric constituents that are in much greater abundance than the CFMs to form the relatively stable hydrated ions H_3O^+ $(H_2O)_n$ and NH_4^+ $(H_2O)_m$, where the degree of hydration, n and m, have values in the range 5 to 7. No reaction has been found to occur in the laboratory between any of the ammonium (NH_4^+) ions and the CFMs. Laboratory measurements indicate that the reactions of the oxonium ions are unlikely to provide tropospheric removal times of CFMs less than 10^3 yr.

C. Stratospheric Processes (Appendix A, Section III)

In the stratosphere, physical removal of the halocarbons does not occur to any extent by deposition on aerosols or any other known process. Chemical processes that destroy halocarbons in the stratosphere lead to products that

decompose ozone and therefore constitute active removal.
An inactive removal process would have to produce a water-
soluble, chlorine-containing product so inert that it would
be transported into the troposphere and washed out before
it could react further to form ClO_x. No such process has
been identified.

The halocarbons are dissociated mainly by uv light at
wavelengths less than 280 nm. Only in the middle and
upper stratosphere are they not shielded from this radiation by ozone. Laboratory measurements have shown that
photolysis releases two chlorine atoms from F-11 and F-12
in their chemically reactive (ClO_x) forms. There is
laboratory evidence that the third chlorine atom is also
released from F-11.

Since $O(^1D)$ atoms have concentrations in the stratosphere about 500 times larger than in the troposphere and
react rapidly with CFMs, they will also contribute to
stratospheric removal of CFMs, amounting to about 1 percent
of that for photolysis in the case of F-11 and 10 percent
in the case of F-12. However, these reactions also lead
to ClO_x formation and therefore will contribute to active
removal. No reactions with measurable rate constants have
been found in the laboratory between the most prevalent
stratospheric ions and CFMs. Even if they did occur, they
would undoubtedly lead to ClO_x formation.

Since the catalytic chain for O_3 destruction involves
the chain carriers Cl and ClO, the efficiency of the chain
depends on the fraction of the "product" chlorine atoms
that are present in either of these forms (ClO_x), i.e., on
the value of the $[ClO_x]/[ClX]$ ratio, where ClX refers to
all the molecular products of halocarbon decomposition
that contain chlorine atoms. The main unreactive forms are
HCl and $ClONO_2$. This latter molecule removes NO_x as well
as ClO_x species, producing nonlinear, additive decreases
in ozone destruction by both catalytic cycles.

The primary uncertainty in this ratio arises from the
uncertainties in the rate constants for the well-established reactions (cf. Chapter 2 and Appendix A), which
largely determine the partitioning among ClX species. The
most uncertain of these rate constants are those for the
reaction between HO and HO_2 and for the photolysis and reactions of $ClONO_2$. Photodissociation of HCl and ClO, and
ion-molecule reactions, are unimportant in affecting the
partitioning.

In laboratory experiments on the reactions of Cl and
ClO in gases at low pressure, it is found that these
species are removed by glass surfaces coated with sulfuric

acid, typically at less than one collision in 10^4. On this basis, reaction with sulfuric acid aerosol will not be a significant sink for Cl or ClO.

Chlorine atoms do combine with NO and NO_2 in slow, three-body processes, but these reactions are much slower than those of Cl with CH_4 or H_2, which form HCl. The combination of ClO with NO_2 has also been included in recent calculations of the ozone reduction. Its earlier omission was based on qualitative estimates of the photolysis cross section of $ClONO_2$, which turned out to be smaller than expected. Further laboratory work is needed to clarify reactions that lead to the formation of higher oxides of chlorine and chlorine oxyacids, but it is unlikely that this will lead to substantial changes in the predicted value of the $[ClO_x]/[ClX]$ ratio.

Finally, mention should be made of potential catalytic ozone cycles by other halogen species. For bromine, the corresponding chain mechanism involving Br and BrO that destroys ozone appears to be fast. The only reaction capable of producing HBr is that between Br and HO_2. Therefore, the catalytic destruction of ozone by BrO_x appears to be more efficient (per molecule) than the ClO_x analog, but there is much less of it being released to the atmosphere. Knowledge of the relevant rate constants is less extensive for the BrO_x system than for ClO_x, and more work is needed.

An analogous FO_x chain will not act to destroy ozone because the steps that form HF are fast, and, once formed, HF does not regenerate FO_x.

III. SUMMARY

Most of the removal of completely halogenated compounds apparently occurs in the stratosphere and not in the troposphere. No significant inactive removal processes have been identified in the stratosphere, where the major removal process is photolysis. Stratospheric removal of CFMs thus invariably leads to products that catalytically destroy ozone, and the principal features of this catalytic cycle are fully supported by all laboratory measurements (see Appendix A).

Three processes have estimated inactive removal times for F-11 and F-12 that are short enough to warrant further, more detailed study. Lower limits of $\sim 10^2$ (70 and 200), 5×10^3, and 10^3 yr have been placed, respectively, on the removal times for solution in the surface waters of the oceans (followed by some unknown degradation process) and by photodissociation and ion-molecule reactions in the troposphere. If each of these processes actively removed F-11 and F-12 in the time corresponding to the *lower limit* set for it, the net effect would be about a 40 percent decrease of the predicted ozone reductions in the absence of such inactive removal.

5 TRANSPORT

I. INTRODUCTION

Transport, simply put, is the motion of parcels of air,
with whatever trace species this air contains, from one
location in the atmosphere to another. This motion is
irregular and occurs on many scales in time and space.
The transport of a species by atmospheric motions is much
faster than molecular diffusion and thus independent of
the molecular weight of that species. Transport contributes importantly to the ultimate disposition of chlorocarbons and, in particular, CFMs released at the surface of
the earth. First, before the chlorocarbons can be photo-
dissociated to form species able to react with ozone they
must be transported from the troposphere to the stratosphere. The time scale for destruction of CFMs by photolysis depends on transport of these chlorocarbons from
regions of very slow photolysis in the lower stratosphere
to regions of fast photolysis in the upper stratosphere.
Transport processes also control the time it takes for the
decomposition products to be moved from the stratosphere
to the troposphere, where they are removed by tropospheric
rain-out.

The term "transport model" refers to a method of representing transport in the mass conservation equation of
chemical species. Three classes of transport models can
in general be distinguished. First, there are *physical*
transport models that actually generate the individual
trajectories of air parcels, either by using observations

of the three-dimensional fields of atmospheric motion or by calculation of these fields in an atmospheric general circulation model (GCM). Second, there are *statistical* transport models that are derived from a physical model by some kind of averaging process. Finally, there are *empirical* transport models that do not utilize atmospheric motions directly but rather are based on observations of the transport-controlled time and spatial distribution of trace constituents. The assumed form of these models has invariably been that of an eddy-mixing model.

Physical transport (3-D) models could in principle be applied to the ozone-chlorocarbon coupling problem. However, the possibilities of such application have not yet been seriously explored, and there is little information that can be gained for the present problem from past studies of such models. Statistical transport modeling for the stratosphere on a global scale has been little explored as yet, and previous attempts have been of doubtful theoretical soundness. For further discussion of these two classes of models, see CIAP Monograph 1, Chapters 6 and 7, and CIAP Monograph 3, Chapter 4.

While there has been essentially no use of physical and statistical transport models for the present problem, it is, nevertheless, worthwhile to discuss briefly their basis to give perspective about the adequacy of the empirical models that acually have been employed. To do this we shall first discuss what is, or can be, measured in the way of atmospheric motions in the stratosphere. Second, we indicate how the transport provided by these motions enters into the continuity equations for atmospheric trace species and how averages over all the details in time and space are obtained.

We then turn to the means by which empirical models are derived from measured distributions of various trace species, thus short-circuiting the need for a detailed knowledge of atmospheric motions. Four possible difficulties with this approach are identified; they are

1. The data used to derive the model may not satisfactorily represent global mean values;

2. A model obtained to reproduce the distribution of chemical species with certain distributions of sources and sinks may not be satisfactory for modeling other species with different distributions of sources and/or sinks;

3. The procedures that have been used for inferring a transport model from a given chemical profile are not unique and appear rather subjective;

4. The chemical loss processes for the chemical species used to infer the transport model may be inadequately known.

Thorough exploration of the first two of these possible difficulties would require more data than presently available. However, an objective procedure for inferring the parameters of a transport model from data on the distribution of a chemical species is useful for estimating the sensitivity of transport models to these possible difficulties. Such a procedure, also necessary to assess the significance of the third and fourth difficulties, has been developed for this study.

Finally the role of transport in determining the distribution of ClX (Cl, ClO, HCl, and ClONO$_2$) from CFMs is discussed. The consequences of the large number of possible scenarios for CFM emission can be approximately evaluated using a few simple concepts established by analytical and numerical considerations given in Appendix B. The time-varying distribution of ClX generated by an instantaneous release is characterized by two time scales: the time required to achieve maximum ClX concentrations at high altitudes (t_1) and the time required for decay from these maximum concentrations (t_2). The variation in ClX concentration in the stratosphere from the time of release can be represented by a simple functional form depending only on these two parameters. It is shown that the effects of CFM releases over time scales of the order of 10 yr or less can be approximated by the effects of instantaneous releases occurring at some average time.

With these concepts, the evolution of ClX can be adequately and readily estimated for essentially all scenarios of interest in studying the effects of CFM emission. Hence, if the potential ozone reduction is known as a function of ClX concentration, it may be estimated for any CFM scenario. Both characteristic time scales are shown to be less sensitive to existing uncertainties in solar flux and CFM photodissociation cross sections than to transport. They may, therefore, be regarded as defined by a given transport model. By evaluating the variation of these time scales with possible uncertainties in the transport parameterization, we can characterize the uncertainties resulting from transport effects in the calculation of ozone reduction due to CFMs.

II. OBSERVED ATMOSPHERIC MOTIONS

Meterologists observe the atmospheric horizontal motions
and temperature variations on a day-to-day basis by track-
ing the horizontal displacement of ascending balloons
launched from several hundred sites around the globe.
These balloon soundings generally extend to an altitude
of approximately 30 km and are taken once or twice a day.
A limited number of rocket soundings (dropping some kind
of wind tracer) provide similar data to twice this altitude. Satellite global temperature observations up to
80 km are now also available to supplement this data base.
Individual observations are analyzed on daily global maps
for altitudes where there are sufficient data or over some
larger time interval, e.g., weekly, where the data are
fewer. These maps at different altitudes are used to obtain a smooth description of the 3-D time-dependent motions of the atmosphere on horizontal spatial scales
larger than 10^3 km and on time scales longer than a few
days. Vertical motions are generally calculated from the
temperatures and the horizontal motions, using dynamic and
thermodynamic constraints.

It must, however, be recognized that the atmosphere
contains a continuum of scales of motion in space and
time. The observation by meteorologists of large-scale
motions has generally been due to the importance of these
motions for such questions as weather forecasting and
aeronautical operations. It is not *a priori* evident that
horizontal motions on a scale smaller than a few hundred
kilometers can be neglected in treating global-scale
vertical transport in the atmosphere. However, the physical transport modeling of Mahlman (e.g., CIAP Monograph
3, Chapter 4, Section 5.1A) provides convincing evidence
that simulated large-scale transport alone can account
for the observed redistribution of ozone and nuclear debris clouds in the *lower* stratosphere.

The motions in the stratosphere on scales too small to
be resolved by the global observational networks have been
studied by sampling, but only in a limited number of locations and for limited times. None of these studies has
pointed to small-scale transport processes as being of
major importance for global transport modeling. At least
in the lower stratosphere, these studies generally indicate that small-scale motions are negligible for global
transport compared with motions on scales of several
hundred kilometers or larger. Prescriptions of the
eddy-mixing variety supposed to represent the effect of

motions with scale small compared with the resolution of
the model are, indeed, used in the GCMs, but their main
role is to smooth out small-scale irregularities rather
than contribute to global-scale transport.

The GCMs are viewed as the ultimate tool for determining stratospheric transport because of their generation
of motions from first principles. Such models should
eventually be able to provide many of the meteorological
feedbacks not present in the simpler models. However, as
far as we know, the uncertainty due to neglect of such
feedbacks as presently included in the GCMs is small compared with the other major uncertainties outlined in this
report. Because of the extreme difficulty of developing
and operating the large models, several years or more are
generally required for their application.

III.   TRANSPORT OF CHEMICALS--AVERAGING METHODOLOGY AND 1-D MODELS

The time evolution of a given chemical species is determined by solution of the mass conservation equation for
that species, allowing for the sources, loss processes,
and transport. Transport is given mathematically by the
product of the atmospheric velocity and the concentration
of the species being transported. As discussed earlier,
this transport is three-dimensional in space and fluctuating in time. If we were to release an inert tracer at the
surface, its subsequent distribution over times of months
or shorter would depend strongly on where it was released.
For example, the same amount of material released at the
equator, in polar northern latitudes, or in the southern
hemisphere would in each case be concentrated for several
months in different latitudes near where it was released.
This result follows from the finite amount of time required
to mix the material within the troposphere. The material
from these various releases would, furthermore, arrive in
the stratosphere with differing latitudinal distributions
and at somewhat different times, in part because of the
pronounced latitudinal variations in large- and small-
scale atmospheric motions. However, on time scales of
several years or longer, an inert gas released anywhere
at the surface becomes essentially uniformly mixed
throughout the troposphere and thus independent of its
initial state. The changes in ozone that are of most
interest here are those of global scale and persisting
for many years. Given the long time scales involved,
transport models that provide descriptions valid only on

a global scale and for long periods in time are believed
to be acceptable for evaluating the effects of CFMs on
ozone.

The empirical eddy-mixing models are of this nature.
Some features of their mathematical formulation are given
in Appendix B. It suffices here to note that (a) they
assume that transport by motions fluctuating in time and
space is determined by averaged gradients of the concentration of the species being transported, and (b) the
quantitative relationship between transport and concentration gradient is determined empirically on the basis of
the model being able to reproduce observed distributions
of chemical species. In principle, averages are made over
longitude to obtain two-dimensional (2-D) eddy-mixing
transport models and further averages over latitude telescope these down to one-dimensional (1-D) eddy-mixing models. Results connected with the CIAP program indicated
that 2-D models did not give very different answers than
1-D models, at that time, in predicting global mean ozone
reduction. Because of their relative lack of validation,
and greater difficulty in development, it was not clear
whether they should yet be considered any more reliable
than the 1-D models for this purpose. This is apparently
still true today. The 2-D models have been, however, useful for estimating latitudinal and seasonal variations of
ozone change. They also, in general, should provide a
more satisfactory means of dealing with latitudinally
varying chemical processes. Only the 1-D models have been
used for a study of the CFM problem, and so only these are
discussed further.

The 1-D model transport description is given by a
vertical-mixing equation with the mixing coefficient defined as a function of the altitude. This latter profile
is chosen to give close agreement between the predicted
and observed globally averaged distribution of some suitable trace species. Questions may be raised regarding
the uncertainties in the transport predicted by such a
parameterization, as mentioned earlier.

The data available for inferring a global 1-D mixing
coefficient in the stratosphere should be global and
seasonally averaged. In practice, they have been obtained
largely over one location in Texas, as discussed in Chapter 6. In the past, it has been suggested (Climatic
Impact Committee, 1975) that data obtained at various times
and locations in the lower stratosphere would probably best
be measured from the local height of the tropopause. This
treatment of the data essentially introduces averaging
over the surfaces of most rapid mixing, which are largely

parallel with the tropopause. Such an averaging procedure is, however, not without disadvantages, for it compli-
cates somewhat the introduction of solar photodissociation
processes. Furthermore, it is not suitable for discussing
transport in the troposphere or upper stratosphere.

For many problems, including those discussed in this report, the relatively slow transport through the lower
stratosphere appears to be a limiting factor and should
be treated as accurately as possible. As discussed further in Appendix B, a modification of the vertical coordinate so that it represents the distance to the tropopause within several kilometers of the tropopause but
represents actual altitude below 10 km and above 20 km
appears to maximize the advantages and minimize the disadvantages of measuring altitude from the height of the
local tropopause.

In the past, the data most suitable for inferring an
eddy-mixing coefficient over a suitable range of altitude
were limited to one average methane stratospheric profile.
Much of our analysis with regard to transport parameterization has been directed toward evaluating the errors that
are likely to be introduced in using these data to determine transport. No objective way is available to us for
determining the precise relationship of these particular
data to the actual long-term global mean profile. However, estimates may be made of uncertainties due to inadequate sampling, peculiarities of the measurement
location, and the limitations of the eddy-mixing concept.
This is done in Appendix B, where we infer from the "raw"
methane data possible "smoothed" and "stretched" profiles,
which we believe to span the possible range of actual profiles allowed by the data. The vertical profile of
nitrous oxide is also sensitive to transport. With the
addition of recent data (cf. Chapter 6 and Appendix C),
there now appears to be sufficient information on the
vertical distribution to N_2O to permit an independent inference of the eddy-mixing coefficient.

Since the eddy-transport parameterizations involve
little physical justification but primarily are simple
"curve fitting," the eddy coefficients most appropriate
for upward versus downward transport are unlikely to be
exactly the same. Unfortunately, no entirely suitable
tracers for studying downward transport are available.
Ozone itself is perhaps most useful, since it is generated
globally in the stratosphere under steady-state conditions.
The complexity of its chemical reactions in the lower
stratosphere, however, discourages its use for an accurate
transport parameterization. Debris from nuclear explosions

can be used, except there is no reason to expect transport parameterization from a local transient source to agree closely with that from a global steady source. Our conclusions do not appear inconsistent with the observed transport of ozone or nuclear debris, and any differences in upward versus downward transport coefficients are not believed to be a major source of error.

The optimum procedure for obtaining an eddy-mixing coefficient from a chemical species profile should be based on objective criteria as to the "goodness" of the agreement between observed and calculated global profiles. Such procedures were not previously available but have been developed, as detailed in Appendix B, to permit a thorough consideration of the uncertainties present in the 1-D transport parameterizations.

Eddy-mixing coefficients are necessarily prescribed for all altitudes, but the available chemical profile data contain only a limited amount of information. It is consequently necessary to assume constraints on the mixing coefficients in order to infer them from the data. Two approaches to such constraints are taken in Appendix B. First, mixing coefficients represented by a limited number of parameters are considered. Second, a smoothing procedure is developed. In applying the first approach, we find that the mean-square deviation between "observed" and calculated data is reduced by increasing the number of parameters, as would be expected. However, also as expected, oscillations in the mixing coefficient profile increase with an increased number of adjustable parameters, probably because of noise in the data. The second approach suppresses such oscillations.

We have obtained best fits to the "raw," "smoothed," and "stretched" methane profiles, using the first approach for a large number of possible parameterizations of the mixing coefficient profile and using the second approach for a wide range of possible smoothings. The range of 1-D mixing coefficients obtained by these two approaches is believed to be as wide as present chemical species data will permit for fixed methane lifetimes. Uncertainties in the methane lifetimes and hence transport parameterization may also be estimated. An additional estimate of the eddy-mixing coefficient with the same procedure and using an estimate of the globally averaged N_2O profile provides another independent approach to the transport parameterization. Thus, the uncertainties in the CFM-ozone problem due to transport can be explored. As further evidence that transport time scales drastically different from those derived here are very unlikely, we note that 3-D model simulations represented by Mahlman (1976) in terms of 1-D mixing coefficients give similar results to those derived in Appendix B.

IV. THE DEPENDENCE OF CFM-OZONE REDUCTION ON TRANSPORT PROCESSES

The primary role of transport in the present context is to determine the distribution in the vertical of the sum (ClX) of the Cl, ClO, HCl, and any other readily inter-converted species such as $ClONO_2$. If this ClX distribution is known, together with the distribution of other relevant species, chemical (and photochemical) processes alone determine the partitioning between the different ClX species and the consequent rates at which the ClO_x (Cl and ClO) catalytic cycle removes ozone.

The crudity with which atmospheric transport processes have been included in modeling the effects of CFMs on ozone has cast some doubt on the accuracy of estimates of ozone reduction. Whereas the Panel finds the transport treatment to be a significant source of uncertainty, the resulting uncertainty in ozone reduction is apparently smaller than that due to the current uncertainty in chemical processes, so that the application of 1-D models is both valid and useful. A detailed analysis of the dependence of the ClX distribution on transport given in Appendix B indicates the degree of uncertainty in estimates of CFM-ozone reduction due to uncertainties in transport parameterization. Crutzen (1975) and Wofsy *et al.* (1975) have previously shown that if CFM releases were abruptly curtailed, the reduction of ozone due to past release would continue to increase to a maximum at some time t_1, a decade or more later, and then would slowly decay over a time t_2 that is an order of magnitude longer.* Our study has led to the interpretation of t_1 as proportional to the time that is required for the Cl atoms in the CFM and the associated ClX to become essentially fully mixed at the altitudes in the stratosphere where ClX determines the amount of ozone reduction.

The calculations reported in Appendix B show that during the time period t_1, only a few percent of the CFM

---

*t_2 is essentially the same quantity as that labeled τ_0 in Chapter 4.

released into the atmosphere at the beginning of that
period is removed by stratospheric photolysis. Furthermore, the CFM concentrations are small compared with ClX
concentrations at levels where ClX significantly influences
ozone concentrations. It follows that the ClX concentrations at t_1 are essentially determined by the complete
mixing of total Cl atoms previously released. Consequently, the maximum ozone reduction after an instantaneous release is quite insensitive to the details of the CFM
transport. All that matters is that the total Cl atoms
become mixed by the transport. (The maximum ozone reduction may still, however, depend on the transport of other
chemical species, especially NO_x in a particular model.
Also, this maximum is not to be confused with the maximum
steady-state reduction, which according to our analysis
is sensitive to transport.)

The times t_1 and t_2 do depend significantly on the
details of the transport. The time t_1 appears to be
controlled largely by the rate at which the CFMs move
upward through the region of relatively slow transport
in the lower stratosphere, and t_2, by the rate at which
ClX moves downward divided by the fraction of total Cl
in the form of ClX. In the 1-D transport models and under
near steady conditions, the upward transport of CFM must
essentially equal the downward transport of ClX. However,
only approximately 10 percent of the CFM released into
the atmosphere need be transported above 16 km during t_1
to achieve a state of nearly complete stratospheric mixing,
simply because only 10 percent of the atmosphere's total
mass lies above this level. On the other hand, the decay
time t_2 is by definition the period required for most of
the total CFM to be transported to the stratosphere, converted to ClX, and then transported back to the troposphere.
Consequently, t_2, the decay time, should be an order of
magnitude longer than t_1, the mixing time. Numerical
studies with instantaneous releases as described in Appendix B show the ration of t_2/t_1 to lie between 5 and 8
for a large number of assumed transport parameterizations.

A very important point to note is that the steady-state
amount of CFMs in the stratosphere is derived from t_2
simply by multiplying the global CFM emissions by t_2
[e.g., Chapter 4 and Eq. (B.27)]. Hence, the steady-state
reduction of ozone by ClX is proportional to t_2 provided
other trace species concentrations, in particular the NO_x,
remain fixed. This is the basis that we have used for
evaluating the uncertainty in steady-state ozone change
due to uncertainty in transport parameterization. Self-
consistent chemical models as discussed in Chapter 7,

which have fixed boundary conditions, are, on the other hand, much less sensitive to transport changes than would be indicated by our analysis. The reason for this behavior appears to involve the sensitivity of calculated ozone change to model NO_x concentration. Ozone change appears to depend approximately on the ratio of ClX to NO_x concentrations, which varies much less with transport than either the ClX or NO_x concentrations individually.

Our task in evaluating the uncertainties in ozone reduction predictions due to transport is greatly simplified by the recognition of the result established here that for long-lived species mostly destroyed in the stratosphere such as the CFMs the role of transport in determining their concentrations can be effectively summarized by specifications of the two time scales referred to above. Our job then is to determine the range of these time scales that is consistent with our present understanding of transport. The approach used has been to evaluate t_1 and t_2 for a large number of eddy-mixing profiles established by fitting in various ways different versions of the observed methane data or by fitting our global average of the available nitrous oxide data (as detailed in Appendix B). The large number of cases considered in Appendix B show that for fixed methane lifetimes there is at most a factor of 2 spread in t_1 and t_2. In view of the comprehensiveness of the cases considered, we believe it extremely unlikely that some further eddy-mixing profile could be found that would still permit satisfactory modeling of the vertical methane distribution but would give time scales for the CFM problem lying outside the range of the cases already considered. There is approximately a factor of 2 uncertainty in the methane lifetimes primarily due to the uncertainties in HO concentrations, the rate at which $O(^1D)$ destroys methane, and HCl concentrations (Chapters 4 and 6 and Appendixes A and C).

The possibility of a large difference between the observed methane and the actual global mean methane profile has been allowed for in our analysis. Some departures from our results might result if global methane were to exhibit more drastic departures from observed values than we have regarded as likely. It is unlikely, however, that our factor of 2 estimate as to the range of possible time scales due to uncertainties in the methane concentrations could be exceeded. For example, an eddy-mixing coefficient taken to be independent of altitude, and chosen to give satisfactory agreement with the methane data at 50 km, would require methane between 15- and 30-km altitude to drop by only 20 percent of the observed drop and yet would give time scales that are no more than 30 percent less than the lowest values inferred by fitting the methane profile. Data on the concentration of N_2O over the equator obtained recently by the National Oceanic and Atmospheric Administration show a remarkably slow decrease with altitude. Adoption of this profile as representative of the global mean gives by our analysis a factor of 3 faster transport than inferred from N_2O data over Texas. However, an estimated global mean profile based on all available data indicates time scales closer to the latter and close to those inferred from methane (cf. Appendix B).

These characteristic times of the CFM problem are also somewhat dependent on the solar flux and CFM absorption cross sections. The solar flux at a given level is sensitive to overlying ozone, which varies with changes in ClX concentration. The uncertainties in these quantities introduce less uncertainty than does transport. The insensitivity to cross sections is evident in comparing F-11 and F-12, which have nearly a factor of 10 difference in photodissociation lifetimes but less than a factor of 2 difference in their characteristic residence times.

The above conclusions were inferred on the basis of a numerical study of instantaneous releases. However, a study of different schedules of CFM release over extended time periods indicates that the effects on ozone of a release occurring on a time scale of 10 yr or less can be approximated by the effects of an instantaneous release. The complete time history of ClX evolution in the stratosphere subsequent to an instantaneous release may be approximated by an analytic expression that depends only on t_1 and t_2 as derived in Appendix B.

The role of these two time scales can be visualized by considering the atmosphere to be two boxes, representing the troposphere and stratosphere, connected by a small pipe. A gaseous substance is transported between these boxes at a rate proportional to the differences in relative concentrations (mixing ratios) in these two boxes. The proportionality constant is determined by the diameter of the pipe, which represents the region of minimum mixing in the lower stratosphere. A CFM released in the troposphere leaks into the stratosphere, where it is destroyed by photodissociation processes proportional to its concentration in the stratosphere. At some time t_1, the mixing ratio of CFM in the stratosphere is brought so close to that in the troposphere by transport that the net rate of supply of CFM from the troposphere has decreased to the rate of loss of the CFM by photodissociation.

As the transport continues to supply CFM to the stratosphere to replenish that lost by photodissociation, the
CFM in both boxes decays in unison on a time scale t_2.
At the same time, a reservoir of ClX is maintained in the
stratospheric box by the products resulting from the continuing destruction of CFM. This amount of ClX is determined by a quasi-steady balance between its production
and its loss to the tropospheric box, where it is washed
away. The ClX concentration thus remains proportional
to the CFM concentration, decaying accordingly while at
the same time participating in the catalytic destruction
of ozone.

In conclusion, the time history of ClX in the stratosphere for the various scenarios of interest in this report can be characterized by two time scales; these
together characterize the times required for mixing and
slow decay of the CFM. Our present estimates of these
times are made from 1-D eddy-mixing models inferred
from observed profiles of methane and nitrous oxide.
Table 5.1 shows our best estimates of most likely values
and uncertainty range for t_2, the time required for the
CFM burden to decay by a factor of 2.7 from its maximum
concentrations in the absence of further sources or a
tropospheric sink. Uncertainties incorporated in Table
5.1 result from the following considerations: errors in
estimating global mean profiles of the observed species
and errors in estimating their stratospheric lifetimes,
the consequences of representing horizontally varying

TABLE 5.1 Lifetimes t_2 in Years, for Factor of 2.7 Reduction of the CFM Burden in the Absence of Sources or
Tropospheric Sinks[a]

	F-11	F-12	Appendix B[b]
Most likely value	54[c]	80	60
Upper limit	80[c]	120	90
Lower limit	27	40	30

[a]The upper and lower limits represent subjective 95 percent confidence limits.
[b]Values for a hypothetical CFM used in most of the calculations of Appendix B.
[c]These most likely values are not significantly different
from the 50- and 90-yr values given in Table 4.2.

quantities by global means without accounting properly for nonlinear chemical terms and of representing the three-dimensional transport processes by a 1-D diffusion expression. With regard to this latter concern, we would expect some differences between upward and downward transport time scales, but these are unlikely to be as much as a factor of 2. Some further relatively small uncertainty results from probable errors in solar fluxes and CFM cross sections.

The concepts introduced in this chapter show how numerical results for a few scenarios can be used to infer the consequences of other scenarios of interest. In particular, the effect of additional CFM releases over the next few years beyond that previously released should vary linearly with the amount. The CFM burden in the atmosphere at the end of 1973 was equivalent to 8 years' release at the 1973 rates. A continuation of 1973 rates for another 8 years before cessation at the end of 1981 would essentially double the maximum decrease of ozone from that which would have occurred after immediate stoppage in 1974. The maximum ozone reduction that would result from other release schedules can be easily estimated by such scaling arguments. Other parameters such as variations in the change of uv light reaching the earth's surface with different CFM release schedules can be estimated by similar arguments.

REFERENCES

CIAP Monograph 1. 1975. *The Natural Stratosphere of 1974*. A. J. Grobecker, ed. Final report, U.S. Dept. of Transportation, DOT-TST-75-51, Washington, D.C.
CIAP Monograph 3. 1975. *The Stratosphere Perturbed by Propulsion Effluents*. A. J. Grobecker, ed. Final report, U.S. Dept. of Transportation, DOT-TST-75-53, Washington, D.C.
Climatic Impact Committee. 1975. *Environmental Impact of Stratospheric Flight: Biological and Climatic Effects of Aircraft Emissions in the Stratosphere*. National Academy of Sciences, Washington, D.C.
Crutzen, P. J. 1975. Testimony before the Committee on Aeronautical and Space Sciences of the U.S. Senate. Updated for the New York Assembly Standing Committee on Environmental Conservation of the Environmental Impact of Chlorofluoromethanes.

Mahlman, J. D.   1976.   Some fundamental limitations of
  simplified transport models as implied by results from
  a three-dimensional general circulation tracer model.
  To appear in Proc. 4th CIAP Conference.
Wofsy, S. C., M. B. McElroy, and N. D. Sze.   1975.   Freon
  consumption: implications for atmospheric ozone,
  Science 187:1165-1167.

6 ATMOSPHERIC MEASUREMENTS

I. INTRODUCTION

There are a number of reasons why the measurement of
minor constituents in the atmosphere plays a central role
in assessing the effects of CFMs on ozone. First, since
most of the CFMs that have been released are located in
the troposphere, accurate measurements at a given time of
their global content in this region of the atmosphere can
be compared with the amounts released. Such comparisons
can, in principle, be used to indicate the presence or
absence of any significant tropospheric removal processes
(sinks), which would lessen the effects of CFMs on the
stratospheric ozone.

Second, atmospheric measurements of some of the minor
atmospheric constituents provide essential input data for
the atmospheric models used to predict the effect of CFMs,
while others provide stringent tests for the validation
and predictions of the models. The most obvious inputs
are the fluxes and the tropospheric concentrations of the
halocarbons from both natural and anthropogenic sources.
However, tropospheric measurements of other gases also fall
into this category. Nitrous oxide, N_2O, measurements are
used to calculate the production rates of the oxides of
nitrogen in the stratosphere, which are intimately involved
in ozone chemistry. Moreover, the transport rate used in
the model is generally calculated from globally averaged,
experimental atmospheric profiles, i.e., the concentrations of chemical species whose chemistry is believed to
be well understood are measured as a function of altitude,

preferably for a wide enough range of latitudes and conditions to provide a suitable global average. Methane and nitrous oxide have been used for this purpose, as discussed in Chapter 5, although the data are limited in extent.

The validity of the models used in this study is best tested by comparing their predictions with measurements, in the stratosphere, of those photochemically related species that determine the amount of ozone in the natural stratosphere. The natural chemical destruction of ozone is dominated by the catalytic cycle involving NO and NO_2 as discussed in Chapter 2 [Reactions (2.10) and (2.11)]. The rate at which ozone is destroyed is, in turn, dependent on the amount of these species produced from N_2O [Reaction (2.13)] and on the fraction of these species that is converted to inactive HNO_3 by HO [Reaction (2.14)]. Existing observations of O_3, O, NO, NO_2, HNO_3, and HO (as summarized below and detailed in Appendix C) confirm our overall understanding of the role of nitrogen oxides in destroying ozone. A more quantitative picture will require considerable improvements in the temporal and spatial extent of the observations.

The predictions of the perturbations caused by an increase in the chlorine content of the stratosphere also require knowledge of the concentrations of the reactive species containing hydrogen (HO_x) or nitrogen (NO_x). The HO_x chemistry couples directly to the chlorine chemistry by conversion of the inactive HCl to the active Cl atoms by reaction with HO [Reaction (2.18)]. The NO_x and chlorine chemistry are coupled through the reaction of ClO with NO [see Appendix A, Section III.D, Eq. (3)] and through the formation of $ClONO_2$ from ClO and NO_2 [Reaction (2.19)]. Thus the comparison between measured and calculated concentrations of the various HO_x and NO_x constituents are important checks on the models.

Finally, the direct measurement of chlorine compounds in the stratosphere provides answers to basic questions of whether the chlorine chemistry is essentially complete or whether any competitive processes have been overlooked. For example, are the products of the processes indeed Cl, ClO, HCl, and $ClONO_2$, and are the amounts of the compounds observed consistent with the amounts predicted by the models? Measurements of these constituents are just beginning to become available.

II. GENERAL METHODS

Measurements of the concentrations of atmospheric constituents provide much of the data base on which our understanding of the atmospheric chemistry important for the present problem is founded. Since the measurements are difficult, sometimes pushing the limits of the technically feasible, we will briefly describe the current techniques and their limitations. Basically, the measurement techniques can be divided into two groups, remote optical sensing and *in situ* measurements. The remote-sensing techniques measure absorption and emission in the infrared region and absorption in the visible (or near uv) region, using the sun, moon, and light scattering from the zenith sky as light sources; emission in the submillimeter region is also used.

To obtain the required sensitivity, the infrared absorption measurements are made during twilight conditions, when long optical path lengths are available. The data therefore always represent an average concentration, possibly over varying atmospheric conditions. Height profiles below the altitude of the balloonborne instrument may be calculated from observations of the total column absorption with increasing solar zenith angle (i.e., from a series of measurements made as the sun sets). Such infrared measurements are capable of high sensitivity but are subject to a number of difficulties in interpretation. High-quality spectral data are needed, such as line shapes and strengths; these are not constant along the viewing path because of inhomogeneities in temperature and pressure. Moreover, care must be taken to prevent interference from other absorbing species. Also, there are computational difficulties in converting the measurements to height profiles. Finally, the need to make measurements during twilight conditions precludes the use of this technique for studying diurnal variations. On the other hand, under twilight conditions the time variations are expected to be greatest, again making interpretation difficult. The emission measurements are subject to many of the same spectral difficulties but can, in principle, be made throughout the day and night. An additional advantage is that height profiles may be obtained from measurements taken as a function of balloon altitudes. Also, the required path length is shorter.

In situ measurements of concentrations are made either directly in the atmosphere or by collecting air samples of the atmosphere, which are later analyzed in the laboratory. They usually do not present interpretational problems, but contamination by the instrument package or platform may pose difficulties that require careful consideration. Interference from other trace gases occasionally presents difficulties. This is especially true in those sampling methods that rely on an unspecific enrichment procedure as, for example, wet scrubbing or sampling with air filters. Measurements involving whole air samples that are subsequently analyzed in the laboratory are faced with problems of possible interaction with the container of the constituents being measured.

The platforms used to carry the instruments for *in situ* measurements at high altitudes include aircraft, balloons, and rockets. Aircraft have a limited altitude range and can only collect data up to about 20 km. They have, however, the great advantage of extended horizontal coverage and have provided latitude-altitude cross sections for a number of trace gases. Balloons have a much higher altitude capability--up to 50 km. Balloons are limited, however, to one vertical profile at a given time. To obtain latitude-altitude cross sections, balloon launches are required at various latitudes, as is the case for rocket launches. Also observational schedules need to reflect longitudinal and seasonal variability, whatever the platform used.

Remote sensing of the atmosphere by satellites is highly promising as a means for continuous global data coverage with good vertical and horizontal resolution. A number of trace gases, notable O_3 and H_2O, can be monitored in this way. So far, only O_3 data have been obtained, and they are not sufficiently accurate. It is likely that remote sensing from satellites will require calibration from local profiles obtained by *in situ* measurements. Remote-sensing measurements can also be made from the ground for certain constituents. The total number of molecules in a vertical column have been obtained for NO_2 and HO. This technique has also been used to derive total column amounts of NO_2 in the stratosphere and in the troposphere separately and to determine the altitude of the maximum concentrations.

III. SUMMARY OF OBSERVATIONS*

A. *Ozone and Atomic Oxygen (Appendix C, Section II)*

Observations of the total ozone column have been carried
out by an extensive surface network of about 60 spectroscopic instruments (largely in the northern hemisphere)
since the International Geophysical Year (1957-1958).
These instruments have also been used to obtain vertical
profiles of limited resolution from backscattered light.
More detailed vertical profiles have been provided by
instruments on balloons and rockets. Furthermore, for
the past several years, satellite instruments have
measured on a global basis the backscattered uv radiation
(partly absorbed by the ozone) and infrared radiation
emitted by the ozone, both of which are related to the
vertical ozone distribution.

The natural distribution of ozone and its fluctuations
in time and space are basic to the assessment of any
changes brought about by man. Most of the measurements
reported so far have come from the ground-based network.
Typical results are shown in Figures C.1 to C.4 and in
Figure 9.1. The measuring stations are irregularly distributed over the earth's surface, and there are considerable difficulties in obtaining reliable records from
different stations over long periods that are on a readily
compared basis. However, over the last 40 yr or so,
a general picture of the variations in the total ozone
column with longitude, latitude, and time has emerged,
along with explanations for the main features of the data.
The daily variation can be of the order of 50 percent;
the seasonal mean variation may be 30 percent. The variability increases greatly with latitude and in higher
latitudes varies with the season. These changes in the
ozone column are due to the passing of the seasons, the
creation of ozone in the tropics, and the general circulation of the atmosphere.

The large localized variations in the ozone column obscure any smaller long-term trends of a global character.
Nonetheless, a number of efforts have been made to average
out the daily, seasonal, and geographical fluctuations;
for example, Figure 9.1 shows the results of one such
average over the northern hemisphere (the southern

---

*Details and references for observations may be found
in the corresponding sections of Appendix C.

hemisphere has too few stations). There is some evidence of a small fluctuation that follows the sunspot cycle (11-yr period) by 1 to 4 yr (Christie, 1973; Angell and Korshover, 1973; Hill and Sheldon, 1975). The results are conflicting, and at the moment we simply call attention to Figure 9.1 in which the long-term variation of the average total ozone in the northern hemisphere is about ±5 percent. Also, there appears to have been an *upward* trend of about 10 percent over the past 15 yr, which now shows signs of reversing (Angell and Korshover, 1975). The fluctuations also covered a 10 percent range in the 1935-1950 period before the CFMs became widely used.

Efforts are under way to identify a human contribution to the long-term variations observed (Parry *et al.*, 1976; Hill *et al.*, 1976). However, there is at present no sound basis for predicting the natural irregularity in the ozone column. Furthermore, as discussed in Chapter 8, the ozone reduction at present by the CFMs is probably no more than 1 percent compared with the natural variation range of 10 percent. Therefore, no independent conclusions can be drawn about how much, if any, of the observed trend in the ozone column is attributable to the CFMs.

Atomic oxygen (O) is closely coupled to ozone chemically in the stratosphere, but it is more difficult to measure. At night, essentially all the O is removed, largely by combination with O_2 to produce O_3 [cf. Reaction (2.2)] but also by reaction with NO_2 [cf. Reaction (2.10)]. During the day, photodissociation processes, i.e., Reactions (2.1), (2.3), (2.12), balance these losses to establish a steady-state concentration of oxygen atoms depending on the photodissociation and reaction rates involved. As a result, measurements of ozone essentially establish the concentration of atomic oxygen. Independent observational confirmation of the oxygen atom is, however, desirable; such a measurement has recently been obtained.

B. Nitrogen Compounds (Appendix C, Section III)

Several vertical profiles have recently been obtained for NO using an *in situ* chemical technique. Other measurements have been made by remote sensing of the solar radiation absorbed by NO. Both techniques employed balloon platforms. Considerable variability of concentrations has been noted, which is probably due in part to conversions between NO and NO_2. The ratio of these two quantities varies especially rapidly near sunset because of the variation of solar photodissociation of NO_2. The species

NO_2 has so far only been measured by various remote-sensing techniques. Considerable NO variability should also be expected due to the lower stratospheric variability of $NO + NO_2 + N_2O_5 + HNO_3$. Such variability is expected to be high because the source of NO is in the middle stratosphere and the sink is below.

Nitric acid (HNO_3) has been measured remotely with infrared techniques and *in situ* by a filter-paper collection technique. The latter method has been employed to determine latitudinal variations, which appear similar to those of ozone.

Nitrous oxide (N_2O) has been measured by a number of techniques in the stratosphere. The most accurate measurements have been obtained from balloons with cryogenic collection of air samples and chemical analysis by mass spectrometer or gas chromatography.

In summary, measurements of NO, NO_2, and N_2O are now adequate to establish approximate vertical profiles of these species. The uncertainties in comparing these profiles with calculated results arise from variations in the observations due to natural fluctuations and instrumental differences. In addition, the available observations very likely are not a representative sample for estimating global mean values. The relative vertical and latitudinal variations of HNO_3 are better established by observation than are those of NO and NO_2. However, there is some question as to the absolute concentrations of HNO_3 because of disagreement between results of the remote and *in situ* techniques. The measurements of all these compounds show that the stratospheric height profiles exhibit both long- and short-term variability. A large number of ground-based, remote-sensing measurements also indicate variability in the total column amount and in the altitude of maximum concentration.

C. Hydrogen Compounds (Appendix C, Section IV)

Measurements of the hydrogen compounds H_2O, CH_4, and H_2 contribute in two ways to our understanding of the present problem. First, the methane data have been used extensively for obtaining the vertical transport rates used in calculating the effects of CFMs on ozone, as discussed in Chapter 5. Second, all three compounds affect the chemical processes by which the CFMs perturb the stratospheric ozone, as described below.

These hydrogen compounds react in the stratosphere with $O(^1D)$ to yield the radicals HO and HO_2. In turn, the latter two species modify the ozone distribution in a number of ways: (a) they themselves are significant although not so important as the oxides of nitrogen in the natural catalytic destruction of ozone [cf. Reactions (2.5) and (2.6)]. (b) HO concentrations determine in part the chemical lifetime of CH_4, whose decrease in the stratosphere provides one of the primary reference points for the empirical determination of transport rates. (c) HO removes NO_x from its active catalytic role in destroying ozone [cf. Reaction (2.14)]. (d) HO converts unreactive HCl back to the ozone-destroying ClO_x species [cf. Reaction (2.18)]. Evidently, if HO concentrations were to decrease, the NO_x compounds would destroy relatively more and the ClO_x relatively less ozone. (e) There is also an indirect dependence on HO since the ClO_x and NO_x cycles are coupled to each other and each depends on the HO concentration in different ways.

Profiles of CH_4 have been established using an infrared technique and *in situ* sampling. The latter is the more accurate of the two procedures and has been carried out a sufficient number of times to provide a good description of the behavior of methane over eastern Texas (32° N). Single measurements of CH_4 at about 46 and 50 km by a rocket "freeze-out" cryogenic sampling technique extend the profile to the top of the stratosphere. Without much information on methane elsewhere, it is necessary to use these results to estimate global mean conditions. It would seem that ~30° is a good latitude to sample because it bisects the global area between the pole and equator. However, there exist at that latitude (at least in the lower stratosphere) unique meteorological conditions that would tend to carry air with lower mixing ratios of methane downward and so distort the vertical profile as compared with global mean values.

Water vapor has been measured extensively at one location up to 30 km (Washington, D.C.) for the past decade. Only a single profile, remotely sensed from a rocket, has been obtained above that level.

Molecular hydrogen has been measured several times over eastern Texas up to 35 km and by a rocketborne cryogenic instrument at higher levels.

In 1971, the first set of HO measurements established a profile extending from 45 to 70 km. Additional measurements have since been made between 30 and 43 km. Since the major sink for HO in the upper stratosphere is the reaction

$$HO + HO_2 \rightarrow H_2O + O_2$$

the concentration of HO in this region is strongly dependent on the rate of this reaction. Past estimates of the value of its rate constant have ranged from 2×10^{-11} to 2×10^{-10} cm^3 sec^{-1}. However, to match the measured values of stratospheric HO concentrations, the model calculations require the lower value for the rate constant (i.e., $\sim 2 \times 10^{-11}$), which we prefer for the independent reasons cited in Appendix A.

The role of HO in the troposphere is also quite important, since reactions of HO with partially hydrogenated halocarbons and unsaturated halocarbons provide a significant sink for these compounds. Tropospheric measurements of HO concentrations close to the ground have been made by two research groups. More recently, measurements have been made in the upper troposphere at 7 and 11.5 km, establishing the fact that HO is present in the troposphere in concentrations close to those predicted by model calculations.

D. Halogen Compounds (Appendix C, Section VI)

A rapidly increasing number of measurements has been made of the chlorinated methanes $CFCl_3$, CF_2Cl_2, CH_3Cl, and CCl_4 during the past few years using gas chromatography with an electron-capture detection method developed by Lovelock. Most of the measurements have been of $CFCl_3$.

The tropospheric measurements, in particular, show that the concentration of F-11 has been increasing. Pack et al. (1976), for example, who have made extensive measurements of F-11 in both the northern and southern hemispheres find that the rate of increase in F-11 since 1970 is 13.2 percent per year in the northern hemisphere and 14.6 percent in the southern hemisphere. This is close to the 13.9 percent per year worldwide growth in F-11 production from 1965-1974 cited in Chapter 3. In addition to the tropospheric measurements, two meridional cross sections (i.e., measurements as a function of latitude) have been obtained in the lower stratosphere, and vertical coverage up to 35 km is now available. There are also several vertical profiles of CF_2Cl_2.

The tropospheric measurements are of most interest in determining whether any inactive removal processes (sinks) for CFMs exist, which would lessen the effects of these compounds on the stratospheric ozone. The stratospheric measurements serve to test the concept that photolysis is the main active removal process in that region of the atmosphere.

Two approaches can be used to obtain indications of the extent to which tropospheric sinks of CFMs exist. One is to compare the rate of increase in the concentration of a particular CFM with the rate predicted from its release rate, assuming that the only loss process is transport into the stratosphere and subsequent photolysis. Any difference between these rates could be indicative of tropospheric sinks. Efforts in this direction have been made recently by Sze and Wu (1976) and by the Manufacturing Chemists Association (1976) for $CFCl_3$. Most of the measurements have been made during the past 2 yr (1974-1975), but some cover the 1970-1973 period. The results of this comparison are inconclusive for a number of reasons. The spread in reported measurements is large for $CFCl_3$ and even larger for CF_2Cl_2, and the accuracy of the measurements is less than ±30 percent *(vide infra)*. It has not yet been established how much of this spread is due to inaccuracies in the measurements and how much is due to fluctuations at the measuring site by localized sources or meteorology. With the existing accuracy, the time base (5 yr) is too small to provide significant comparisons between measured and predicted growth curves. Any differences between these curves will increase with time so that either increased accuracy or longer observation times are required. Moreover, the curves are directly dependent on the choice of eddy-diffusion coefficients used to parameterize the transport of the tropospheric CFMs into the stratosphere.

There have also been two stratospheric measurements of F-11 and F-12 obtained by a remote-sensing infrared technique (Williams *et al.*, 1976a), one in 1968 and one in 1975. The increased concentration between these two dates (20 ppt in 1968, 49 ppt in 1975 at ~19 km) are not inconsistent with the increase in release rates and with stratospheric photolysis providing the only significant removal process. But the accuracy is insufficient to rule out the presence of significant tropospheric sinks.

The second and perhaps preferable approach to determination of the extent to which tropospheric sinks exist is to compare the global atmospheric inventory with the total release up to the time of the inventory. This approach is considered in detail in Appendix E. Globally averaged measurements of the CFMs in the troposphere are essential for this purpose. The magnitude of the observational task is suggested in Figure 6.1, which presents contour maps showing the dependence on latitude and altitude measured for F-11 in early and late 1974. Globally

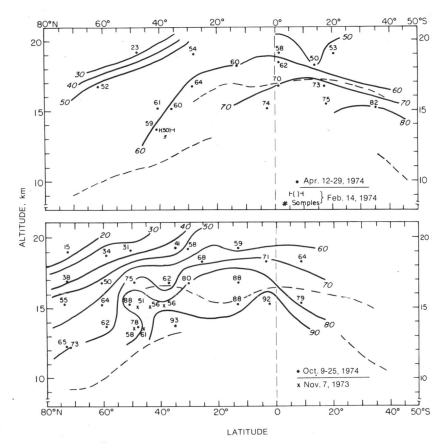

FIGURE 6.1 Contour maps of the concentration of CFCl₃
in parts per trillion by volume, measured for different
altitudes and latitudes. Dashed lines denote the average
observed positions of the tropopause during the sampling
periods. The contour lines are subjectively drawn between
the observed points (after Krey and Lagomarsino, 1975).

averaged measurements of CH₃Cl and CCl₄ are desirable in
order to assess the total chlorine flux into the stratosphere and hence determine the percentage increase to
be expected from continued release of CFMs.

Unfortunately, the sampling, analytical and calibration
uncertainties, and inadequate geographic coverage prevent
an accurate determination at this time of the global mean
tropospheric content of CFCl₃ and CF₂Cl₂. Several attempts have been made recently to compare the analyses of

the same samples by different groups engaged in atmospheric halocarbon measurements (Workshop on Halocarbon Analyses and Measurements, Boulder, Colorado, 1976). The results indicate that the agreement is about ±30 percent for $CFCl_3$ and CF_2Cl_2, ±50 percent for CH_3Cl, and a fourfold range in the case of CCl_4. The question of absolute standards for the measurements has yet to be resolved, but it is likely that this will not appreciably increase the range of uncertainties given above.

Rasmussen (1975) and co-workers have made the most extensive measurements of these halocarbons in the troposphere, and their results have been used to obtain the approximate global averages discussed in Appendix E and the average tropospheric profiles shown in Figures 6.2 to 6.5. The latter give the impression of rather uniform distributions; however, individual profiles do show considerable scatter, somewhat below ±10 percent in the case of Rasmussen's data. Although much of that scatter is caused by instrumental error or calibration problems, the rest of it must be due to variations caused by the highly localized release and nonequilibrium distribution of these man-made compounds. Furthermore, the CFMs are released predominantly in the northern hemisphere, which generates large systematic latitudinal variations and increases the importance of having good measurements for the southern as well as the northern hemisphere. In addition, more profiles in the middle and upper troposphere, where mixing is more extensive would help to better define the global average. With such wider-scale measurements it should be possible to reduce the uncertainty in the total tropospheric burden of halocarbons to less than 10 percent.

As of September 1975, Rasmussen (1975), who has made a large number of halocarbon measurements in remote areas, reports tropospheric averages for the northern hemisphere of 123 ppt for $CFCl_3$, 208 for CF_2Cl_2, 111 for CCl_4, and 449 for CH_3Cl. Measurements of F-11 in the southern hemisphere have been much more limited. Rasmussen (1975) has found, in measurements off the coast of Peru and Equador, 20 percent less F-11 than in the northern hemisphere. This is in good agreement with Lovelock (1975), who has made one set of southern hemispheric measurements and found that they are 23 percent lower than northern hemispheric values. In addition, measurements of radioactive krypton released in the northern hemisphere indicate that the southern hemispheric concentrations are also 20 percent lower (Telegadas and Ferber, 1975). Using the

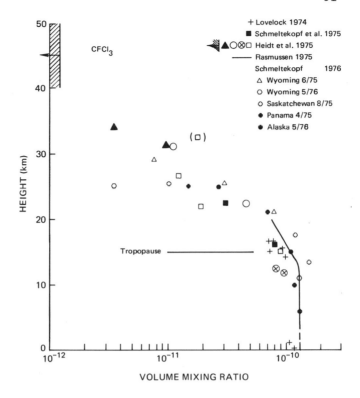

FIGURE 6.2 Vertical distribution of the CFCl$_3$
volume mixing ratio. The data are from various authors and collected in different years
and locations (see Appendix C, Section VI.A).
The sampling altitudes are plotted relative
to the tropopause (assumed height 15 km) to
facilitate comparison. For clarity, the 1975
data of Rasmussen are indicated by a line denoting the average of several measurements.

above information, the concentration of F-11 in the
southern hemisphere is (0.78)(123) = 96 ppt. Following
Rowland and Molina (1975), one may *estimate* a global
average that gives

$$\frac{123 + 96}{2} = 110 \text{ ppt}$$

The Manufacturing Chemists Association (1976) has used
a somewhat different method for estimating global inven-
tories of F-11. They use the data obtained from a cruise

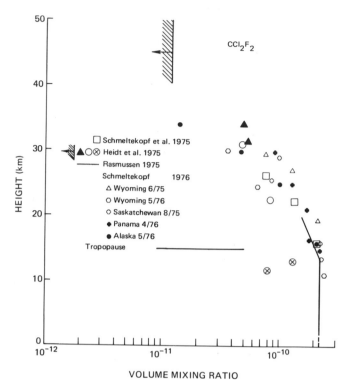

FIGURE 6.3 Vertical distribution of the CF_2Cl_2 volume mixing ratio. The data are from various authors and collected in different years and locations (see Appendix C, Section VI.B). The sampling altitudes are plotted relative to the tropopause (assumed height 15 km) to facilitate comparison. For clarity the 1975 data of Rasmussen are indicated by a line denoting the average of several measurements.

made in 1972 between latitude 65° S and 60° N to derive latitude conversion factors. They then apply these conversion factors to measurements reported by Pack *et al.* (1976) at 51° N and at 38° S to derive 1975 global tropospheric burdens of F-11. The value obtained by this procedure agrees with that given above within 17 percent, well within the uncertainty of the measurements.

Stratospheric profiles of $CFCl_3$ and CF_2Cl_2 (Figures 6.2 and 6.3) show some variation, but sufficient

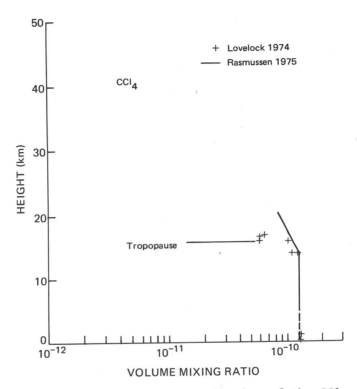

FIGURE 6.4 Vertical distribution of the CCl₄
volume mixing ratio. So far stratospheric
data are restricted to the lower stratosphere.
The sampling altitudes are plotted relative
to the tropopause (assumed height 15 km) to
facilitate comparison. For clarity the 1975
data of Rasmussen are indicated by a line denoting the average of several measurements.

measurements have not been made to give clear indications
of whether these variations are due to seasonal or latitudinal effects. Since other minor constituents show such
variations in the stratosphere, these measurements can also
be construed as showing nonuniformity of the stratosphere
and the effect of stratospheric motions. The stratospheric
profiles are consistent with photodissociation occurring
in the upper and mid stratosphere as discussed in Chapter
7 (compare Figures 7.12 and 7.13). The observed mixing
ratios drop off more rapidly in the stratosphere than is
predicted on the basis of release data and transport

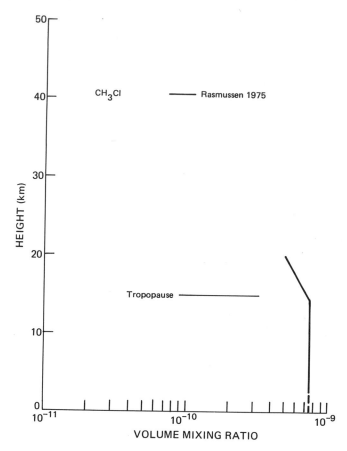

FIGURE 6.5 Vertical distribution of CH₃Cl
volume mixing ratio. For clarity the 1975
data of Rasmussen are indicated by a line denoting the average of several measurements.

alone. Furthermore, CFCl₃ drops off more rapidly than
CF₂Cl₂. This behavior is consistent with the larger
absorption cross section and faster photodissociation of
CFCl₃. Thus the observations support the basic concept
that the CFMs are transported to the stratosphere where
they are photodissociated.

The present measurements of halocarbons in the atmosphere indicate that industrial and natural sources of
chlorine atoms reaching the stratosphere are now of comparable magnitude. Since the natural chlorine species
are presumably already in steady state, industrial

chlorine compounds can be expected soon to exceed natural chlorine (except possibly for localized injections by special volcanic episodes).

1. HCl Hydrogen chloride (HCl) profiles have been de- termined both by remote-sensing infrared measurements and by *in situ* collection on filter papers. In the latter case, the chlorine content of solid particles was shown to be negligible. The absence of interference by ClO and $ClONO_2$ in the filter-paper measurements is not yet com- pletely certain. The available data, Figure 6.6, show that the HCl concentrations increase with altitude above the tropopause. This behavior is consistent with the model predictions that the CFMs are photolyzed in the mid and upper stratosphere, where the products are partially converted to HCl, some of which is transported downward to be removed by rain-out in the troposphere.

There is some indication from one *in situ* profile and one remote-sensing measurement that the HCl concentration decreases at the highest altitudes of the measurements, a feature not predicted by all the models (cf. Figure 7.15). However, too much significance should not be attached to this result at the present time. The remote- sensing profile shows the behavior only for the highest altitude, which has the greatest uncertainty. There are too few profiles available to date to consider them to be representative of the stratosphere in view of the great

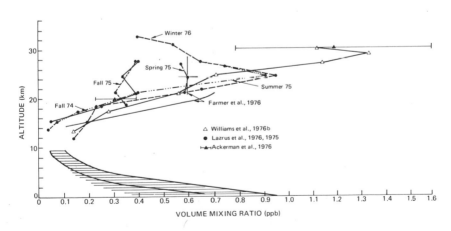

FIGURE 6.6 Vertical distribution of HCl volume mixing ratio (ppb). The hatched area indicates uncertainty limits for the tropospheric concentration (see Appendix C, Sec- tion VI.C).

variability shown by other minor constituents. If further measurements confirm the drop-off at higher altitudes, it will be necessary to establish whether it is due to horizontal transport effects or to some other inadequacy in the 1-D model.

2. *ClO* Since ClO is directly involved in the catalytic destruction of ozone by chlorine [Reactions (2.15) and (2.16)], the measurement of this compound in the stratosphere provides one of the most important tests of the models. Considerable effort is being expended in making such measurements, and reliable results are expected shortly. At the time of writing, three attempts have been reported, two involving remote-sensing techniques and one *in situ* measurement.

Carlson (1976) attempted to observe the solar absorption of ClO in the uv region at 303.5 nm using the solar telescope at Kitt Peak Observatory in Arizona. Some absorption was indicated in this spectral region, although barely above the statistical noise level. The author does not claim to have made a positive identification of ClO, but the upper limit of 0.5 percent absorption permits him to place an upper limit of 2×10^{15} molecules of ClO in a cm^2 column, which corresponds to a mixing ratio of 2 ppbv at 30 km.

Ekstrom *et al.* (1976) have made a tentative identification of ClO in the microwave region at 93 GHz using the Kitt Peak radio telescope. Both solar absorption and emission from the zenith sky was utilized. There was considerable variability in the baseline and indication of interference from other species. Estimation of the ClO concentration also involved assumptions of the temperature distribution in the stratosphere. The indication that the ClO concentration is about 50 times the value calculated from the model must therefore be considered to be highly tentative.

The *in situ* measurement was obtained by Anderson (1976) in June 1976 at Palestine, Texas, using a balloon-lofted, parachuteborne flow-tube method. Nitric oxide, NO, was added to the sampled air to convert ClO to Cl atoms, which were then detected by resonance fluorescence. ClO concentrations were obtained over the height range from 25 to 45 km, which were 20 to 100 times the value calculated from the models. The measured concentrations are also inconsistent with the HCl measurements and with the limits placed on the $ClONO_2$ concentration. Most disturbing, however, is that the amount of ClO measured exceeds the total Cl content of all the chlorine-containing constituents believed to be present in the stratosphere. Obviously, considerably more work must be done to ensure that no experimental errors are involved in the measuring technique and to determine whether this single measurement is representative of the normal stratosphere.

3. $ClONO_2$ Because of the possible role of chlorine nitrate in stratospheric chlorine chemistry, the measurement of this compound has assumed considerable significance. Earlier infrared solar spectra obtained by Migeotte *et al.* (1971) at 10,000 ft in the Swiss Alps can be compared with laboratory spectra of $ClONO_2$. A weak absorption feature in the solar spectrum at 808.7 cm^{-1} corresponds to a sharp line from $ClONO_2$ and should be relatively free from interference from other atmospheric constituents. If this absorption is attributed to $ClONO_2$, a crude upper limit of a few parts per billion can be placed on the stratospheric mixing ratio of $ClONO_2$.

A more significant limit can be derived from the solar absorption spectra obtained by Murcray (1976) taken from a balloon in September 1975. He observed absorption features near the 781 cm^{-1} absorption feature of $ClONO_2$, which increased with increasing solar path length through the atmosphere. However, the resolution was not sufficient to rule out interference from other species such as O_3. Hanst (1976) has also analyzed the same data and concluded that the $HNO_3/ClONO_2$ ratio is greater than 50, and, therefore, the $ClONO_2$ is not important in stratospheric chemistry. Murcray believes this conclusion to be unwarranted since the analysis of Hanst did not take into account the difference in altitude at which HNO_3 and $ClONO_2$ have their maximum concentrations. Murcray's own analysis places an upper limit of 1 ppbv on the $ClONO_2$ at 25 km.

Murcray has also analyzed spectra taken in 1967, which included the spectral region at 1292.4 cm^{-1} and which correspond to the center of the relatively strong Q branch of $ClONO_2$. He could find no evidence for an absorption feature. However, there is an uncertainty of a factor of 4 in the absorption coefficients of $ClONO_2$ at this wavelength, which makes the upper limit for the 1967 mixing ratio at 25 km lie in the range 2×10^{-10} to 8×10^{-10}.

Since the models predict the mixing ratio of $ClONO_2$ in 1976 to be of the order of 5×10^{-10}, the upper limits given above are not definitive in assessing the significance of $ClONO_2$ in stratospheric chemistry. Additional laboratory measurements of the absorption coefficients at 1292.4 cm^{-1} and stratospheric measurements in this spectral region are planned for the near future and should provide data with sufficient sensitivity and accuracy to make such an assessment possible.

4. Other Fluorine-Containing Compounds Measurements of fluorine-containing compounds in the stratosphere would be useful in determining the amount of CFMs dissociated in the stratosphere. The results may not be clear-cut in the case of HF since some injection of this compound (1.4×10^3 metric tons/yr) into the stratosphere may result from volcanic eruptions (Cadle, 1975). The phosgene derivatives F_2CO and $ClFCO$, on the other hand, are probably formed only from CFMs. Apparently no attempts have been made to measure the latter two compounds, but measurements of HF are currently under way, and reliable data should be available shortly.

IV. SUMMARY

The atmospheric measurements outlined in this chapter and described in detail in Appendix C are summarized below and some general conclusions drawn.

Tropospheric measurements of the CFMs have not been made with sufficient geographic coverage or accuracy to permit unequivocal conclusions about whether or not significant tropospheric sinks for these compounds exist. The rates of increase of CFMs with time are consistent with the increase in the amounts released, but again the uncertainties are too large to infer an accurate tropospheric lifetime for these compounds. The accuracy with which the global tropospheric contents of CH_3Cl and CCl_4 are known is much lower than for the CFMs, and therefore the relative contributions of natural and anthropogenic chlorine to the stratosphere cannot be inferred with any certainty, although the indications are that they are of approximately equal magnitude at the present time.

Tropospheric N_2O and CH_4 are used as input data to the models, both as sources for stratospheric NO_x and HO_x, respectively, and for determining the vertical transport coefficients. Both appear to be well mixed in the troposphere, above the boundary layer (i.e., above about 1 km from the surface), and the average global concentrations are probably not known to any accuracy of better than ±35 percent.

The major conclusion that can be drawn from the stratospheric measurements is the inhomogeneity of this region of the atmosphere. Height profiles of the concentrations of stratospheric constituents show considerable variability, both on long (month) and short (less than 1 day) time scales. This variability has been observed both for relatively inert constituents, such as N_2O, as well as for the short-lived, reactive constituents, such as NO. Variability has also been observed in the total amount of some compounds (HO, NO_2, NO, O_3) in a vertical column and in the height of maximum concentration.

The lesson to be learned from these observations is that extreme caution should be used in inferring any general stratospheric condition from single measurements or height profiles obtained at one time and place.

The one-dimensional models used in this study are inherently incapable of treating such variability since they are designed to represent globally averaged conditions. The question of whether this inability seriously challenges the validity of the models must be answered on the basis of the purpose to which the models are applied and how the measurements are used in the models. In this study, the models are indeed used to provide predictions of an average global nature. The measurements that provide input data for these models and the resulting uncertainty in the predictions are discussed in Chapter 8. The tests of the models against the measurements of other trace constituents (e.g., NO_x) are discussed in Chapter 7 and show that the calculated concentrations fall within the range of variability of the measurements.

Finally, the measurements of the chlorine-containing compounds in the stratosphere provide some, but not complete, validation of the models' ability to describe the CFM-ozone interactions. The stratospheric measurements of F-11 and F-12 show that these compounds are transported into the stratosphere. The rates at which their concentrations decrease with altitude are also consistent with photolysis being the dominant loss process for CFMs in the stratosphere. The measured concentrations of HCl in the stratosphere are consistent with the calculated values, although there have been too few measurements of height profiles to permit definitive conclusions. Upper limits have been placed on chlorine nitrate concentrations from infrared measurements, but these are marginal for determining the role of this compound in stratospheric chemistry. Measurements of ClO, Cl, $ClONO_2$, and HF are under way, and results are expected in the near future that should greatly enhance future evaluation of the problem.

REFERENCES

Ackerman, M., D. Frimout, A. Girard, M. Gottignier, and
C. Muller. 1976. Stratospheric HCl from infrared
spectra, *Aeron. Acta* A158.

Anderson, J. G. 1976. University of Michigan. Private
communication.

Angell, J. K., and J. Korshover. 1973. Quasi-biennial
and long-term fluctuations in total ozone, *Mon. Weather
Rev.* 101:426-443.

Angell, J. K., and J. Korshover. 1975. Global analysis
of recent total-ozone fluctuations. Air Pressures
Laboratory, NOAA, Washington, D.C.

Cadle, R. D. 1975. Volcanic emissions of halides and
sulfur compounds to the troposphere and stratosphere,
J. Geophys. Res. 80:1650-1652.

Carlson, R. W. 1976. Investigation of atmospheric chlorine
oxide through solar absorption spectroscopy. Preprint.

Christie, A. D. 1973. Secular or cyclic change in ozone,
Pure Appl. Geophys. 106-108:1000-1009.

Ekstrom, P. A., K. C. Davis, F. O. Clark, and R. A. Stokes.
1976. Stratospheric ClO abundance: a tentative microwave emission measurement. Preprint.

Farmer, C. B., O. F. Raper, and R. H. Norton. 1976.
Spectroscopic detection and vertical distribution of
HCl in the troposphere and stratosphere, *Geophys. Res.
Lett.* 3:13-16.

Hanst, P. E. 1976. Evidence against chlorine nitrate being a major sink for stratospheric chlorine. Part II.
Examination of stratospheric infrared spectra to establish limits for the $HNO_3/ClNO_3$ ratio. Preprint.

Heidt, L. E., R. Lueb, W. Pollock, and D. H. Ehhalt. 1975.
Stratospheric profiles of CCl_3F and CCl_2F_2, *Geophys.
Res. Lett.* 2:445-447.

Hill, W. J., and P. N. Sheldon. 1975. Statistical modeling of total ozone measurements with an example using
data from Arosa, Switzerland, *Geophys. Res. Lett.*
2:541-544.

Hill, W. J., P. N. Sheldon, and J. J. Tiede. 1976.
Analyzing worldwide total ozone for trends. Paper pre-
sented at the San Francisco meeting of the American
Chemical Society, August 29-September 3.

Krey, P. W., and R. J. Lagomarsino. 1975. Stratospheric
concentrations of SF_6 and CCl_3F. Health and Safety
Laboratory, *Environ. Quart.* HASL-294.

Lazrus, A. L., B. W. Gandrud, R. N. Woodard, and W. A.
Sedlacek. 1975. Stratospheric halogen measurements,
Geophys. Res. Lett. 2:439-441.

Lazrus, A. L., B. W. Gandrud, R. N. Woodard, and W. A. Sedlacek. 1976. National Center for Atmospheric Research. Private communication.

Lovelock, J. E. 1974. Atmospheric halocarbons and stratospheric ozone, *Nature* 252:292-294.

Lovelock, J. E. 1975. Natural halocarbons in the air and in the sea, *Nature* 256:193-194.

Manufacturing Chemists Association, Technical Panel on Fluorocarbon Research. March 1, 1976. The effect of fluorocarbons on the concentration of atmospheric ozone. Preliminary data from this report supplied to the Panel on Atmospheric Chemistry by R. L. McCarthy, December 15, 1975, and revised data February 9, 1976.

Migeotte, M., L. Nevin, and J. Swensson. 1971. The solar spectrum from 2.8 to 23.7 microns, Part I, *Photometric Atlas*. University of Liège. (See *Handbook of Lasers*, R. J. Pressley, ed. The Chemical Rubber Co., Cleveland, Ohio, 1971, pp. 91, 112.)

Murcray, D. G. 1976. University of Denver. Private communcation.

Pack, D. H., J. E. Lovelock, G. Cotton, and C. Curthoys. 1976. Halocarbon behavior from a long time series. To be published in *Atmospheric Environment*.

Parry, H. D., R. W. Parry, and G. R. Kelly. 1976. Study of the integrity of the ozone layer. Ecosystems, Inc., Westgate Research Park, McLean, Va., Jan. 15.

Rasmussen, R. A. 1975. Washington State University. Private communication.

Rowland, F. S., and M. J. Molina. 1975. University of California at Irvine. Private communication.

Schmeltekopf, A. L. 1976. National Oceanic and Atmospheric Administration, Boulder, Colo. Private communication.

Schmeltekopf, A. L., P. D. Goldan, W. R. Henderson, W. J. Harrop, T. L. Thompson, F. C. Fehsenfeld, H. I. Schiff, P. J. Crutzen, I. S. A. Isaksen, and E. E. Ferguson. 1975. Measurements of stratospheric $CFCl_3$, CF_2Cl_2 and N_2O, *Geophys. Res. Lett.* 2:393-396.

Sze, N. D., and M. F. Wu. 1976. Measurements of fluorocarbons 11 and 12 and model validation: an assessment, *Atmospheric Environment*. In press.

Telegadas, K., and G. J. Ferber. 1975. Atmospheric concentrations and inventory of krypton-85 in 1973, *Science* 190:882-883.

Williams, W. J., J. J. Kosters, A. Goldman, and D. G. Murcray. 1976a. Measurements of stratospheric fluorocarbon distributions using infrared techniques. To be published.

Williams, W. J., J. J. Kosters, A. Goldman, and D. G.
  Murcray. 1976b. Measurement of the stratospheric
  mixing ratio of HCl using infrared absorption techniques,
  Geophys. Res. Lett. To be published.

7 ONE-DIMENSIONAL MODELS OF THE STRATOSPHERE

I. INTRODUCTION

As was outlined in Chapter 2, calculations of stratospheric properties are usually based on a set of mathematical equations describing the stratosphere's principal
physical and chemical processes. In Chapters 3 and 4 and
Appendix A, the relevant chemical processes are discussed,
and a detailed analysis is made of the source and sinks of
the halogen compounds. Sources and sinks of other chemical species have been discussed previously in the report
of the Climatic Impact Committee (1975) and in CIAP Monographs 1 and 3 (1975). Chapter 5 describes the transport
of materials into the stratosphere. In particular, the
representation of the net vertical motion of gases through
one-dimensional (1-D) transport parameterization has been
analyzed in detail (Appendix B). In the present chapter,
we consider how the mathematical treatment of transport is
extended to include the chemical and photochemical processes that govern the amount of stratospheric ozone.

The nature of the ozone calculation depends not only on
our current knowledge of the stratosphere but also on the
tools available for analyzing the complex set of mathematical equations involved. Although multidimensional
descriptions of the stratosphere should, in principle,
give more detailed and more accurate results than the 1-D
models, such efforts are more time-consuming, require more
atmospheric observations for their testing and application,
and have not yet been developed or employed as extensively

as the 1-D models.* In fact, much of the current knowledge about the chemical structure and the coupling processes of the stratosphere has been derived with the help
of 1-D calculations. As is true of all theoretical models
and perhaps more so than in most cases, the usefulness of
the 1-D approximation must be established by comparison
of results derived from the model with measurements in the
atmosphere, mainly with the distributions of trace species.
The simplest, most valid comparison would use measurements
averaged in space and time to correspond to the assumptions of the particular 1-D calculation. However, this
requires extensive measurements with global and time coverage far beyond what is available in most cases. For
this reason, the comparisons possible at this time generally fall short of the ideal, and their interpretation
requires an understanding of both possible observational
variabilities and model limitations.

In Section II, the structure of 1-D models is presented,
including a discussion of the various input parameters.
In Section III, the general validity and usefulness of the
1-D models are analyzed through quantitative and qualitative comparisons with the atmospheric measurements (Chapter 6 and Appendix C).

II. GENERAL DESCRIPTION OF 1-D MODELS

In 1-D models of the stratosphere, the behavior of a specific trace species c_i, averaged over latitude and longitude, is described through the continuity equation

$$\frac{\partial c_i}{\partial t} = \frac{\partial}{\partial z}\left[K\rho\,\frac{\partial}{\partial z}\,(c_i/\rho)\right] + P_i - L_i c_i + S_i \qquad (7.1)$$

where c_i is the concentration (number density) of the ith
constituent at time t and altitude z, P_i and $L_i c_i$ are the
production and loss rates of c_i due to photochemical and
chemical interactions, K is the vertical transport or mixing coefficient, S_i is the net production or loss rate due
to any other possible processes, and ρ is the air density.

---

*There are several research groups working with multidimen-
sional models of the stratosphere, but few of the results
available so far are directly applicable to the present
problem.

In general, all of these quantities have been averaged over latitude and longitude, leaving them as functions of both time and altitude. In addition, depending on circumstances, averages over time may also be employed.

It is well known that, on a global scale, the variation in local concentrations of trace species is principally in the vertical direction (cf. the measurements in Chapter 6 and Appendix C). The concentrations of a given chemical species may vary by as much as several orders of magnitude with altitude but only a factor of 2 or 3 with latitude and longitude. Consequently, by taking into account the vertical transport and vertical distribution of the trace species, the 1-D model includes the most dominant aspect of their spatial variability. Furthermore, inasmuch as horizontal mixing occurs in the troposphere on a time scale of weeks to months, even a localized surface source (such as CFMs) would in effect be a uniform source to the stratosphere on a time scale of years to decades. Also it is shown in Appendix B that the basically phenomenological, average vertical transport (i.e., as given by the eddy-mixing coefficient K) can be prescribed by minimizing the disagreement between observed and calculated profiles of natural tracers such as CH_4 and N_2O. Thus, it is clear that 1-D models are physically reasonable and can approximate the averaged long-range behavior of the trace species in the stratosphere. Hence, 1-D models have been used to study the interactive coupling of vertical transport and the chemical and photochemical processes that determine the stratospheric role of pollutants such as the CFMs.

The elements central to all 1-D calculations are the sets of input variables used to describe the various processes involved. These include a definition of the physical domain and boundary conditions, the vertical transport coefficient (K) with its associated approximations, the system of chemical reactions, and the photodissociation processes. A brief discussion of these input data serves to characterize the calculations and provide the perspective needed to evaluate the results.

A. *Physical Domain and Boundary Conditions*

Consideration of the dominant photochemical processes and the sources and sinks of the relevant trace species will determine the optimal physical domain for analysis. For example, although the catalytic destruction of ozone by

ClO_x is most pronounced in the limited altitude range of 30 to 40 km, it is still necessary to include the atmosphere from the surface to about 50 or 60 km. The sources of chlorine, such as CFMs, CH_3Cl, or CCl_4, are at the surface. The precise placement of the upper boundary is not important as long as it is well above the region where the photolytic and chemical reactions are all much faster than transport so that a steady state is established among the species involved (a total photochemical equilibrium). If this is done, the results are relatively insensitive to the upper boundary conditions. Any uncertainty in the latter, such as in the local fluxes or concentrations, will then be compensated by the photochemical reactions in the region just below the boundary. For studying the effects of the NO_x, ClO_x, and HO_x catalytic cycles on stratospheric ozone, an upper boundary at or above 50 km is acceptable, and the concentrations or fluxes derived for local photochemical equilibrium serve sufficiently well as boundary conditions.

At the surface, the measured concentrations of naturally occurring trace species are used as boundary conditions, e.g., the concentrations of CH_4, N_2O, O_3, HCl. For well-documented trace species released by man, such as CF_2Cl_2 and $CFCl_3$, the estimated net fluxes (i.e., atmospheric release rate minus any surface destruction rate) are the best boundary conditions. For other trace species, either the estimated surface flux (e.g., CCl_4) or concentration (e.g., HO, H_2O_2, HO_2, NO) is used. The choice of boundary conditions for water-soluble and reactive species has little effect on the solution for their concentrations in the stratosphere (Chang, 1974).

In order to study the effect of the atmospheric release of CFMs, it is necessary to start with the stratosphere before the introduction of CFMs. Usually the initial stratosphere is calculated by means of the model (without the CFMs) and verified by comparison with known concentrations. The initial chlorine content of the stratosphere is obtained from the measured average surface concentrations of HCl, CH_3Cl, and CCl_4, assuming the distributions of these trace species are at (or near) steady state. Starting from this "natural" condition, calculations can then be made to examine the time-dependent effect of CFMs on ozone.

B.   1-D Vertical Transport Coefficient

As was pointed out in Chapter 5, vertical transport in the 1-D model is parameterized through the so-called eddy-mixing coefficient *K* in Eq. (7.1). In Appendix B, it is shown that there is no unique approach for deriving a suitable set of eddy-mixing coefficients. The uncertainty in *K* for a given altitude might be as large as a factor of 10 or more. However, Appendix B points out that the overall uncertainty in the average *K*, as reflected in transport time scales, is probably only a range of 2 to 3. In Figure 7.1, several representative choices of *K* are shown. Profiles A, B, an earlier version of C, and

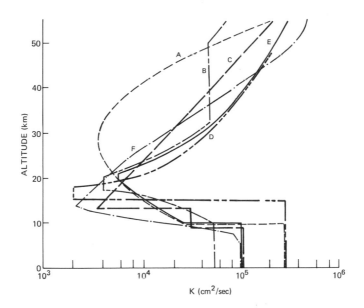

FIGURE 7.1 Some typical 1-D vertical eddy-mixing profiles used in the models (A, Chang, 1974; B, Crutzen and Isaksen, 1975; C, Wofsy, 1975; D, Wofsy *et al.*, 1972; E, Chang, this study; F, Liu and Cicerone, 1976). Profile B is derived from a profile first suggested by Ehhalt (see Crutzen and Isaksen, 1975); profile C is a modified version of a profile attributed to Hunten (Climatic Impact Committee, 1975); and profile E corresponds essentially to the best estimate in Appendix B.

D have been used to analyze the effect of supersonic transport effluents on ozone. The most recent results of this nature are those of Chang (1974), Crutzen (1974), the Climatic Impact Committee (1975), and McElroy *et al.* (1974). Since then, profiles A, B, and C have been used to study the effects of CFMs on ozone (Chang *et al.*, 1975; Crutzen, 1975; Wofsy, 1975).

Most profiles of K are similar in their essential characteristics, having a rather large value in the troposphere ($\sim 10^5$ cm^2/sec), followed by a low-value region ($\sim 10^3$-10^4 cm^2/sec) near the tropopause, and then a fast rise to high values ($\sim 10^5$ cm^2/sec) in the upper stratosphere. In the following detailed analysis only profile E is considered. (Ozone reductions for profiles A, B, and E are compared in Chapter 8.) Profile E is a choice "in-between" profiles A and B in terms of time scales; in fact, it is almost exactly a smoothed version based on fitting global N_2O data of the K presented in the last column of Table B.5. Since the ozone reduction by CFMs occurs primarily in the photochemical equilibrium region (above 30 km), and thus is not strongly dependent on the detailed local representation of transport, analyses based on these K profiles should represent the full range of possible responses of the stratosphere to the continued introduction of CFMs.

C. *The System of Chemical Reactions*

The chemical reactions that establish the concentration of ozone have been discussed in Chapter 2 and Appendix A. Of all models that incorporate some type of transport, 1-D calculations are generally the most complete in considering photochemical kinetics. Most 1-D models now include the Chapman cycle, the NO_x, HO_x, and ClO_x cycles, and many of the transfer reactions among these basic cycles. The number of reactions considered ranges from 60 to more than 100 (Crutzen and Isaksen, 1975; Wofsy and McElroy, 1974; Chang *et al.*, 1975; Turco and Whitten, 1975). The more extensive reaction sets used in some models include many minor reactions, but a comparison of the various publications establishes that the chemical kinetics systems used are in essential agreement. However, the same reaction rate coefficients have not been used for the same reaction in all calculations. Table 8.1 illustrates some of the significant differences. The consequences of such

differences are discussed both in Section III of this
chapter (the natural stratosphere) and in Chapter 8 (the
perturbed stratosphere). A typical detailed set of reactions is listed in Appendix D, Table D.2.

It should be pointed out that, although many of the reaction rate constants are functions of temperature and
the reaction rates are functions of air density, the sensitivity of the results to local variations in these variables is small. Of course, for some reactions local
deviations from the average could be quite large, perhaps
as large as factors of 2 but usually only 30 percent or
less (Callis et al., 1976). The cumulative effect of
such variations on the average, therefore, is likely to
be small; however, direct quantitative verification is
not within the scope of 1-D models. Such uncertainties
can best be evaluated by multidimensional models in the
future. In some recent calculations, average temperatures
were derived for different altitudes in a self-consistent
manner from the distributions of the trace species; they
are in good agreement with and thus lend support to the
commonly employed standard temperature distributions
(Callis et al., 1975; Luther et al., 1976).

In general, the qualitative aspects of stratospheric
chemistry are now well understood, and many of its features can be described in quantitative terms. There are,
however, several significant reactions for which the rate
constants have not yet been determined to a desirable
level of accuracy. They are listed in Table 8.1, and
their effects are analyzed in detail in Chapter 8. Also,
it should be noted that an approximation is introduced in
the use of space- and time-averaged concentrations to calculate the reaction rates that determine the ozone reduction. Most of the reactions are bimolecular with rates of
the form $k_{ij}c_i(v,t)c_j(v,t)$, where k_{ij} is the rate constant
for reaction of species i with species j, at concentrations
c_i and c_j in a particular small volume v at time t. The
approximation arises because one does not obtain the same
average rate by (a) first averaging the concentrations in
different volumes and then multiplying as is found by
(b) first multiplying and then averaging. The latter (b)
gives the right answer, but it is a theoretical ideal probably not achievable even with a three-dimensional (3-D)
model; the former (a) is a simplifying approximation that
can be handled with the 1-D model. Similar considerations
apply to the time-averaging (diurnal and seasonal) of the
photodissociation processes that generate many of the most
reactive species.

D. *Photodissociation Processes*

Photodissociation processes in the atmosphere are often the most important mechanisms for the production and destruction of chemical species. The photodissociation rate $J(z,\Theta,\Psi,t,c_i,\ldots,c_j)$ is a complicated function of the spatial coordinates (z,Θ,Ψ), time (t), the absorption coefficients (implicit constants), and the concentrations $c_i,\ldots,c_j$ of the reactants, which again are functions of z,Θ,Ψ, and t (CIAP Monograph 3). For a given species and process, the photolysis rate is proportional to $I(\nu,v,t)c_i(v,t)$, where I is the intensity of radiation at frequency ν incident upon a particular small volume v at time t. As in the case of bimolecular reactions, the separate averaging of $I(\nu,v,t)$ and $c_i(v,t)$, followed by their multiplication, is an approximation inherent to 1-D calculations. Its effects are particularly important in averaging the relatively large amplitude diurnal and seasonal changes in $I(\nu,v,t)$, for which a variety of methods have been used. At this time, there exists no detailed evaluation of their relative merits. However, the possible limitations due to this approximation are known.

For example, the effects of prior averaging will be greatest for species such as $O(^1D)$, $O(^3P)$, NO, and HO, whose concentrations are most strongly dependent on solar radiation. In such cases, comparisons of calculated and observed concentrations should be made with great care, keeping in mind the possible importance of the approximations that have been made (cf. Section III). Furthermore, the diurnal or seasonal variations that do occur will not be reproduced in kind by calculations in which diurnal or seasonal averages are used for the individual photodissociation rates (J's). Nonetheless, the results of such calculations can be a reasonable approximation to the average of the actual variations. If the bias introduced by this averaging approximation is only a first-order effect, it will tend to cancel and be small when the same time-averaging is used for the natural state of the stratosphere as for its perturbation by the CFMs. Studies of this question indicate that the predicted perturbations will usually be accurate to within a factor of 0.8 to 1.2 (Wuebbles and Chang, 1975). Nonetheless, care must be exercised in handling this question.

There are circumstances for which the use of a diurnally averaged photodissociation rate (a constant sun at half the solar flux) will produce significant error for a species. Johnston (1976) has pointed out that this problem

has long been recognized by photochemists as the difference
between applying to a photochemical system all the light
half the time versus half the light all the time. The
two options are equivalent if all effects [terms in the
solution of Eq. (7.1)] are linear in light intensity (I)
but differ if any effects are nonlinear in I. The importance of the difference depends not only on the relative magnitudes of linear versus nonlinear terms but also
on the time constants of the nonlinear terms compared to
the length of the day. For example, two photochemical
processes limit the amount of $ClONO_2$ in the stratosphere.
One controls its destruction; the other controls its formation by dissociating NO_2 with which ClO reacts to form
$ClONO_2$. Both processes have time constants in the 0.5-
to 5-hour range. This makes the $ClONO_2$ concentration
inversely proportional to I^2. Thus a diurnally varying
sun should be used for the accurate determination of its
effects.

In most current models of the stratosphere, including
the reference model for this report (Appendix D), coefficients describing the absorption of solar radiation are
computed in a self-consistent manner from the concentration profiles of trace species predicted by the calculations. For example, consider the photoproduction of O_3,
which is involved in the often-discussed partial "self-
healing" mechanisms of ozone. Ozone is produced by the
combination of O and O_2 in a three-body collision process
(Chapter 2). The required O atoms are produced by photodissociation of O_2 at wavelengths shorter than 240 nm.
In this spectral region, both O_3 and O_2 absorb solar radiation. Consequently, if for any reason there is a decrease in the local O_3 concentration above a given
altitude, z_0, then more radiation in this spectral region
will reach below z_0 and produce more O and hence more O_3
than before. Therefore, a reduction of O_3 at higher
altitudes (such as by an increase in NO_x or ClO_x) will
increase O_3 production at lower altitudes. However, the
amount is relatively slight, because most absorption by
O_3 takes place at wavelengths where photodissociation of
O_2 (and production of O_3) does not occur. Detailed calculations of the reduction in stratospheric ozone now always
include the effect of such partial "self-healing" processes.
However, care must be exercised to ensure that they are
treated properly (Chameides and Walker, 1975).

There are some other sources of uncertainty, such as
the amount of or the variation in the natural solar flux
or the effects of multiple scattering and albedo in the

calculation of photodissociation coefficients. Any such uncertainties in the solar flux will affect calculations of the "natural" amount of stratospheric ozone as well as those of its perturbation by man-made pollutants. Hence, the relative effects of the latter will be unchanged to first order; the second-order effects will tend to be small, as in the case of diurnal and seasonal variations. Two studies of the effect of multiple scattering and albedo on local photodissociation coefficients (Callis *et al.*, 1975; Luther and Gelinas, 1975) indicate it to be mostly localized around 20 km. Consequently, these processes would be expected to have little direct influence on the ClO_x cycle, which is most active above 30 km.

E. *Other Physical Data*

Other input data for 1-D calculations include such parameters as the natural background air density and, at various times, the vertical distributions of such constituents as H_2O, H_2, CO_2, and CO. The data required depend on the nature of the problem and the purpose of the study. Comparatively little is known about the sensitivity of 1-D perturbation calculations to these atmospheric quantities. Because of the similarity in the role of these variables to that of temperature, it seems reasonable to expect that their effects are relatively minor, although more direct tests should be available in the near future.

Another important input parameter to 1-D models is the removal rate by precipitation of the water-soluble species such as HNO_3, NO_2, and HCl. The detailed mechanisms and their effects are described in Chapter 4 and Appendix A. Although this rate is not always accurately known, any reasonable estimate gives virtually the same reduction in ozone and seems to affect the natural tropospheric distributions of only a few species.

III. COMPARISONS WITH ATMOSPHERIC MEASUREMENTS

The results of a 1-D calculation can be evaluated by comparison with suitably averaged atmospheric measurements. Several basic types of information have proved to be useful for this purpose: height profiles of individual species, partitioning of related species, diurnal and seasonal variations, and height profiles for groups of trace species. Below, each of these is described briefly, after which detailed comparisons are made for those species most directly involved in the CFM problem.

Height profiles giving the concentrations of individual trace species as a function of altitude are the most obvious and direct results from calculations and atmospheric measurements. Because of the local and discrete nature of all measurements, there is significant variability in the experimental data. Similarly, different calculations often yield highly variable results because of basic differences in the input parameters (Chang, 1974). Nevertheless, a comparison of height profiles is usually the first and a valuable step in evaluating a model.

The concentrations of some trace species are strongly coupled by chemical and/or photochemical processes, for example, the active and inactive species in catalytic cycles such as NO_y (NO, NO_2, HNO_3), HO_y (HO, HO_2, H_2O_2), and ClX (Cl, ClO, HCl, $ClONO_2$,...). In these cases, the partitioning of the total amount among the individual species provides perhaps the best detailed information about the catalytic cycles. This partition, in effect, determines the strength of the catalytic cycle in the stratosphere, so it is most useful in establishing the predictive capability of a calculation.

The diurnal and seasonal variations of individual or groups of trace species are most illuminating in studying the photodissociation process. Analyses of these variations are useful not only in establishing the detailed photochemistry but also in evaluating the accuracy of the averaged parameters employed to describe the photochemistry.

The photochemical processes are such that members of several selected groups of trace species, for example NO_y, can be considered collectively as an almost inert tracer in the stratosphere. Therefore, the height profiles for the NO_y species can be combined and used to test directly nonchemical aspects of the model, such as the 1-D vertical-mixing coefficients. Seasonal trends of these "neutral" tracers can also tell much about their global production and destruction rates. At present, this kind of information is quite limited; but when available, it has been useful.

The study of the role of CFMs in the atmosphere requires an understanding of the upward transport of CFMs from the surface, the conversion of CFMs to the catalytically active chlorine species in the stratosphere, and the downward transport of these active species (Chapters 2, 4, and 5). In the remainder of this chapter, we examine the accuracy and usefulness of 1-D calculations for studying these processes, using the general procedures outlined above. Most of the calculations are similar in scope to the one described in Appendix D, although there are differences in the choice of the particular input parameters as described previously.

A. O_x Species

The stratospheric ozone distribution is highly variable both in time and in space (see Figures C.1-C.4). Most of this variability is averaged out in the 1-D calculations. The uncertainties due to this averaging are not easily evaluated. All models can predict the midlatitude vertical profiles of ozone (for example, McElroy et al., 1974; Shimazaki and Whitten, 1976). A somewhat indirect but more meaningful test of the 1-D calculations can be obtained through comparisons such as those in Figures 7.2 and 7.3 of the results from 2-D and 3-D models with atmospheric measurements. Although on a limited basis, the similarity of comparable results from 1-D, 2-D, and 3-D models reported in the CIAP study demonstrates the extent to which 1-D calculations are capable of describing the

FIGURE 7.2 The ozone column in units of 10^{-3} cm at STP (Dobson units) in the northern hemisphere as a function of season and latitude: (a) as determined by Dütsch (1971) and (b) as predicted by the 3-D model developed by Alyea et al. (1975).

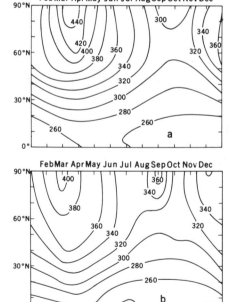

FIGURE 7.3 Some comparisons of calculated (———) (2-D model of Widhopf, 1975) and observed vertical profiles of ozone at selected latitudes.

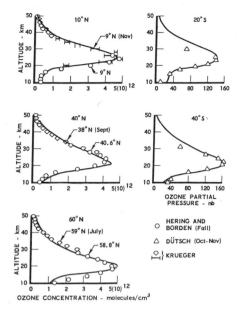

average stratospheric ozone content and its perturbations (CIAP Monograph 3; Climatic Impact Committee, 1975). Furthermore, the available 2-D and 3-D results have provided confidence that no unexpected effects are introduced by the multidimensionality of the problem.

Both $O(^3P)$ and $O(^1D)$ are produced by photodissociation and are directly dependent on the local O_3 concentrations. As a result, in the region of photochemical equilibrium (largely above 30 km) when the photochemical processes are fast compared with transport, the concentrations of O_3 and hence of $O(^3P)$ and $O(^1D)$ are independent of transport for fixed NO_y and HO_y concentrations. Below 30 km, transport processes can be expected to affect directly the local concentrations of the total "odd oxygen" (O and O_3). The 1-D models are not capable of describing local variations due to horizontal motions, so good agreement with any local measurement of $O(^3P)$ in the absence of simultaneous O_3 data cannot be expected in the region below 30 km. Above 30 km, however, the theoretical and experimental results for O atom concentration would be expected to be in fairly good agreement. A typical $O(^3P)$ profile calculated for midlatitude conditions is included in Figure 7.4, and the comparison of it with the measurements shows agreement within a factor of 2. Furthermore, this agreement is not sensitive to variations in the parameters employed for the calculation.

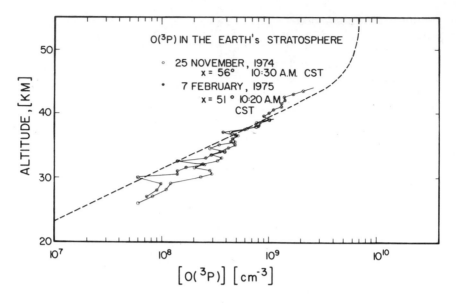

FIGURE 7.4 Comparison of measured (————O————) and calculated (---------) vertical distributions of $O(^3P)$. Calculations are based on the 1-D model using the eddy-mixing profile E of Figure 7.1. Measurements are from Anderson (1975).

B. *NO_Y Species*

Next to ozone, the species NO, NO_2, and HNO_3 are the most intensively studied trace species in the stratosphere. In recent years, a large body of data has been collected, for example, as shown in Figures 7.5-7.7. The wide variability in the observations for a single species of the three has been discussed in Chapter 6. The partition-ing between NO and NO_2 in the upper stratosphere is gov-erned mainly by the reactions

$$NO_2 + h\nu \xrightarrow{J_1} NO + O$$

$$NO_2 + O \xrightarrow{k_1} NO + O_2$$

$$NO + O_3 \xrightarrow{k_2} NO_2 + O_2$$

The relative concentrations of NO and NO_2 maintain a quasi-steady-state ratio of approximately

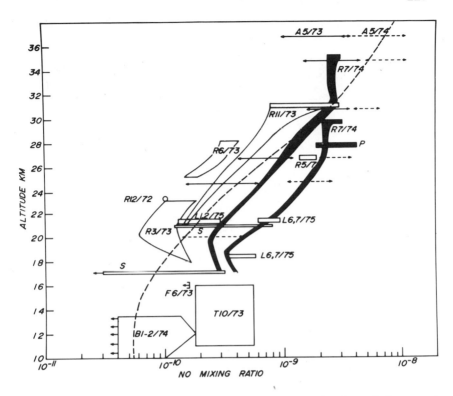

FIGURE 7.5 Comparison of measured and calculated (------)
vertical distributions of NO. Calculations are based on
the 1-D model using the eddy-mixing profile E of Figure
7.1. For references, see Appendix C.
Bl-2/74, Briehl *et al.* (1975); 25-37° N; chemiluminescence.
T10/73, Toth *et al.* (1973); 43-51° N; solar absorption.
S, Lowenstein *et al.* (1974); 25-49° N; chemiluminescence.
L6, 7/75, Lowenstein *et al.* (1975); 25-49° N; chemilumi-
  nescence.
A5/73, Ackerman *et al.* (1973a); 44° N; solar absorption.
A5/74, Ackerman *et al.* (1974); 44° N; solar absorption.
P, Patel *et al.* (1974), Burkhardt *et al.* (1975); 33° N;
  spin-flip laser absorption.
F6/73, Fontanella *et al.* (1974); 43-51° N; solar absorp-
  tion.
R12/72, R3/73, Ridley *et al.* (1973, 1974); 33° N; chemi-
  luminescence.
R6/73, R11/73, Ridley *et al.* (1975); 33° N; chemilumines-
  cence.
R7/74, Ridley *et al.* (1976a); 58° N; chemiluminescence.
R5/75, Ridley *et al.* (1976b); 33° N; chemiluminescence.

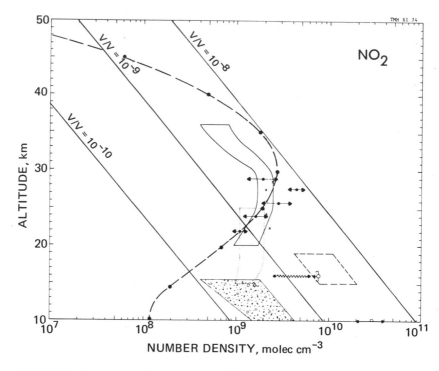

FIGURE 7.6 Comparison of measured and calculated (------)
vertical distributions of NO$_2$. Calculations are based on
the 1-D model using the eddy-mixing profile E of Figure
7.1. For references, see Appendix C.

Ackerman *et al.* (1974); balloon 44° N; sunset V 74;
solar absorption at 1597 and 1600 cm^{-1}; grille spectrometer; authors' uncertainty in measuring equivalent widths as amplified by inversion.

o Ackerman and Muller (1972a, 1973); balloon 44° N; X 70;
2850-2925 cm^{-1} solar absorption; grating spectrometer;
authors' uncertainty at 12.5 km, upper limit at 16.1
km.

• Ackerman and Muller (1972a, 1973); based on spectra of
Goldman *et al.* (1970); balloon 33° N; sunset XII 67;
1612-1616 cm^{-1} solar absorption; grating spectrometer;
authors' uncertainty.

□ Brewer *et al.* (1973); aircraft 44° N; VIII 73; 430-450
nm absorption in scattered sunlight; noon value with
authors' uncertainty.

Brewer *et al.* (1973, 1974); aircraft 43-51° N; sunrise
and sunsets VI, X, XI 73; 430-450 nm absorption in
scattered sunlight; envelope of several profiles.

$$\frac{[NO]}{[NO_2]} = \frac{J_1 + k_1 \, [O]}{k_2 \, [O_3]}$$

The ratio responds with a time constant of about 100 sec to changes in J_1, $[O]$, or $[O_3]$. The rate constant k_2 has a large temperature coefficient; therefore, the ratio will be a sensitive function of the temperature T as well as of the O_3 concentration, both of which are known to have considerable local variability. Consequently, this ratio would not be a good parameter to test 1-D or even 2-D calculations. Above 35 km, the sum of NO and NO_2 should be a quantity that is fairly well conserved because there is very little HNO_3 above this altitude. Below 35 km, HNO_3 constitutes a significant, and even principal, fraction of the total NO_y.

From a 2-D study (Widhopf, 1975) we expect only mild latitudinal variations in total NO_y. Also, the source function (in the form of N_2O) is expected to be essentially uniform in time and space because of rapid tropospheric mixing, so 1-D models are expected to be reasonably accurate at representing total NO_y. Recently, a simultaneous measurement of NO, NO_2, and HNO_3 became available (Evans et al., 1975). Prior to that, the best available data were a set of bounds derived from many measurements of HNO_3 (see Appendix C) and one simultaneous measurement of NO and NO_2 (Ackerman et al., 1974). These bounds by nature of their derivation are only an indication of variability of NO_y in the middle stratosphere. Considering them along with the data of Evans et al., one has a

---

Farmer et al. (1974); aircraft 43-51° N; sunsets VI, IX, X 73; 2850-2925 cm^{-1} solar absorption; interferometer; authors' uncertainty.

Fontanella et al. (1974); aircraft 43-51° N; sunset VII 73; 1604-1606 cm^{-1} solar absorption; interferometer; authors' uncertainty.

Harries et al. (1974a); aircraft 65-75° N; V 73; thermal emission at 38 cm^{-1}; interferometer; authors' uncertainty.

x Murcray et al. (1974); based on spectra of Goldman et al. (1970) and Murcray et al. (1969); balloon 33° N; sunset XII 67; 1604-1616 cm^{-1} solar absorption; grating spectrometer.

Δ Murcray et al. (1974); based on spectra of Ackerman et al. (1973b); aircraft 43-51° N; sunset VII 73; 1604-1616 cm^{-1} solar absorption; grille spectrometer.

reasonably solid although still minimal basis for asses-
ing the qualitative and quantitative accuracy of the 1-D
models.*

A comparison of vertical profiles calculated for the
three species, with the experimental data (Figures 7.5-
7.7), shows that their general qualitative features are
the same. The differences between predictions and measurements of NO and NO_2 in the lower stratosphere are to
be expected because of the important influence of horizontal transport on NO_y. This would influence the partition of total NO_y into HNO_3, NO, and NO_2, so that the
quasi-steady-state partition ratios are a function of the
delicate balance between horizontal transport and photochemistry. In Figure 7.8 there is quantitative consistency between theoretical predictions and experimental
measurements of total NO_y above 15 km. Below 15 km the
theoretical results are highly sensitive to the assumed
rain-out mechanism, and the agreement becomes poor. The
calculated stratospheric burden of NO_y is, however, not
very sensitive to the calculated tropospheric values.

An additional verification of the accuracy of the photochemical aspects of the calculations comes from a comparison of *in situ* measurements of the changes in stratospheric
NO at sunrise and sunset (Figures C.7 and C.8) with the
prediction of a time-dependent 1-D model. The magnitude
of these diurnal changes, relative to the concentrations
at noon, provides a convincing demonstration that the
relationship between NO, NO_2, and O_3 can be calculated.

Oxidation of N_2O by $O(^1D)$ is the principal source of
stratospheric NO_y. The calculated vertical distribution
of NO_y scales linearly with the tropospheric mixing ratio
assumed for N_2O. In the range of 15 to 30 km, the distribution of NO_x is relatively insensitive to the 1-D vertical-mixing coefficient (Chang, 1974). The comparison with
high-altitude data for N_2O (Figure 7.9) and CH_4 (Figure
7.10), at particular latitudes, suggests a preference for
high values of the eddy-mixing coefficients above 30 km.
However, it is apparently impossible to simultaneously
fit the CH_4 and N_2O data at midstratosphere, the reason
being the rapid destruction of CH_4 by HO in concentrations
similar to those recently measured (see next section).

---

*It should be noted that because of instrumentation dif-
ferences, the measurements of the three species are not
for identical parcels of air, although they are close
enough on a global scale for our purpose.

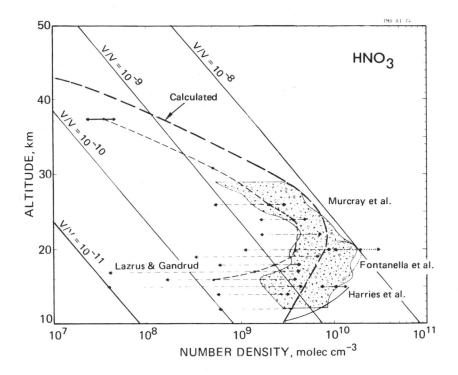

FIGURE 7.7 Comparison of measured and calculated (------)
vertical distributions of HNO₃. Calculations are based
on the 1-D model using the eddy-mixing profile E of
Figure 7.1. For references, see Appendix C.
···· Fontanella *et al.* (1974); aircraft 43-51° N; VII 73;
    solar absorption at 1326 cm⁻¹; grille spectrometer;
    authors' uncertainty.
— Harries *et al.* (1974a, 1974b); aircraft 65-70° N;
    V 73; thermal emission between 9 and 31 cm⁻¹;
    interferometer.
-- Lazrus and Gandrud (1974); aircraft and balloon 75° N-
    51° S; 71-73; paper filter capture. Dashed profile
    above 16 km is average of 33° N and 34° S; IV-VI 73;
    dashed arrows indicate extremes over all latitudes
    and seasons.
  Murcray *et al.* (1973, 1974); balloon 33° N; V-XI 70;
    and 64° N; IX 71 and IX 72; 810-955 cm⁻¹ thermal
    emission; filter radiometer; hatched area is en-
    velope of eight profiles. Concentrations at 64° N
    are generally lower in altitude and larger than
    at 33° N.

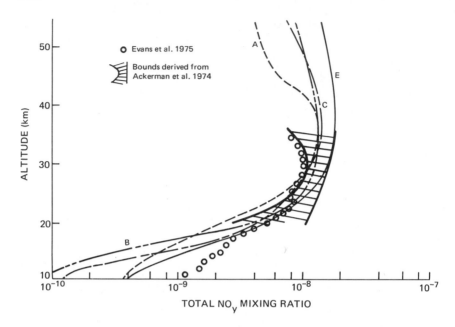

FIGURE 7.8 Comparison of the total NO_y measurements with vertical profiles calculated for several 1-D eddy-mixing profiles of Figure 7.1 (dashed lines). The open circles are from Evans *et al.* (1975), and the shaded region represents bounds from Ackerman *et al.* (1975).

Furthermore, in order to fit the CH_4 data closely everywhere, the eddy-mixing coefficient required in the 20-25 km range has to be so large that it distorts the O_3 profile beyond reason. This difficulty may be due to the assumed surface concentration of CH_4. But, on the average, the calculated CH_4 profiles differ from the measurements only at a few altitudes and within an acceptable range. The fact that N_2O and the CFMs both have largely photodissociative sinks with therefore similar latitudinal and vertical distributions supports the choice of adjusting vertical transport to N_2O for CFM predictions. The primary difficulty with now using N_2O to infer transport is the sparseness of observations above 30 km and the sensitivity of the inferred transport to high level N_2O concentrations at those altitudes (see Appendix B).

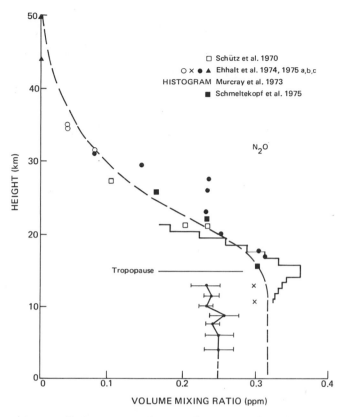

FIGURE 7.9 Comparison of measured and calculated (--------) vertical distributions of N₂O. Calculations are based on the 1-D model using the eddy-mixing profile E of Figure 7.1. The stratospheric data were obtained by various authors in different years and locations (see Appendix C, Section III.E). They are normalized to the tropopause (assumed height 15 km) to facilitate comparison. The tropospheric profile is an average over six individual profiles (Ehhalt *et al.*, 1975b). See Appendix C for references.

C. *HO$_y$ Species*

At present, the HO$_y$ species (HO, HO$_2$, H$_2$O$_2$) are the least well understood of the major trace species. Most of our knowledge derives from theoretical considerations.

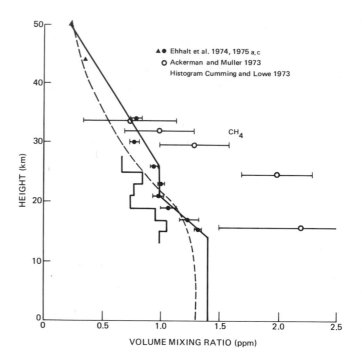

FIGURE 7.10 Comparison of measured and calculated (---------) vertical distributions of CH_4. Calculations are based on the 1-D model using the eddy-mixing profile E of Figure 7.1. The heavy line and full dots represent the average of nine profiles from nine different balloon flights at 32° latitude. The error bars represent the standard deviation of the data and are mainly due to the natural variability in CH_4. The other data are individual profiles; the error bars represent the experimental uncertainties. The data are normalized to the tropopause (assumed height 15 km) to facilitate comparison. See Appendix C for references.

Calculations show that the concentration profiles of HO, HO_2, and H_2O_2, unlike those of NO_y, apparently are not sensitive to changes in the vertical transport rate. This is due to the dominance of HO production by the reaction

$$O(^1D) + H_2O \xrightarrow{k_3} 2HO$$

The H_2O concentrations are obtained from observations, and the $O(^1D)$ from calculated concentrations of O_3. The major uncertainties in the production rate are then the H_2O and $O(^1D)$ concentrations, and the rate coefficient k_3. The HO is converted back to H_2O largely by the loss mechanism

$$HO + HO_2 \xrightarrow{k_4} H_2O + O_2$$

The principal uncertainty in the loss rate is the rate constant k_4 (cf. Chapter 4 and Appendix A). For the ten-fold range $2 \times 10^{-11} \leq k_4 \leq 2 \times 10^{-10}$ the resulting differences in HO are quite significant. These are shown by the profiles in Figure 7.11, calculated with the 1-D model described in Appendix D (for a constant sun at one half the actual solar flux). It is seen that the uncertainty in k_4 causes the profile to vary by a factor of 2 to 3 in the altitude range of 20 to 55 km. The HO concentration has an important effect on the chlorine chemistry because

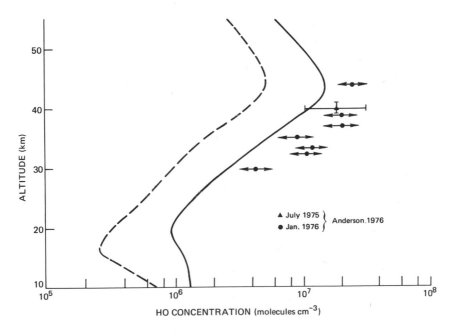

FIGURE 7.11 Effect of changing a particular chemical rate constant (HO + HO$_2$ $\xrightarrow{k}$ H$_2$O + O$_2$) on the stratospheric vertical profiles of HO. $k = 2 \times 10^{-11}$(————); $k = 2 \times 10^{-10}$(--------).

of the reaction

$$HO + HCl \xrightarrow{k_5} Cl + H_2O$$

This reaction is a major process in determining the partitioning of chlorine species.

The recent *in situ* measurements of HO by Anderson (1976) suggest that a value of 2×10^{-10} would be much too high for k_4; the HO profile corresponding to this value is almost an order of magnitude less than Anderson's data. Even 2×10^{-11} is a bit high (Figure 7.11), giving a profile that is nearly a factor of 2 too small. An analysis of results from diurnal models indicates that this discrepancy is not attributable to the use of a constant sun. At the corresponding zenith angle ($\chi = 80°$) 1-D diurnal calculations still yield HO profiles a factor of 2 less than the measurements.

Calculations using a rate constant for $O + HO_2 \rightarrow OH + O_2$ of 3×10^{-11} cm^3 sec^{-1} give results in much better agreement with the experimental HO data than is the case with the presently adopted lower value (Table D.2). This higher value would not appear to be ruled out by preliminary results of recent studies.

The tropospheric concentration of HO is determined by the surface chemistry and rain-out and, possibly, by some heterogeneous reactions. Measurements of HO close to the ground have been made by two research groups. More recently measurements have been made in the upper troposphere at 7 and 11.5 km (see Chapter 6 and Appendix C). When compared with the theoretical profile of HO, the agreement is reassuring but not conclusive. Therefore, in the analysis of the effect of CFMs on ozone, the sensitivity of ozone reduction to k_4 is studied in detail both individually and in conjunction with the effects of other rate constants.

D. *Halocarbons*

The mixing ratio profiles of CF_2Cl_2 and $CFCl_3$ were calculated for late 1975, based on the past industrial production rates (McCarthy, 1974) of $CFCl_3$ and CF_2Cl_2 and assuming a 90 percent atmospheric release coefficient. In this, it was assumed that the only loss mechanism is photodissociation and reaction with $O(^1D)$. The actual photodissociation coefficients used are global averages (e.g., Rowland and Molina, 1975). Profiles calculated with and without photodissociation are compared with the recent observations in Figures 7.12 and 7.13. The profiles *with* photodissociation are in much better agreement with the observations than the profiles without photodissociation. The comparison confirms the basic validity of the postulated photodissociation of chlorofluoromethanes in the stratosphere. However, the comparison has little bearing on the existence of tropospheric sinks, because

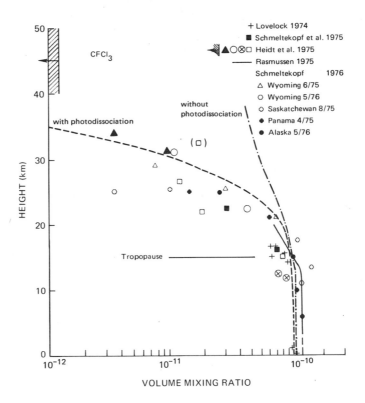

FIGURE 7.12 Comparison of measured and calculated (---------) vertical distributions of $CFCl_3$. Calculations are based on the 1-D model using the eddy-mixing profile E of Figure 7.1. The data are from various authors and collected in different years and locations (see Chapter 6, Section III.D). They are normalized to the tropopause (assumed height 15 km) to facilitate comparison. See Chapter 6 for references.

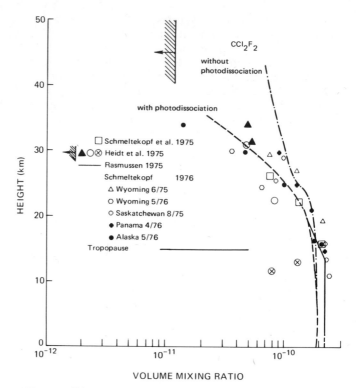

FIGURE 7.13 Comparison of measured and calculated (--------) vertical distributions of CF_2Cl_2. Calculations are based on the 1-D model using the eddy-mixing profile E of Figure 7.1. The data are from various authors and collected in different years and locations (see test). They are normalized to the tropopause (assumed height 15 km) to facilitate comparison. See Chapter 6 for references.

the data are not accurate or extensive enough for the purpose (cf. Appendix E).

Some measurements of CCl_4 have been made near the surface and just above and below the tropopause. As may be seen in Figure 7.14, the measurements do not extend to high enough altitudes to provide much evidence about photodissociation of the compound in the stratosphere.

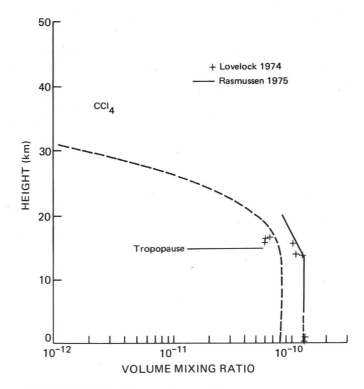

FIGURE 7.14 Comparison of measured and calculated (---------) vertical distributions of
CCl_4. Calculations are based on the 1-D
model using the eddy-mixing profile E of
Figure 7.1. So far, stratospheric data are
restricted to the lower stratosphere. The
data are normalized to the tropopause (assumed height 15 km) to facilitate comparison. For references see Chapter 6.

E. ClX Species

Although the ClO_x catalytic cycle governs the effect of
CFMs on ozone, current experimental data on atmospheric
concentrations of the ClX species (Cl, ClO, HCl, $ClONO_2$)
are still limited to HCl except for several preliminary
measurements. Even for HCl, the measurements are relatively few in number (Chapter 6 and Appendix C). Nevertheless,
using the most current information, 1-D calculations for
the stratosphere give a concentraion profile of HCl that

is consistent with the available measurements, which are limited to the lower stratosphere (Figure 7.15). For the lower stratosphere, the calculated concentrations of ClX are insensitive to the choice of vertical transport coefficients considered in this report. With other factors the same, all the mixing profiles shown in Figure 7.1 will predict the same changes in the ozone concentration profile up to 30 km, due to ClX. Above this level the more pronounced differences in ozone (in smaller concentrations) have but little effect on total ozone concentration.

For the troposphere, the predicted vertical distribution of HCl is directly dependent on the assumed rain-out rate and the tropospheric transport coefficients. However, tropospheric HCl contributes only a small fraction of the stratospheric ClX, so uncertainties in the former have a negligible effect on the calculations of stratospheric ClX and the resultant reductions in ozone.

The vertical distribution of ClX in the stratosphere does depend on the source strengths of Cl (and thence ClO, HCl, and ClONO$_2$) from the various halocarbons such as CCl$_4$, CH$_3$Cl, CF$_2$Cl$_2$, and CFCl$_3$ and their altitude.

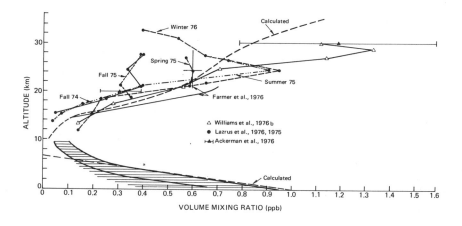

FIGURE 7.15 Comparison of measured and calculated (------) vertical distributions of HCl. Calculations are based on the 1-D model using the eddy-mixing profile E of Figure 7.1. The data are from various authors (see Appendix C, Section VI.C, and Chapter 6, Section III.D). The hatched area indicates uncertainty limits for the tropospheric concentration (see Appendix C). For references see Chapter 6.

In the case of CH_3Cl and CCl_4 there are some uncertainties in the strengths of natural sources (Chapters 3 and 6). But these uncertainties should have little effect on the uncertainties in the reductions in stratospheric ozone by additional ClX from man-made sources. The contributions to the total ClX from the various surface sources are linearly additive, and the ozone reduction by them is also linearly dependent on total ClX, except for reductions larger than 20 percent or more. Hence, the reduction in ozone by ClX from human sources is in addition to that from natural sources.

The chlorine species, $ClONO_2$, has proved to be less significant than estimated early this year (cf. Chapters 4 and 8). Nevertheless it may be an important link between the ClX and NO_y species. Through the reaction $ClO + NO_2 + M \to ClONO_2 + M$, an increase in ClO could lead to a decrease in NO_2, and vice versa. Consequently, there is a negative feedback mechanism between these two catalytic cycles that diminishes their collective effect on ozone, by storing the reactive species in the noncatalytic $ClONO_2$ reservoir. As yet, the only stratospheric measurement of $ClONO_2$ is the upper bound of 1 ppb at 25 to 30 km placed upon its concentration by solar infrared spectra (Chapter 6). Based on present model results (Figure 7.16) it can best be detected at about 23-28 km, where its concentration should be a few tenths of a ppb. Below or above this altitude range the $ClONO_2$ concentration decreases by an order of magnitude within approximatley 5 km.

The direct measurement in the stratosphere of ClO (or of Cl if it becomes feasible) would confirm the stratospheric photodissociation of halocarbons, the occurrence of which is demonstrated by the concentration profiles observed and calculated for $CFCl_3$ (Figure 7.12) and CF_2Cl_2 (Figure 7.13). Stratospheric observations of ClO and Cl could also serve as another check on the accuracy of the predictions (Figure 7.16). Preliminary measurements of ClO are being attempted (Chapter 6 and Appendix C).

F. Concluding Comments

In this chapter (Figures 7.2-7.11), most of the comparisons made are related to the "natural" ozone balance, for which a sound understanding is needed in order to predict the effects of pollutants on it. An impressive demonstration of the present extent of this understanding has been provided by the recent analysis of solar proton events.

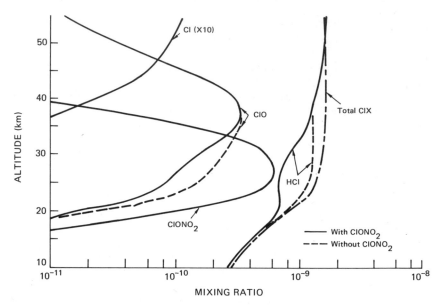

FIGURE 7.16 Calculated vertical profiles of various Cl*X* species calculated for late 1975 on the basis of past releases of F-11 and F-12 plus estimates of Cl from HCl, CH_3Cl, and CCl_4 as discussed in Section II.A.

Crutzen *et al.* (1975) predicted that high-energy protons produced by solar flares could generate enough nitric oxide at high latitudes to produce substantial, predictable, local decreases in stratospheric ozone by the catalytic NO_x cycle that dominates the natural ozone balance. Such solar proton events may be considered as natural experiments, in which the magnitude of the perturbation is determined by the proton flux and energy distribution. The predictions have been verified by subsequent analysis (Crutzen *et al.*, 1976) of satellite data characterizing a major proton event that occurred near the north pole in August 1972. A localized ozone decrease of about 16 percent was calculated for it, which is to be compared with a decrease of the order of 20 percent that was actually observed in that region at that time by the Nimbus 4 polar-orbiting satellite. The analysis is stated to constitute "a strong validation of the current photochemical theory of stratospheric ozone above 30 km."

Most of the species important to the reduction of stratospheric ozone by CFMs have been observed, at least to some extent. In this chapter, we have compared

measured concentrations with the results of 1-D calculations based on our present understanding of the chemical
and photochemical processes involved. The atmospheric
measurements are often limited by poor coverage in space
and time, as well as by experimental sensitivity and
accuracy. The 1-D models have clearly defined and well-
understood limitations, not all of which have been enu-
merated in this report. Nonetheless, 1-D calculations
are useful for studying the balance in global ozone. As
"engineering" and empirically adjusted approximations,
these calculations have a demonstrable validity and accuracy. They are reasonably successful in describing the
vertical transport of trace species. The general level
and extent of agreement between the observed and calculated concentrations are within the known limitations
of both.

REFERENCES

Ackerman, M., J. C. Fontanella, D. Frimout, A. Girard, N.
Louisnard, and C. Muller. 1974. Simultaneous measure-
ments of NO and NO_2 in the stratosphere, *Proc. of the
Anglo-French Symposium on the Meteorological Effects of
Stratospheric Aircraft*, Vol. 1, pp. 1-23.

Alyea, F. N., D. M. Cunnold, and R. G. Prinn. 1975.
Stratospheric ozone destruction by aircraft-induced
nitrogen oxides, *Science* 188:117-121.

Anderson, J. G. 1975. The absolute concentrations of
$O(^3P)$ in the earth's stratosphere, *Geophys. Res. Lett.*
2:231-234.

Anderson, J. G. 1976. The absolute concentrations of
$OH(X^2\pi)$ in the earth's stratosphere, *Geophys. Res. Lett.*
3:165-168.

Callis, L. B., R. E. Boughner, and V. Ramanathan. 1976.
The importance of coupled stratospheric phenomena:
temperature, chemistry, solar scattering, and vertical
transport, preprint.

Callis, L. B., V. Ramanathan, R. E. Boughner, and B. R.
Barkstrom. 1975. The stratosphere: scattering effects,
a coupled 1-D model, and thermal balance effects, *Proc.
of the Fourth Conf. on the Climatic Impact Assessment
Program*, U.S. Dept. of Transportation. To be
published.

Chameides, W. L., and J. C. G. Walker. 1975. Stratospheric ozone: the possible effects of tropospheric-
stratospheric feedback, *Science* 190:1294-1295.

Chang, J. S. 1974. Simulations, perturbations and interpretations, *Proc. of the Third Conf. on the Climatic Impact Assessment Program,* U.S. Dept. of Transportation, DOT-TSC-OST-74-15, pp. 330-341.

Chang, J. S., D. J. Wuebbles, and W. H. Duewer. 1975. Sensitivity to parameter uncertainties for ozone reduction from chlorofluoromethanes, Lawrence Livermore Laboratory Rep. UCRL-77432.

CIAP Monograph 1. 1975. *The Natural Stratosphere of 1974.* A. J. Grobecker, ed. Final report, U.S. Dept. of Transportation, DOT-TST-75-51, Washington, D.C.

CIAP Monograph 3. 1975. A. J. Grobecker, ed. *The Stratosphere Perturbed by Propulsion Effluents,* U.S. Dept. of Transportation, DOT-TST-75-53, Washington, D.C.

Climatic Impact Committee. 1975. *Environmental Impact of Stratospheric Flight: Biological and Climatic Effects of Aircraft Emissions in the Stratosphere.* National Academy of Sciences, Washington, D.C.

Crutzen, P. J. 1974. Estimates of possible variations in total ozone due to natural causes and human activities, *Ambio* 3:201-210.

Crutzen, P. J. 1975. As reported in Figure III.3, p. 27 of the IMOS (1975) report.

Crutzen, P. J., D. F. Heath, and A. J. Krueger. 1976. Influence of a solar proton event on stratospheric ozone. Preprint.

Crutzen, P. J., and I. S. A. Isaksen. 1975. The impact of the chlorocarbon industry on the ozone layer. Preprint.

Crutzen, P. J., I. S. A. Isaksen, and G. C. Reid. 1975. Solar proton events: stratospheric sources of nitric oxide, *Science* 189:457-459.

Dütsch, H. U. 1971. Photochemistry of atmospheric ozone, *Advan. Geophys.* 15:219-322.

Evans, W. F., J. B. Kerr, and D. I. Wardle. 1975. The AES stratospheric balloon measurements project: preliminary results, Atmospheric Environment Service, Canada, Internal Rep. No. APRB-30-X-2.

Hering, W. S., and T. R. Borden, Jr. 1964, 1965, 1967. Ozonesonde observations over North America, Environmental Research Papers No. 38, AFCRL-64-30 (II); No. 133, AFCRL-64-30 (III); and No. 279, AFCRL-64-30 (IV).

Johnston, H. S. 1976. University of California at Berkeley. Personal communication.

Krueger, A. J. 1973. The mean ozone distribution from several series of rocket soundings, *Rev. Pure Appl. Geophys.* 106-108:1272-1280.

Liu, S. C., and R. J. Cicerone. 1976. University of
Michigan. Private communication.

Luther, F. M., and R. J. Gelinas. 1975. Effect of
molecular multiple scatterings and surface albedo on
atmospheric photodissociation rates, Lawrence Livermore
Laboratory Rep. UCRL-75160.

Luther, F. M., J. S. Chang, and D. J. Wuebbles. 1976.
Radiation feedback effects in a one-dimensional strato-
spheric model, Lawrence Livermore Laboratory Rep.
UCRL-77298.

McCarthy, R. L. 1974. Fluorocarbons in the environment.
Paper presented at the American Geophysical Union Meeting, San Francisco, Calif., December 13.

McElroy, M. B., S. C. Wofsy, J. E. Penner, and J. C.
McConnell. 1974. Atmospheric ozone: possible impact
of stratospheric aviation, *J. Atmos. Sci.* 31:287-303.

Rowland, F. S., and M. J. Molina. 1975. Chlorofluoromethanes in the environment, *Rev. Geophys. Space Phys.*
13:1-35.

Shimazaki, T., and R. C. Whitten. 1976. A comparison of
one-dimensional theoretical models of stratospheric
minor constituents, *Rev. Geophys. Space Phys.* 14(1):1-12.

Turco, R. P., and R. C. Whitten. 1975. Chlorofluoromethanes in the stratosphere and some possible consequences for ozone, *Atmos. Environ.* 4:1045-1061.

Widhopf, G. F. 1975. Meridional distributions of trace
species in the stratosphere and the effect of SST pollutants. Preprint.

Wofsy, S. 1975. Interactions of CH_4 and CO in the
earth's atmosphere. Preprint.

Wofsy, S. C., J. C. McConnell, and M. B. McElroy. 1972.
Atmospheric CH_4, CO, and CO_2, *J. Geophys. Res.* 77:
4477-4493.

Wofsy, S. C., and M. B. McElroy. 1974. HO_x, NO_x and
ClO_x: their role in atmospheric photochemistry, *Can.
J. Chem.* 52:1582-1591.

Wuebbles, D. J., and J. S. Chang. 1975. Sensitivity of
time varying parameters in stratospheric modeling,
J. Geophys. Res. 80:2637-2642.

8 EXPECTATIONS FOR
OZONE REDUCTION
BY THE CFMs F-11 and F-12

I. INTRODUCTION

Earlier chapters have discussed our knowledge of the
chemistry and physics of stratospheric ozone, both its
natural stability and its susceptibility to perturbation.
This knowledge has risen substantially in the last 4
yr through the scientific effort stimulated by the
recognition of possible stratospheric disturbance by
emissions from supersonic transports (SST's) and then by
concern about possible similar effects due to the Space
Shuttle. Even though there remain aspects of the stratospheric structure that are not fully understood, this
intensive effort* has greatly improved our ability to
analyze and predict perturbations to the ozone shield.

   With this immediately relevant background, the
numerous recent studies of possible stratospheric perturbation by the CFMs provide a base for assessment of
their possible long-range effects on the stratospheric
ozone, both the magnitude and the time dependence. As
an important part of that assessment we can attempt to
assign to these expectations uncertainty ranges that
indicate the current limitations of our knowledge.

   Several premises can be stated with certainty.

---

*The CIAP program alone had a budget of $21 million over
the period 1971-1974.

*-- Most of the CFMs released have been accumulating in
the troposphere, in particular, CFCl$_3$ (F-11) and CF$_2$Cl$_2$
(F-12).*

This accumulation in the troposphere is known through
direct atmospheric measurements. It is clearly attributable to the relative chemical inertness of these com-
pounds--the very property that has given the CFMs so many
practical applications.

*-- The CFMs are slowly transported into the stratosphere,
where the chlorine they contain is converted, at least
in part, to a chemically active form (Cl, ClO) through
absorption of light in a spectral range that does not
penetrate to the troposphere.*

The CFMs have been directly measured in the stratosphere.
The photolytic sensitivity of the CFMs to irradiation in
the solar spectral range encountered in the stratosphere
is well known from laboratory measurements. Furthermore,
the stratospheric concentrations of CFCl$_3$ and CF$_2$Cl$_2$ have
been observed to decrease above 20 km by different amounts
consistent with their predicted photolysis and transport
rates.

*-- If stratospheric Cl or ClO is added (e.g., through
photolysis of CFMs) to that already present from natural sources it will perturb, to some extent, the amount
of stratospheric ozone.*

Laboratory studies have shown that the chemical reactions
of Cl and ClO with other stratospheric ingredients do
occur, including the catalytic reactions by which ozone
is destroyed.

The reduction of stratospheric ozone has been identified as a potential hazard. Since the atmospheric CFM
content is rising every year, its effect on the ozone
shield is increasing. Hence, we need the best possible
quantitative estimates of the *magnitude* of the disturbance, its projected *growth,* and its *recovery period.*
Such estimates are presented here on the basis of three
possible schedules of CFM release. But first the sources
and magnitudes of the uncertainties that accompany these
predictions are reviewed.

II. UNCERTAINTIES IN THE PREDICTIONS

Discussions of the various sources of uncertainty in estimating the reduction of stratospheric ozone by the CFMs

are scattered throughout our report in connection with the topics where the uncertainties occur. Our intent here is to summarize the most significant sources and to relate them to the particular uncertainties that are treated in detail in this chapter. As a definition of uncertainty range we have adopted 95 percent confidence limits. This would correspond to twice the standard deviation (2σ) if the uncertainties had a Gaussian error distribution, which most of them probably do not. In any case, there is a 2.5 percent chance that the actual value of the uncertain quantity is higher than the upper end of the range and likewise a 2.5 percent chance that it lies below the lower limit. The most probable value should be centrally located between the two limits.

It is difficult to place numerical values on a number of the more serious possible sources of error, and there are many subjective elements in our uncertainty analyses. Our numerical estimates correspond to the odds the panel would find reasonable in betting on a particular outcome. We have categorized the various sources of error under release rates, transport, stratospheric chemistry, and other factors, each of which is considered below.

A. *Release Rates*

As discussed in Chapter 3, the uncertainties in past production and release of F-11 and F-12 are relatively small (±5 percent). They are important primarly in the materials balance method for determining the extent of any inactive removal of the CFMs from the atmosphere (Appendix E). Future releases are highly uncertain. This may be seen in Figures 3.1 and 3.2, where it is shown that after nearly two decades of 10 percent per year exponential growth, production dropped by 15 percent in 1975. Therefore, in our calculations we have assumed particular release schedules for the three major types of situations: continued growth, constant release, and interrupted release. The "1973" release rates that we have used for the constant release estimates of ozone reduction are close to the actual values for 1975 (within 1 percent for the total). (They are 3 percent less than the more recent 1973 values given in Chapter 3.)

B. Transport

Two types of approximation can be identified in 1-D calculations of stratospheric chemistry: (1) the approximations inherent in replacing the actual three-dimensional
processes with *any* one-dimensional (1-D) model and (2)
the choice of the eddy-mixing or vertical-transport profile K for a particular calculation (Chapters 5 and 7,
Appendix B). As yet, insofar as we know, no objective
method of demonstrated reliability has been employed to
determine the inherent approximations. Furthermore, in
practice the choice of K depends in some measure on the
other approximations. We *estimate* that these approximations in the transport model cause a combined uncertainty
in the predicted ozone reduction by a factor of 1.7 in
either direction (i.e., a total range of 3). Fortunately,
the CFM problem is quite a bit less sensitive to transport
uncertainty than is the SST problem.

The choice of K has been analyzed in detail in Appendix
B. In addition, some of the calculations reported in this
chapter were repeated with three different K's covering
a range of values considered to be reasonable. The corresponding results for the ozone reduction have a spread
of no more than ±30 percent (a twofold range) compared
with the threefold range adopted for the combined uncertainties.

C. Stratospheric Chemistry

Factors related specifically to the *particular* reaction
scheme used to predict the ozone reduction include the
chemical reaction rate constants, the solar flux, photolysis rates, the temperature distribution in the stratosphere, and the concentrations (or source and sink strengths)
of trace species in the unperturbed atmosphere (Chapter 7
and Appendix D). Of these, the reaction rate constants
are the largest source of uncertainty that we have identified. The detailed analysis given later in this chapter
indicates that the cumulative uncertainty in ozone reduction due to experimental error in the rate constants is
about a fivefold range.

Lesser but appreciable uncertainties are associated
with the photochemical processes and with the concentrations of natural species. These uncertainties tend to be
interdependent and are more difficult to evaluate. We expect each to be no more than a twofold range, which would

140

combine with the fivefold range for rate constants to give an overall sixfold range for the particular reaction scheme employed (Appendix D).

Photochemical Processes Variations and uncertainties in the solar flux should affect the natural ozone cycles as well as the CFM perturbation and therefore have little effect on the predicted ozone reduction, except for reactants whose concentrations are not linear in the solar flux (Chapter 7). Similarly, uncertainties in photolysis rates of species in the natural cycle should also affect the CFM perturbation, and thereby tend to cancel. As for the species that are a direct part of the CFM perturbation, the photolysis of ClO and HCl were shown to be negligible (Appendix A). The most significant processes are the photodissociation of the CFMs and $ClONO_2$. The Cl production rate by CFM photolysis is known within considerably better than a factor of 2; this uncertainty has a smaller effect on the ozone reduction because of its dominance by transport processes. The case of $ClONO_2$ warrants further study.

Concentrations of Natural Species The two most important families of chemical compounds whose concentrations are determined in part by modeling and in part by observations are the NO_x and the HO_x species. Recent analyses have revealed that the reduction in stratospheric ozone by the CFMs is sensitive to NO_x concentrations. A doubling of NO_x apparently could decrease the ozone reduction by the CFMs by as much as a half under the prevailing circumstances (Chapter 9). As discussed in Chapters 6 and 7, the NO_x concentrations are constrained by modeling and observation to within better than a factor of 2, which would give a corresponding uncertainty range in the ozone reduction. The uncertainty in HO_x concentrations has up to now been determined by the large uncertainty of the rate constant for HO + HO_2. Thus, it is simpler to include the effects of its uncertainty in the analysis of rate constants, which has been done.

D. Other Factors

Included in this category are inactive removal processes (Chapter 4 and Appendix A), feedback mechanisms (Chapter 7 and 9), and the completeness of the reaction scheme employed (Appendix A). The only inactive removal process of consequence that has been identified so far is solution in the oceans, followed by some unknown degradation mechanism. The evidence for such oceanic removal is limited; it might decrease the ozone reductions by about 1/5 (20 percent) of the values otherwise predicted.

The predictions are based on a particular reaction scheme and will be inaccurate if the scheme is incomplete. Until early this year, it was thought that the formation of $ClONO_2$ would have little or no effect on the results, so it was not included. Extensive laboratory studies since then have shown that it can be significant, and the effects of its inclusion are considered later in the chapter.

There is a possibility that in spite of the thorough study already made some as yet unidentified factor, such as an inactive removal process or addition to the reaction scheme, could affect the predictions. But there is no basis for assigning probabilities to such unidentified events. However, the probability of a future discovery should, of course, decrease with increasing length and intensity of the search that has already been completed.

III. PRELIMINARY ESTIMATES OF MAGNITUDE

Before turning to the detailed, computer-based calculations of the ozone reduction, it is helpful to examine the problem through "short-cut" or "rule-of-thumb" estimation procedures. Such procedures achieve simplicity by focusing on a separable part of the problem, by aspiring to limited predictive aims, and by incorporating experimental tie-points that bypass difficult computational areas. These estimation procedures can sometimes provide assurance about the magnitude of a quantity of interest without the opacity of an intricate computer program; they "show us what is going on" and they give us a "ballpark estimate" of the result. Thus, they help protect against a gross error hidden by elaborate computations. At the same time, these estimation procedures may reveal the need for the more ambitious, and more informative, computer-based analyses.

One such estimation procedure is that devised by Johnston (1975). It attempts to define the magnitude of the ozone reduction that would ultimately accompany long-continued CFM release at a constant rate, for a particular reaction scheme. This limited predictive goal, the "steady-state" ozone reduction, obviates the need to calculate the atmospheric transport that carries the CFMs from the point of release, through the troposphere, and into the stratosphere. In the Johnston approach, the CFM interaction with the ozone through the Cl/ClO catalytic cycle is compared with the analogous nitric oxide interaction. This takes advantage of our considerable understanding of the NO/NO_2 catalytic cycle and simplifies the treatment of some reaction sequences that involve molecular species whose stratospheric concentrations are difficult to measure. Such an analysis is described in Section III.D of Appendix A. The magnitude of the estimates, with allowance for the probable role of $ClONO_2$, confirms the need for careful and detailed scrutiny of the matter.

IV.   REDUCTIONS PROJECTED FOR THREE RELEASE SCHEDULES

Calculations of the stratospheric impact on CFMs have been made by a number of different groups of investigators. We need not discuss every attempt that has been made to model the atmospheric behavior of the CFMs. The recent, detailed studies performed by three groups, P. J. Crutzen and I. S. A. Isaksen (NOAA and NCAR), S. C. Wofsy and M. B. McElroy (Harvard University), and J. S. Chang, D. J. Wuebbels, and W. H. Duewer (Lawrence Livermore Laboratory) provide a sufficient range of computer programs to minimize the possibility of a program-induced, spurious result and to permit evaluation of the sensitivity of 1-D predictions to the choice of the eddy-mixing profile K. A number of earlier, less detailed projections have aided in defining the CFM problem; they include studies by Cicerone et al. (1974), Crutzen (1974), Rowland and Molina (1975), Johnston (1975), and Whitten et al. (1975).

Table 8.1 lists the key reactions and the numerical magnitudes of the rate constants used by each of the three groups (Chang et al., 1975; Crutzen, 1975; Wofsy et al., 1975). Of great interest are the choices of the rate constant for the $HO + HO_2$ reaction, for which the uncertainty range is large. Similarly, Figure 7.1 (see Chapter 7) compares the eddy-mixing profiles, for which marked differences are seen. This variability has significance to our analysis. It displays the extent to which "parametric freedom" is available and it shows how different scientists have exercised this freedom as they seek a meaningful description of the atmosphere. Finally, it is of particular importance to note that the last reaction in Table 8.1 between ClO and NO_2 to form chlorine nitrate,

TABLE 8.1 Rate Constant Values Used for Key Reactions by Three Different Groups in 1-D Calculations of Ozone Reduction by ClO_x (cm^3 molecule^{-1} sec^{-1})

Reaction	Crutzen[a]	Wofsy[b]	Best Current Value[c]
$Cl + O_3 \rightarrow$ $ClO + O_2$	$4.23 \times 10^{-11} \exp\left(-\dfrac{358}{T}\right)$	1.8×10^{-11}	$2.97 \times 10^{-11} \exp\left(-\dfrac{243}{T}\right)$
$ClO + O \rightarrow$ $Cl + O_2$	5.7×10^{-11}	5.3×10^{-11}	$3.38 \times 10^{-11} \exp\left(+\dfrac{75}{T}\right)$
$ClO + NO \rightarrow$ $Cl + NO_2$	1.7×10^{-11}	1.7×10^{-11}	$1.13 \times 10^{-11} \exp\left(\dfrac{200}{T}\right)$
$HO + HCl \rightarrow$ $H_2O + Cl$	$2.1 \times 10^{-12} \exp\left(-\dfrac{340}{T}\right)$	$2.1 \times 10^{-11} \exp\left(-\dfrac{1037}{T}\right)$	$2.0 \times 10^{-12} \exp\left(-\dfrac{310}{T}\right)$
$Cl + CH_4 \rightarrow$ $HCl + CH_3$	$4.76 \times 10^{-12} \exp\left(-\dfrac{1104}{T}\right)$	$5.0 \times 10^{-11} \exp\left(-\dfrac{1791}{T}\right)$	$5.4 \times 10^{-12} \exp\left(-\dfrac{1133}{T}\right)$
$HO + HO_2 \rightarrow$ $H_2O + O_2$	6.0×10^{-11}	7.0×10^{-11}	2.0×10^{-11}
$ClO + NO_2 +$ $M \rightarrow ClONO_2$ $+ M^c$	--	--	$\dfrac{1.38 \times 10^{-14} \exp\left(\dfrac{880}{T}\right)}{1.166 \times 10^{18} \exp\left(\dfrac{220}{T}\right) + M}$

[a] P. J. Crutzen and I. S. A. Isaksen. The impact of the chlorocarbon industry on the ozone layer. Submitted for publication.

[b] S. C. Wofsy and M. B. McElroy. 1974. HO_x, NO_x, and ClO_x: their role in atmospheric photochemistry. Can. J. Chem. 52:1582-1591.

[c] See Appendix A for discussion and literature references leading to our choice of these values.

$ClONO_2$, was not included in any of the three studies to be compared. All three studies also assume that there is no inactive removal of the CFMs.

In these studies, three types of schedule for future CFM release into the atmosphere (scenarios) have been examined:

-- *Continued Growth:* Continued CFM release at an increasing rate (10 percent yr^{-1})
-- *Constant Release:* Continued CFM release at a constant rate (1973)
-- *Interrupted Release:* Continued CFM release at an increasing rate (10 percent yr $^{-1}$) up to a sharp cutoff in 1978.

A. Continued Growth

Over the past decade or two, until a 15 percent drop in 1975, the annual production of CFMs has increased by about 10 percent per year, a 7-yr doubling rate. Projections based on such a continued growth pattern are presented in Figure 8.1, as reported for the three versions of the 1-D approximation. Just a glance at the figure indicates that the models agree qualitatively: *continued growth of CFM releases at 10 percent per year will have a major effect on stratospheric ozone within a few decades.* How soon this is predicted to happen differs somewhat; the three projections give a 10 percent reduction of stratospheric ozone in 23 to 28 yr, a range of ±20 percent in the time required.

B. Constant Rate of Release

Another possible scenario is that the CFM release rate might continue to be constant, as it has been since 1974. Chang *et al.* (1975) and Wofsy *et al.* (1975) have made calculations for constant release, although based on different assumptions about the release rate. Chang used an early estimate by McCarthy (Chapter 3) of the 1973 world-wide CFM production (3.14×10^5 and 4.70×10^5 metric tons for F-11 and F-12, respectively), whereas Wofsy took a constant production rate about two thirds as large (2×10^5 and 3×10^5 metric tons, roughly the 1969 world-wide production rate). Their results are shown in Figure 8.2.

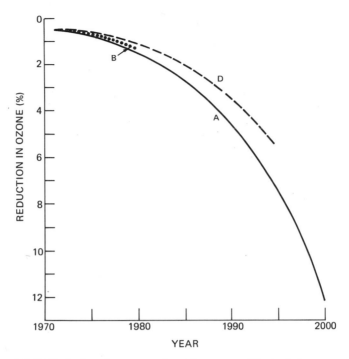

FIGURE 8.1 Ozone reductions predicted by
three different research groups using 1-D
calculations for continued 10 percent growth
per year in release of F-11 and F-12, for
various periods of time. Curve A is by
Chang; B, Crutzen, and D, Wofsy. The calculations do not include the effects of ClONO$_2$
formation; also it is assumed that there is
no inactive removal of the CFMs.

    Obviously, the curves in Figure 8.2 differ in form from
those in Figure 8.1. Both calculations show that releasing CFMs at a constant rate leads, ultimately, to a steady
state, after which the ozone reduction no longer increases,
which is intuitively reasonable. Thus, the "runaway"
quality of the *continued growth* scenario does not develop.
What matters now is the limiting ozone reduction that will
ultimately occur and how rapidly it will be reached. With
allowance for the smaller release rate assumed by Wofsy,
the reductions predicted at steady state (14 and 10 percent) agree within a range of ±20 percent. These results
indicate that *a limited reduction of stratospheric ozone*

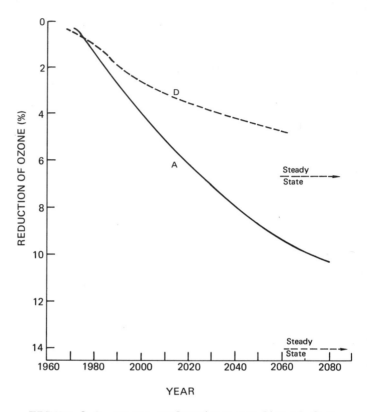

FIGURE 8.2 Ozone reductions predicted for
constant release rates of F-11 and F-12, by
two different research groups. Curve A
(Chang) is for 3.19×10^5 and 4.70×10^5
metric tons/yr of F-11 and F-12, respectively, while curve D (Wofsy) is for 2×10^5 and
3×10^5 metric tons/yr. The steady-state
limit for A is about 14 percent and for B,
6.7 percent. The calculations do not include
the effects of $ClONO_2$ formation; also it is
assumed that there is no inactive removal of
the CFMs.

*will occur if CFMs are released at a constant rate, the
maximum extent of the reduction being directly proportional to the magnitude of the constant release rate. The
time required to reach one half of the maximum (steady-
state) reduction is about half a century.*

C. Interrupted Release

There remains the scenario in which CFM release is stopped at some point in time or sharply reduced. Figure 8.3 shows the three projections by Chang, Crutzen, and Wofsy for ozone reduction based on continued CFM release with a 10 percent increase per year until 1978, at which time release is abruptly halted. Once again, the curves differ in form from those produced by the other types of scenario.

When release is stopped, the CFM already in the tropospheric "reservoir" has not had time to attain equilibrium with the stratosphere. Therefore, after release is stopped, the amount of CFM in the stratosphere and the ozone reduction continue to increase for some years before beginning to slowly decay. The results indicate that *even after a sharp cutoff in releases the perturbation of the stratospheric ozone by CFMs would continue to grow for about 10 yr and the subsequent recovery would be extremely slow.*

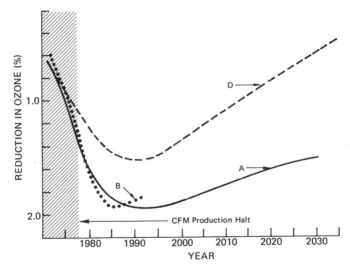

FIGURE 8.3 Ozone reductions predicted by three different research groups for 10 percent per year growth in release rates of F-11 and F-12 until 1978, when all release is stopped. Curve A is by Chang; B, Crutzen; and D, Wofsy. The calculations do not include the effects of $ClONO_2$ formation; also it is assumed that there is no inactive removal of the CFMs.

*The peak perturbation would be about 1.5 times that at
cutoff, and about 50 yr would be required after the
peak for the perturbation to return to its value at cutoff.*

In Chapter 1, it was pointed out that inactive removal
of the CFMs will decrease the steady-state ozone reduction
in proportion to the fraction of CFM removed by such processes. In addition, inactive removal will reduce the
extent of ozone reduction under any scenario of release,
and it will also modify the time scales involved. For
example, it would decrease the time required for the
stratospheric ozone to recover to a given extent after a
cutback in release rates.

V.   UNCERTAINTY DUE TO CHOICE OF *K*

All the factors that influence the uncertainty due to
transport in 1-D calculations have been discussed in
Chapters 5 and 7. Also, as noted earlier in this chapter, their combined effects are considered to lead to
at most a threefold uncertainty range in the prediction
of ozone reduction by CFMs at steady-state conditions.
Here we shall examine the sensitivity of the results to
the choice of eddy-mixing profile (A, B, and E of Figure
7.1) within our standard 1-D model, i.e., with all other
input variables held fixed. As we shall see, the actual
results so obtained span ranges in values for the predicted ozone reductions that are significantly smaller
than the total uncertainty range inferred from Chapters
5 and 7 for transport, and so provide further assurance
that these uncertainties can be bounded.

The Chang calculation program and reaction scheme
(Appendix D) were used together with the rate constants
listed in the last column of Table 8.1, still excluding,
for the purposes of this comparison, the last reaction
involving formation of chlorine nitrate. With these
uniform input variables, Figures 8.4 and 8.5 show the
variability caused by the eddy-mixing profiles A, B, and
E. The results for profile E are given, rather than those
for profile D as in Figures 8.1-8.3, and will be used here-
after to maintain consistency with earlier chapters. The
three profiles are typical of the range that has been
used for *K*.

Figure 8.4 illustrates the constant production scenario.
The shaded area shows the range of projections obtained

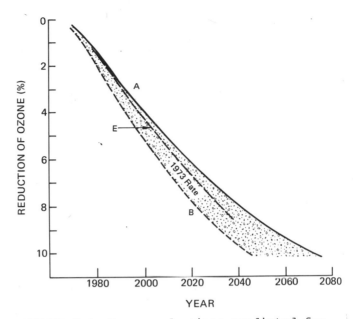

FIGURE 8.4 Ozone reductions predicted for
release rates (1973) obtained in calculations
differing only in the eddy-mixing profile.
Curve A is for the profile of Chang; B, that
of Crutzen; and E, a composite curve from
Chapter 7 (see Figure 7.1). The 1973 release
rates were taken to be 3.14×10^5 and 4.70
$\times 10^5$ metric tons/yr for F-11 and F-12,
respectively. The calculations do not include the effects of $ClONO_2$ formation; also
it is assumed that there is no inactive removal of the CFMs.

when the 1973 production rates are maintained.* The ultimate (steady-state) ozone reduction ranges from 13 percent (Chang) to 15 percent (Crutzen). The time needed to

---

*In this calculation earlier production estimates (IMOS,
1975) of F-11 (3.14×10^5 metric tons) and F-12 ($4.70 \times$
10^5 metric tons) were used instead of the most recent
estimates by the Manufacturing Chemists Association (3.68
$\times 10^5$ and 4.41×10^5 metric tons, respectively). Use of
the latter would increase the calculated values by about
3 percent.

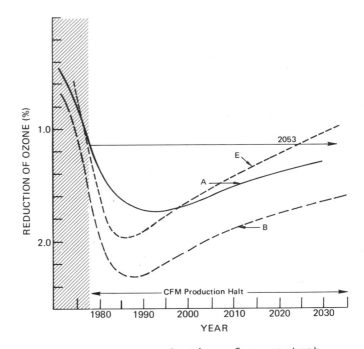

FIGURE 8.5 Ozone reductions for constant release rate (1973) until 1978, when release is stopped, obtained in calculations differing only in eddy-mixing profile. Curve A is for the profile of Chang; B, that of Crutzen; and E, a composite curve from Chapter 7 (see Figure 7.1). The calculations do not include the effects of $ClONO_2$ formation; also it is assumed that there is no inactive removal of the CFMs.

reach one half the steady-state reduction (measured from 1973) ranges from 50 yr for the Chang profile A to 42 yr for the Crutzen profile B.

The results for recovery of the ozone reduction after CFM release is stopped are given in Figure 8.5. As before (Figure 8.3), the ozone reduction continues to grow for a number of years after release of CFMs is completely stopped, and then more gradually recovers from the perturbation. Hence, there are several quantities to consider as we compare the results for the three profiles:

1. How many years are required (after cessation of CFM
release) for ozone reduction to reach its maximum value,*
and what is the peak value?

2. At this maximum value, how much does the ozone
reduction exceed the value it held when CFM release was
stopped?

3. How many years are required (after cessation of
CFM release) for the stratosphere to recover from its
perturbation by the CFMs?†

These quantities are compared in Table 8.2. The last
column gives the central value and the actual range for
the three values. It is predicted that if CFM release
were to be completely halted in 1978, ozone reduction due
to CFMs would continue to rise for a decade (8 to 13
yr) to a peak reduction of 2 percent (1.7 to 2.3 percent) that is 1.5 times larger (1.4 to 1.6) than the 1978
value, and it would require 55 yr (44 to 75 yr) to
return to the 1978 value.

The results are comparable with the different curves
in Figures 8.2 and 8.3 for the same scenarios but with
different rate constants as well as different vertical
mixing profiles. The analysis of Appendix B covers a
wider range of profiles, including some that are unlikely,
and gives a larger range in the values found for the rise
time (±5 yr) and recovery time (±30 yr) but indicates a
smaller spread in the peak values after release is
stopped.

Thus, *the choice of the eddy-mixing profile K in 1-D
calculations leads to predictions of the ozone reduction
by CFMs that fall generally within a twofold range of
values, whatever the assumed schedule of release.*

VI.   ROLE OF CHLORINE NITRATE ($ClONO_2$)‡

The formation of chlorine nitrate by the reaction ClO +
NO_2, with stabilization by a third body M, was included
by Rowland and Molina (1975) in their summary table of

---

*This is the time called t_1 in Chapter 5 and Table B.8.
†This is related to the time called t_2 in Chapter 5 and
Table B.8.
‡$ClONO_2$ is structurally similar to nitric acid (HNO_3) in
which the hydrogen is also bonded to one of the oxygens,
$HONO_2$.

TABLE 8.2 The Ozone Reduction Predicted for Interrupted Release of F-11 and F-12 and Its Dependence on the Choice of the Vertical Transport Profile[a]

	Crutzen B	Chang A	Profile E	Central Value and Range[b]
Year of peak reduction	1986	1991	1986	1988 ± 3
Peak reduction	-2.31%	-1.75%	-1.96%	-2.0 ± 0.3%
Time to peak reduction	8 yr	13 yr	8 yr	10 ± 3 yr
(Peak/1978) reduction	1.43	1.52	1.63	1.5 ± 0.15
Year of return to 1978 reduction	2033	2053	2022	2033 ± 10
Time to return to 1978 reduction	55 yr	75 yr	44 yr	55 ± 10 yr

[a]The release schedule assumes past releases to 1973, then constant release at the 1973 rates until 1978, when all release is stopped. The 1973 rates are taken to be 3.14×10^5 and 4.70×10^5 metric tons/yr for F-11 and F-12, respectively.
[b]This is the actual range from the calculations, not confidence limits.

possible chain termination reactions. It was not regarded at that time by them nor, until quite recently, by any others to be an important reaction. Early this year, however, Rowland et al. (1976) reopened the question of the possible role of $ClONO_2$. Insofar as $ClONO_2$ is formed and not rapidly reconverted to active species, it provides an inactive reservoir of *both* ClO and NO_2, reducing the destruction of ozone by the NO_x catalytic cycle as well as by the ClO_x cycle. Thus, the significance of $ClONO_2$ formation is twofold compared with the removal of ClO alone. The effects of its formation depend on three factors:

-- The rate of the reaction $ClO + NO_2 + M$
-- The photolytic dissociation rate of $ClONO_2$
-- Any other chemical and physical processes that convert $ClONO_2$ back into the active ClO and NO_2 forms or remove the $ClONO_2$ from the stratosphere

In considering the role of $ClONO_2$, Rowland et al. (1976) found that much of its supposed reactivity and short photochemical lifetime reported by earlier workers was due to impurities. They redetermined its ultraviolvet spectrum, extended it to shorter wavelengths, and made preliminary estimates of chlorine nitrate's possible effect on calculations of ozone reductions. In the estimates, they assumed that the rate constant for the $HO + NO_2 + M$ reaction could be taken as an upper limit for the formation of $ClONO_2$. Their preliminary results drew attention to the fact that chlorine nitrate formation *must* be considered in the reaction scheme that describes the effect of CFMs on stratospheric ozone.

Because of the sharp focus of attention that is now placed on the CFM problem, several groups quickly made laboratory measurements of the rate constants needed to predict the effect of $ClONO_2$, while others made 1-D calculations analyzing its role. The considerable data now available on the chemistry of $ClONO_2$ are described in Section III.D of Appendix A. We have used them to assess the effect of adding $ClONO_2$ formation to the CFM reaction scheme (Chapter 7). The evidence on the rate constant for the $ClO + NO_2 + M$ reaction places it an order of magnitude less than the analogous $HO + NO_2 + M$ reaction (with a negative activation energy, near 950 cal/mole) (see Tables 8.1 and A.3). Rowland's photolytic cross sections (1976) were used with the assumption that the photolysis returns both chlorine and nitrogen to active form (ClO_x and NO_x) with unit quantum efficiency. The only other $ClONO_2$ destruction process included was the reaction with oxygen atoms, $O(^3P)$, using the rate constant measured at 245 K, $k = 2 \times 10^{-13}$ cm^3 sec^{-1} (Table A.3), and again assuming that both chlorine and nitrogen are returned to active form (Appendix D).

The effect of adding this chlorine nitrate chemistry was examined by 1-D calculations with the eddy-mixing profile E for the same three CFM release schedules employed previously. Like the behavior displayed in Figure 8.1, continued exponential growth of CFM release still causes large perturbations in stratospheric ozone, although the time development is slower and the nature of these perturbations differs somewhat. Initially, the ClO_x and NO_x catalytic processes are linked by the $ClONO_2$, the increase in one being offset by a decrease in the other. Ultimately, sufficiently high CFM concentrations are reached so that the amount of ozone becomes dominated by the chlorine chemistry. The final result is still a major reduction in stratospheric ozone.

The predictions for constant and interrupted release schedules with $ClONO_2$ in the reaction scheme have virtually the same evolution in time as those given in Figures 8.4 and 8.5 without it. However, the magnitudes of the steady-state and peak ozone reductions are decreased from 14.0 to 7.5 percent and from 2.0 to 1.1 percent, respectively, with eddy-mixing profile E. Thus, unless there is some presently unknown process that quickly returns $ClONO_2$ into active ClO_x and NO_x species, *the effect of chlorine nitrate formation is to decrease the projected ozone reductions by about a factor of 1.85 compared with the values calculated for the CFMs without this reaction.*

This predicted effect is large enough that further evidence about its magnitude is desirable. Therefore, efforts are being made to observe the presence of $ClONO_2$ in the stratosphere by spectroscopic methods (Chapter 6). So far, infrared data have placed *an upper bound* of about 1 ppb on the amount of $ClONO_2$ that could be present at altitudes of 15 to 21 km. The value calculated to be in that region at present is about 0.5 ppb (Figure 7.16); therefore, more sensitive measurements should resolve the issue.

VII. CHANGES IN THE OZONE PROFILE

Thus far, attention has been focused on the expected reduction in the total ozone column. This quantity, of course, determines the intensity of the ultraviolet light transmitted to the earth's surface through the atmosphere. There is also the possibility that at particular altitides the local change in ozone is much larger or smaller than the average for the total column. Large perturbations in the ozone profile might have their own consequences; for example, they could alter the stratospheric temperature profile somewhat, conceivably enough to be a noticeable factor in climate or in the transport of water vapor through the tropopause (Chapter 9).

Changes in the ozone profile tend to be overestimated unless special care is taken to include all feedback mechanisms between the CFMs, ClX, and O_3. In Figure 8.6 we show three ozone profiles calculated for the stratosphere with the standard 1-D model (Appendix D), which includes such mechanisms. The curve for the unperturbed atmosphere is a reference profile; the other two are the steady-state perturbations by constant CFM release at the 1973 rate, with and without $ClONO_2$ in the reaction scheme. The ratios of the perturbed profiles to the reference are

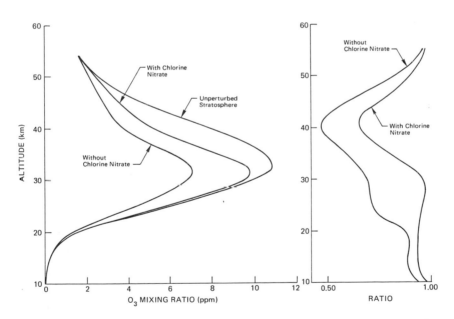

FIGURE 8.6 Steady-state ozone profiles calculated for the unperturbed stratosphere and for constant CFM release at the 1973 rate, with and without $ClONO_2$ formation. The ratios of the perturbed profiles to the reference profile are given at the right.

also given. In both instances, the maximum perturbation
is in the 35- to 45-km region, where it is about four times
larger than the average reduction. However, the perturbation without $ClONO_2$ in the reaction scheme is also relatively large at lower altitudes, down to 25 km. These
results suggest that the shape of the ozone profile might
be a more sensitive indicator of CFM perturbations than
the total column density.

VIII.   SENSITIVITY OF OZONE PREDICTIONS TO REACTION RATE
        CONSTANTS

There remains the question of how sensitive the projections are to the uncertainties in rate constants for
the reaction scheme employed. This can be explored by
making calculations of the ozone reduction (R_0) for
standardized conditions (with the preferred set of parameters), by chaning a particular rate constant from its
preferred value k_0, and then recalculating the ozone reduction. The ratio of the fractional change in ozone reduction to that in the rate constant is a measure of the
reduction's sensitivity to the rate constant, where

$$r = \frac{(R - R_0)/R_0}{(k - k_0)/k_0} = (\Delta R/R_0)/(\Delta k/k_0) \qquad (8.1)$$

gives the sign as well as the magnitude of the dependence.
This type of sensitivity test has been conducted for seven
rate constants that have been identified as having an appreciable effect on the projections.
    The reactions and our presently preferred values for
their rate constants are listed in Table 8.1. These values
are based on the recent determinations and references giv-
en in Table A.2 and either are the average of the values
cited or were chosen for the reasons summarized in Section
III.D of Appendix A. The sensitivities calculated via
Eq. (8.1) for each rate constant are listed in Table 8.3.
For the HO + HCl and HO + HO_2 reactions, r was found to
depend somewhat on whether $ClONO_2$ is included in the reaction scheme. In both cases, the sensitivity is greater
with $ClONO_2$ included, presumably because of feedback from
its effect on the NO_x cycle. For HO + HO_2, the ozone
reductions is not linearly related to k, so the sensitivity
for the reaction was obtained from the slope at k_0 of a
ln k versus ln R plot at k_0.

TABLE 8.3 The Uncertainty in Ozone Reduction Projections due to Reaction Rate Constants, with and (without) Inclusion of $ClONO_2$ in the Reaction Scheme[a]

Reaction	f	r	$(r \ln f)^2$
$Cl + O_3$	1.20	0.290	0.003
$ClO + O$	1.35	0.256	0.006
$ClO + NO$	1.40	-0.194	0.004
$HO + HCl$	1.30	1.30 (0.724)	0.116 (0.036)
$Cl + CH_4$	1.50	-0.324	0.017
$HO + HO_2$	3.00	-0.592 (-0.506)	0.423 (0.309)
$ClO + NO_2$	1.50	-0.766	0.096

Cumulative uncertainty *with* $ClONO_2$:

$$\varepsilon = [\Sigma (r \ln f)^2]^{1/2} = 0.815 \qquad e^{+0.815} = 2.26$$

$$e^{-0.815} = 0.442$$

Cumulative uncertainty *without* $ClONO_2$:

$$\varepsilon = [\Sigma (r \ln f)^2]^{1/2} = 0.686 \qquad e^{+0.686} = 1.99$$

$$e^{-0.686} = 0.50$$

[a]The range of uncertainty for the rate constant of the reaction is defined by the factor f; the sensitivity of the predicted ozone reduction to the value for the rate constant is given by r, the ratio of the fractional change in the ozone reduction to the corresponding fractional change in the rate constant.

IX. UNCERTAINTY DUE TO REACTION RATE CONSTANTS

Having explored the sensitivity of the ozone reduction to experimental error in individual rate constants, we can consider the uncertainty that they generate as a whole. While the data are not susceptible to a rigorous error analysis, it is possible to make a useful estimate of the cumulative effect. In order to do this, we need to know whether the effects of the rate constants on the ozone reduction are independent of one another (additive), at least over the ranges of uncertainty in question. It is conceivable that concerted changes in two or more of the rate constants could affect the ozone reduction by an amount differing from the sum of the effects taken individually. To search for such a possibility, the rate constants were simultaneously altered in the directions that (taken singly) would give r values of the same sign (+ or -), using the reaction scheme that includes $ClONO_2$. In both cases, the ozone reduction that resulted was nearly the sum of the changes calculated individually.

This makes it reasonable to treat the uncertainties as independent and additive. In combining them, allowance must be made for the fact that some are likely to contribute in the opposite sense to others. This dictates a form based on the square root of a sum of squares. Second, the manner in which experimental difficulties may be reflected in the magnitude of a rate constant must be considered. The experimental data are interpreted through a logarithmic plot. Hence, the uncertainty in the logarithm of the rate constant is more directly connected to the data, so we have used logarithmic terms in our analysis.

The contribution of each rate constant to the uncertainty in the predicted ozone reduction depends on the uncertainty factor f for the constant as well as on the sensitivity r. The value of f is chosen so that the 95 percent confidence limits (2σ) are k_0/f and fk_0. A detailed discussion is given in Appendix A (Section III.D) of the value of f for the HO + HO_2 reaction. For ClO + NO_2 + M, the rate constant has uncertainties (2σ) in the A factor and in the activation energy that correspond to ±14 percent ($f = 1.14$) at 250 K. However, there are indications that inclusion of $ClONO_2$ in the reaction scheme increases the dependence of the ozone reduction on the choice of eddy-mixing profile. To allow for this, we increase the f for the reaction to 1.50.

For the other five reactions, f was calculated by taking the square root of the sum of the squares of the following quantities: twice the standard deviation of the recent experimental data, the deviation of the most divergent experimental result from the mean value, and the estimated systematic error of a typical experimental study. The resulting values for f are included in Table 8.3 along with those for r. The last column in Table 8.3 is the square of the value of (r ln f) and, at the bottom, the square root of the sum of these squares, $\varepsilon = [\Sigma(r \text{ ln } f)^2]^{1/2}$. The quantity exp($\pm\varepsilon$) expresses the limits of the uncertainty range due to these rate constants.

Inspection of the results shows that *the large uncertainty in the HO + HO₂ rate constant (a sixfold range) is at present the largest single source of uncertainty identified in predicting ozone reductions by the CFMs. It produces a threefold range of uncertainty in the projections. The overall uncertainty produced by the seven reactions is a fivefold range with inclusion of ClONO₂ in the reaction scheme and a fourfold range without it.*

Although this analysis for the rate constants was made in terms of the ozone reduction calculated for 1975, it applies as well to the steady-state ozone reduction in a constant-production scenario. The curves for the latter (e.g., Figure 8.4) are well described by an exponential expression of the form $A(1-e^{-\lambda t})$, in which λ represents the total removal rate for the CFMs. The value of λ depends predominantly on the eddy-mixing profile (which describes the rate of CFM injection into and HCl removal from the stratosphere) and on the rate of CFM photolysis. For ozone reductions less that 15 percent, the chemical rate constants have very little effect on λ or the time dependence. On the other hand, the rate constants strongly affect the catalytic chain length for ozone destruction, which appears in the multiplicative constant A. This implies that a change in a rate constant produces, to a good approximation, a linear scaling of the entire time-dependent curve. Thus, the relative uncertainties estimated for the 1975 ozone reductions are applicable to the steady-state limit of a constant production scenario.

From the above, it is apparent that improvement in the accuracy of the rate constants is called for. The first four reactions in Table 8.3 are known fairly accurately, with standard deviations (±1σ) of ±20 to ±25 percent. The f factors in Table 8.3, however, were chosen conservatively to assure 95 percent confidence limits (2σ). Three of the reactions, and especially HO + HO₂, account for over two thirds of the uncertainty that the rate constants contribute to the ozone reduction. At least two additional studies of the Cl + CH₄ reaction are under way, and within the next 6 to 12 months there is reason to expect a reduction of its f factor from 1.50 to about 1.30. For the experimentally much more difficult HO + HO₂ reaction, further work is planned or under way in three or four laboratories, and considerable improvement may be expected in 1 to 2 yr.* If its f factor is reduced to, say 1.5 and those of the other reactions to 1.3, the sum of these improvements would lead to a substantial re-

duction of the uncertainties in the ozone projections.
The contribution of these rate constants would then have
a range of about 2.5 instead of 4 to 5.

X. OVERALL PROJECTIONS OF OZONE REDUCTION BY CFMs

We can now combine the previous analyses to give an estimate of the ozone reduction that seems likely to result
from a constant release schedule of F-11 and F-12. *At
present there is no reason not to include ClONO$_2$ in the
reaction scheme. With its inclusion, the steady-state
reduction in stratospheric ozone is estimated to be
7.5 percent for the 1973 release rates. However, this
value could be decreased to about 6 percent (by 1/5) by
an oceanic sink, if more detailed study confirms the preliminary indication of its presence.*
    The sources of uncertainty on which we have been able
to place numerical factors (*f*) are 1.05 for release rates,
1.7 for transport (a threefold range), and 2.4 for atmospheric chemistry (a sixfold range), the latter being
largely due to the reaction rate constants (a fivefold
range). Their effects are generally multiplicative in
nature, so we combine them as

$$u = [\Sigma (\ln f)^2]^{1/2} \qquad\qquad (8.2)$$

which gives overall uncertainty limits of exp(−u) and
exp(+u). The resultant uncertainty factor is 2.8. *Thus,
the combined uncertainties, for which numerical estimates
could be made, correspond to an eightfold range with
limiting factors of 0.36 and 2.80. Application of these
limits to the likely value of 6 to 7.5 percent for the
ozone reduction gives an overall range of from 2 to 20
percent.*

---

*At present, the large uncertainty in this rate constant
dominates the other smaller sources of uncertainty in
stratospheric HO$_x$ concentrations. An improved value for
it, especially if on the low side of the range, would increase the need for narrowing the other uncertainties.

XI.   UNCERTAINTIES IN TIME DEPENDENCES

In the discussion so far, emphasis has been placed on the
steady-state projections, the simplest, most direct description of the consequences of CFM release. However,
one should not overlook the importance of the time required for the CFMs to build up in the stratosphere or to
decay after a cutback in release rates. As pointed out
in Chapter 5 and Appendix B (see Table 5.1) and as shown
by the discussion of $ClONO_2$ earlier in this chapter, the
time constants are more sensitive to uncertainties in
transport than to other parameters that have a major effect on the ozone reduction. Furthermore, the integrated
amount of ozone reduction can be much the same for a fast
buildup and recovery as for a slow buildup and recovery.
In any case, *we estimate the uncertainties in the time
dependence of the ozone reduction to have a threefold
overall range.*

XII.   SUMMARY

The detailed considerations in this and previous chapters
permit extensions of the premises stated at the beginning
of this chapter. These extensions provide the current
scientific basis that we have for planning. They refer
to projections of ozone reduction, which must be coupled
with expectations about the extent and seriousness of the
impact of ozone reduction on man and the environment.
   Taking into account all the presently known processes
by which CFMs can reach and perturb the stratospheric
chemistry, it is clear that

-- *Release of CFMs at a rate growing in the exponential
   pattern of the last decade could, within a few decades,
   cause a reduction of stratospheric ozone much larger
   than natural variations.*
-- *Release of CFMs at a constant rate will perturb the
   ozone by an increasing amount that will ultimately
   reach a steady state, the limiting reduction depending
   on the worldwide release rate. The 1973 CFM production rate would be expected to reduce the ozone by
   about 6 to 7.5 percent at steady state (with an uncertainty range extending from as little as 2 percent to
   about 20 percent). If $ClONO_2$ proves to be less important than indicated by the present data, the ozone
   reduction could be larger than this by a factor of up*

to 1.85. The time required to reach half of the steady-state reduction would be about 40 to 50 yr.

-- *If CFM release were suddenly and completely terminated,*
   - *Ozone reduction would be expected to continue to rise for about a decade (± about 4 yr),*
   - *Ozone reduction would reach a peak about one and one-half times larger than the ozone reduction at the time of termination,*
   - *50 to 75 yr would be needed to return the ozone reduction level to one half the peak disturbance.*

-- *The possibility of an oceanic inactive removal process should be vigorously investigated, and evidence bearing on the existence of other tropospheric sinks should continue to be sought.*

-- *There are several important areas in which improved data can be expected within the next 1 or 2 yr, most particularly in atmospheric compositional monitoring, in the HO + HO_2 rate constant, in the oceanic sink, and in the chemistry of chlorine nitrate. Since the presently anticipated CFM effect on the stratospheric chemistry is substantial (although it carries large uncertainty limits), a re-evaluation of the question should be made no more than 2 yr hence, whether immediate regulatory action is judged to be appropriate or not. It is essential that the most crucial research called for in Appendix F be encouraged with this time span in mind.*

REFERENCES

Chang, J. S., D. J. Wuebbels, and W. H. Duewer. 1975. Lawrence Livermore Laboratory. Private communciation.

Cicerone, R. J., R. S. Stolarski, and S. Walters. 1974. Stratospheric ozone destruction by man-made chloro-fluoromethanes, *Science* 185:1165-1167.

Crutzen, P. J. 1974. Estimates of possible future ozone reductions from continued use of fluoro-chloromethanes, *Geophys. Res. Lett.* 1:205-220.

Crutzen, P. J. 1975. As reported in Figure III-3, p. 27 of the IMOS (1975) report.

IMOS. 1975. Report of Federal Task Force on Inadvertent Modification of the Stratosphere. *Fluorocarbons and the Environment.* Council on Environmental Quality, Federal Council for Science and Technology.

Johnston, H. S. 1975. Pollution of the stratosphere, *Ann. Rev. Phys. Chem.* 26:315.

Rowland, F. S., and M. J. Molina. 1975. Chlorofluoro-methanes in the environment, *Rev. Geophys. Space Phys.* 13:1-35.

Rowland, F. S., J. E. Spencer, and M. J. Molina. 1976. University of California at Irvine. Private communication.

Whitten, R. C., W. J. Broucki, I. G. Popoff, and R. P. Turco. 1975. Preliminary assessment of the potential impact of solid-fueled rocket engines in the stratosphere, *J. Atmos. Sci.* 32:613-619.

Wofsy, S. C., M. B. McElroy, and N. D. Sze. 1975. Freon consumption: implications for atmospheric ozone, *Science* 187:1165-1167. See also Figure III-2, p. 28 of the IMOS (1975) report.

9 THE SPACE SHUTTLE AND OTHER CONSIDERATIONS

I. INTRODUCTION

The preceding chapters discuss the central aspects of the CFM problem: release, removal processes, stratospheric chemistry, transport, atmospheric measurements, and the probable effect of CFMs on stratospheric ozone. There are, in addition, several topics that seem relevant to our overall charge but that do not fit comfortably under any of those headings. We have elected to collect them together in this final chapter.

II. THE EFFECTS OF THE SPACE SHUTTLE

As presently planned, the Space Shuttle will use a solid fuel consisting of ammonium perchlorate in a matrix of powdered aluminum. The combustion products consist of hydrochloric acid gas and solid particulates of aluminum oxide (Al_2O_3) with traces of aluminum chloride and ferric chloride. The emission rates of the exhaust products are estimated to be of the order of 120 metric tons of Al_2O_3 and 80 metric tons of HCl per flight.

Whitten *et al.* (1975) have estimated the effect on stratospheric ozone of the HCl from the postulated Space Shuttle flight schedule of 50 flights per year, all launched at the latitude of Cape Canaveral, 30° N. They used an eddy-mixing profile that resembles qualitatively that of Chang (profile A in Figure 7.1) but has a much faster vertical mixing rate in the lower stratosphere.

164

Two assumptions about the latitudinal distribution of
Space Shuttle exhaust products were examined. In the
first case, the exhaust products were evenly distributed
over a 1000-km latitudinal band centered at 30° N; in
the second case, the products were evenly distributed
over the entire northern hemisphere. In both cases, no
further horizontal mixing was considered. The altitude
of exhaust release was assumed to be 10 to 42 km. They
found that the HCl released would reduce stratospheric
ozone by 0.3 to 1 percent, the range within these limits
depending on the latitudinal distribution assumed.

These estimates should be brought up to date for two
reasons. First, the rate constants employed include for
the Cl + O_3 reaction the value 1.84×10^{-11}, used by
Wofsy and McElroy (1974) in their earliest published work
on the CFMs (see Table 8.1, column 3). The "best current
value" listed in the last column of Table 8.1 is lower by
a factor between 1.5 to 2 (depending on the temperature).
Second, the eddy-mixing profile used by Whitten *et al.*
(1975) at that time has a vertical transport rate at 20
km that is as much a factor of 4 larger than presently
seems reasonable (Chapter 5 and Appendix B).

With the current best set of rate constants, global
mixing, and the same Space Shuttle launch schedule and
altitude of exhaust release, the ozone perturbations have
been recalculated using three of the eddy-mixing profiles
shown in Figure 7.1. The ozone reduction (without inclusion of $ClONO_2$) falls in the range 0.29 percent (Crutzen
profile) to 0.34 percent (Chang and Wofsy profile). Applying the adjustment factor for $ClONO_2$ formation and uncertainty limits adapted from Chapter 8, *we conclude that
the effect on the ozone column of HCl from the proposed
Space Shuttle launch schedule (50 flights per year) would
be a reduction of about 0.15 percent, with an uncertainty
range of roughly 0.05 to 0.45 percent.* Tropospheric sinks
would not affect the projected reduction; however, there
is a new factor--the extent to which the Space Shuttle
effects are localized rather than global.

In addition to the effects of HCl, there is the question of possible effects from the particulate matter.
Various laboratories within NASA and the Jet Propulsion
Laboratory (JPL) have performed studies aimed at assessing these effects. The results of these studies have or
will be published in reports from the respective organi-
zations. A brief summary of the findings is given here.

The particles of aluminum oxide emitted into the stratosphere are found to have a size distribution that decreases

steeply with size over the radius range 0.1-1.0 μm and is relatively level in the vicinity of 0.02 μm. Corresponding to this crudely determined size distribution and a steady-state loading of the stratosphere from 50 flights per year, the change in optical thickness of the stratosphere was estimated as $<10^{-4}$. Effects of this magnitude are not considered to be significant for climate alteration.

Laboratory studies at JPL sought to determine whether Al_2O_3 can affect reaction rates leading to destruction of various gases, most notably ozone. The results were negative with the maximum effect on ozone being roughly estimated to be about a factor of 10^6 too slow to be of significance (Keyser, 1976).

When exhaust gases and particles were held in a container, about 5 percent of the HCl was removed as $AlCl_3$. This effect is likely to be even smaller in the open atmosphere, where HCl concentrations are much lower. Thus it appears unlikely that Al_2O_3 can significantly alter the ClO_x chemical interactions in the stratosphere.

Generally speaking, the above findings (albeit somewhat sketchy) suggest that particulate matter from projected Space Shuttle operations would not significantly affect the radiation and ozone chemistry in the stratosphere or the climate at the surface. The possibility of a significant effect due to an unsuspected mechanism cannot be ruled out absolutely.

III. FEEDBACK AND COUPLING MECHANISMS

There are several feedback and coupling mechanisms that might modify the effect of CFMs on stratospheric ozone or alter the consequences of a given reduction in it. One such case, the partial "self-healing" of the stratosphere, has been considered in Chapter 7. The others that have come to our attention are discussed in this section.

A. *Temperature and Water Vapor Feedback*

Liu *et al.* (1976b) have pointed out that reduction in stratospheric ozone by CFMs would lead to enhanced solar heating of the lower stratosphere and the tropopause, which might increase the flux of water vapor into the stratosphere. The added water vapor would increase the production of hydroxyl radicals by the reaction

$$H_2O + O(^1D) \rightarrow HO + HO$$

and, in turn, the HO would react with HCl to produce more Cl atoms

$$HO + HCl \rightarrow H_2O + Cl$$

Finally, the regenerated Cl atoms would increase the catalytic destruction of ozone by ClO_x. In particular, Liu et al. (1976b) find that with a doubling of the stratospheric water vapor concentration, the amount of ozone reduction produced by adding a given amount of stratospheric chlorine also nearly doubles.

They argue that the temperature of the tropical tropopause acts as a cold trap to maintain stratospheric humidity at the saturation mixing ratio for that temperature. Observations of water vapor in the stratosphere (Mastenbrook, 1971) provide some suggestion of such a relationship. However, a better understanding of the transfer of water vapor between troposphere and stratosphere than now available would be needed before this relationship can be regarded as established. To evaluate the possible importance of this effect, it is helpful to note that the saturation water vapor mixing ratio at the tropical tropopause would double with a 4° increase of temperature at that altitude. Hence, significant changes of stratospheric water vapor might be expected if the temperature of the tropical tropopause were to change by a few kelvins or more.

We have attempted to examine the known physical processes that might affect the temperature of the tropical tropopause. One process consists of the radiative effects due to changes in the ozone profile itself. A global mean climate model developed at NCAR by Coakley (1976) indicates a 1 to 2° warming of the tropopause using preliminary calculations by Crutzen (1976a) of the ozone reduction due to steady-state CFM release (1973 rates), including the chlorine nitrate species. As shown in Figure 8.6, the CFM perturbation causes a larger fraction of the reduced total ozone column to be at lower altitudes.

This estimate of the change in temperature at the tropical tropopause must be regarded as highly tentative. In reality, the temperature of the tropical tropopause depends on local dynamic and radiative balances as well as on the global mean equilibrium. Above 35 km, where the large ozone descreases are predicted (Figure 8.6), the climate model shows a temperature decrease of several tens of degrees. Significant changes in dynamics and transport processes in the upper stratosphere would also likely result from the predicted large decreases in heating rates.

Another warming effect of approximately the same magnitude is provided by the CFMs themselves through their absorption of terrestrial thermal radiation in the 10-μm window region (Ramanathan, 1975). The steady-state concentrations of CFMs for 1973 release rates are estimated to warm the temperature of the tropical tropopause by an additional 1-2 K.

In conclusion, *a preliminary evaluation of the direct and indirect effects of CFMs on the temperature of the tropical tropopause indicates that,* provided the "cold-trap" hypothesis is valid, *there could well result significant increases of stratospheric water vapor concentrations, and corresponding increases in the ozone reduction. This problem needs further attention using more rigorous modeling approaches.*

B. Interaction between ClO_x and NO_x Catalysis

Another instance of feedback is the already recognized interaction between the NO_x and the ClO_x catalytic cycles. As first pointed out by Wofsy (1976), if there were to be a major change in stratospheric NO_x concentrations in the future (as might conceivably result from greatly increased use of nitrogenous fertilizers) the effect on ozone by a given amount of CFM would be appreciably less than if the NO_x concentrations remained at the present values. The model of Chang (Appendix D) applied to the constant 1973 CFM production rate provides a quantitative example.

With the present N_2O injection rate (which determines the NO_x concentrations), continued release of CFMs is predicted to cause a steady-state reduction in stratospheric ozone of 14 percent, without inclusion of the $ClONO_2$ chemistry. If now the N_2O injection rate is doubled, the stratospheric ozone *increases* slightly to a net reduction of 13 percent. This same final state can be reached by first doubling the N_2O (without CFMs), which causes an ozone reduction of 6 percent, and then adding the CFMs, which cause a 7 percent reduction. Thus, *if one considers only the CFMs, their effect on stratospheric ozone is inversely proportional to the concentration of NO_x. The numerical relationships are modified by inclusion of $ClONO_2$ in the reaction scheme.*

C. Speculations

The three feedback loops considered so far involve processes within the stratosphere. There are other less
direct and more conjectural ways in which there might be
some compensation for or amplification of a reduction in
stratospheric ozone. For feedback to occur, the ozone
reduction or a process associated with it must cause
other changes that serve to ameliorate or enhance the effects that would occur in the absence of the feedback
mechanism.

One possibility involves the N_2O produced by bacterial
action in the soil and surface waters, which leads to most
of the ozone destruction in the natural cycle. If stratospheric ozone is reduced, the amount of biologically harmful uv (290-320 nm, uv-B) that reaches the surface is increased. This *conceivably* could affect plant life so as
to decrease bacterial production of N_2O and thereby compensate for *some* of the ozone reduction. The compensation
would never be complete because there must be at least
some net ozone reduction left to maintain the mechanism.
Furthermore, because of the interaction between the ClO_x
and NO_x cycles just described, a decrease in N_2O would
make a given amount of CFM more effective in reducing
ozone. Thus, in order to restore in large part a given
percentage of ozone reduction, the bacterial production
of N_2O would have to decrease by at least that percentage,
with major agricultural and ecological implications.

Therefore, even if there is some relation between uv-B
level and bacterial production of N_2O, it is not necessarily beneficial. Similar arguments apply to the possibility that an ozone reduction would increase cloud coverage,
which would then shield the earth's surface from some of
the extra uv radiation. If any such mechanisms exist and
are powerful enough to affect the ozone destruction rate
significantly, or to modify the consequence of an appreciable reduction in stratospheric ozone, they could themselves have a major impact on life.

IV. NATURAL VARIATIONS IN THE OZONE COLUMN

Chapter 6 and Appendix C have summarized the extensive
measurements carried out over the past four decades of
the total amount of ozone present in the vertical column
of atmosphere above various points on the earth's surface.
Such measurements of the total ozone column give results

that vary considerably not only with latitude but also with the time of day and the season of the year. These latitudinal, daily, and seasonal changes are relatively large and regular in character, and their origins are generally well understood. In addition to them, longer-term, less-regular natural fluctuations have been observed in the annual averages.

These are apparent in Figure 9.1, taken from the report of the Climatic Impact Committee (1975), which gives the ozone column averaged over the northern hemisphere, for 1934-1970. The seasonal changes in the top part have been averaged out at the bottom. The smoothed annual values that result exhibit irregular fluctuations covering a range of about ±5 percent. Two likely causes of the fluctuations are sporadic or periodic changes in solar activity such as flares or the sun spot cycle and year-to-year variations in stratospheric circulation, although other factors may contribute (e.g., the atmospheric nuclear explosions of the 1960's). Mechanisms producing the fluctuations could include the variable generation (or removal) of chemically active species in the stratosphere (Ruderman *et al.*, 1976) and variable transport of the ozone from "fast chemistry" regions to slow ones.

If there were a detailed understanding of the mechanism(s) causing the long-term fluctuations, one might be able to predict and subtract them from the observations and look for further (residual) changes that might be

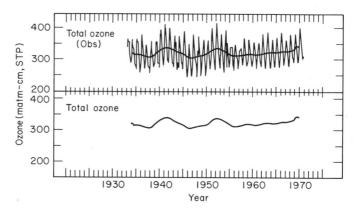

FIGURE 9.1 Long-term variation in total ozone in the northern hemisphere. At the bottom, the seasonal variations have been averaged out.

attributed to the reduction predicted for the CFMs. However, this is not yet the case. As a consequence, the irregular natural variations increase materially the difficulty of identifying man-made perturbations of stratospheric ozone. One is limited to statistical analyses, based on continuation of recent "natural trends," that search for an "abnormal decrease" (by the CFMs), which becomes larger with time in a particular way (see, e.g., Hill *et al.*, 1976).

Because of the large natural fluctuations, the question has been asked whether an additional, sustained reduction of smaller magnitude would actually be of any significance. In terms of possible biological consequences of increased uv-B radiation, the answer is clearly *yes*, since these are, in general, cumulative effects of total radiation received, with natural fluctuations averaged out. The effect is much like the relation between mean temperature and changes in the weather. Temperature at a given location may vary by as much as 30°C on a daily or 70°C on an annual basis, but a 3°C change in the annual mean temperature amounts to a profound change in climate. Similarly, the existence of local natural fluctuations in stratospheric ozone does not necessarily have any bearing on the possible significance of man-made changes in the average amount, changes that would persist for the major part of a century.

V. N_2O FROM FERTILIZERS

The highly likely further increase in the use of nitrogenous fertilizer has been identified as a potential problem, initially by Crutzen (1974, 1976b), subsequently and more extensively by McElroy *et al.* (1976), and also by Sze and Rice (1976). It has long been known that some small fraction of the nitrogen in fertilizer is converted into N_2O by processes in the soil (Hardy and Havelka, 1975) and oceans (Hahn, 1974) and released into the troposphere. From there it eventually contributes to the destruction of ozone by conversion to NO_x in the stratosphere. Our *quantitative* understanding of the processes is unsatisfactory for the soil and poorer for the sea (see, e.g., Liu *et al.*, 1976a, Table 1). Furthermore, there are many gaps in our knowledge of how much of the fertilizer is incorporated in the crops; how much stays in the soil as inorganic nitrogen; how much is released from the soil as N_2, NH_3, or N_2O; and how much is lost as runoff to the oceans, adding to their enormous store of inorganic and organic nitrogen compounds. There is also uncertainty about the fraction of tropospheric N_2O from the oceans versus that from continental sources, as well as about their total amount.

Because of these limitations in our knowledge, it is difficult to project how much of an effect fertilizers may have on stratospheric ozone. The problem is compounded by the coupling between the ClO_x and NO_x cycles described earlier in this chapter. In any event, fertilizer usage has been increasing about 6 percent per year, and continued increase at this rate appears needed for an indefinite period to feed the growing population of the world (Hardy and Havelka, 1975). Such growth would increase worldwide fertilizer application from 40 million metric tons/yr in 1974 to 200 in the year 2000. There are controversial differences in estimates of the corresponding effects on stratospheric ozone because of the gaps in our knowledge. The present assessment of the problem appears to be that there is no excuse for complacency--and, equally, no cause for immediate alarm. But *it does seem desirable to plan to reduce greatly our uncertainties about the sources and sinks of N_2O and the potential role of fertilizers as a source of N_2O so that within 5 yr or so (by 1981) we can make a definitive evaluation of the potential effects on stratospheric ozone of continued growth in the use of nitrogen fertilizers.*

VI. OZONE REDUCTION AND ULTRAVIOLET TRANSMISSION

The most immediate effect of a reduction in stratospheric ozone is an increase in the transmission of solar radiation by the atmosphere. Stratospheric ozone acts as a highly effective screen to prevent short-wavelength uv light from reaching the earth's surface, the cutoff occurring rather sharply between 320 and 290 nm, in the so-called uv-B region of the spectrum. It is, accordingly, in this wavelength region that changes in ozone concentration have their largest effect. For small changes in ozone concentration, a 1 percent decrease in stratospheric ozone permits approximately 2 percent additional uv-B radiation to reach the earth's surface. For larger reductions, this ratio rises above 2, as given in Figure 9.2.

In turn, an increase in uv-B radiation could have harmful biological effects, and this feature of ozone reduction

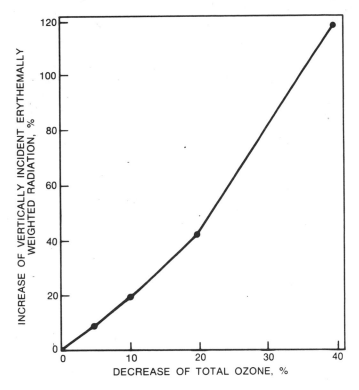

FIGURE 9.2 Percentage of change in dose of vertically incident erythemally weighted radiation versus percentage of change in ozone column. (After Schulze, 1973.)

has received intensive study and discussion (cf. Climatic Impact Committee, 1975). The most clearly identified consequence stems from the apparent relationship between the amount of uv-B radiation and the incidence of skin cancer. Beyond this, a marked increase in uv-B radiation could have a detrimental effect on a whole range of biochemical processes in the plant and animal world. A related possibility is the more rapid deterioration of polymeric materials (plastics, paints, elastomers, textiles).

Furthermore, the absorption of less radiant energy in the stratosphere would be expected to reduce its temperature and allow more radiation to reach the troposphere, both of which might alter global climate. A different but related effect is the contribution by the infrared absorption of the CFMs and ozone to the "greenhouse effect," warming the troposphere. These critical features of the overall problem are beyond the scope of this report, except to note their possible importance, which is being investigated and reported on separately by the Committee on Impacts of Stratospheric Change.

VII. SUPPLY OF FLUOROSPAR FOR FUTURE CFM PRODUCTION

The present large-scale and increasing uses of CFMs for the long periods of time considered in Chapter 8 require that the world's fluorine resources be adequate for the purpose. If not, deplenishing reserves might provide a natural limit on CFM release in the future. This, how-ever, does not appear to be the case (Bradbury, 1975).

Roughly 20 percent of the total U.S. consumption of fluorospar in 1973 was used in the production of CFMs. Applying this factor to estimated world consumption of fluorospar (4,164,000 tons in 1970) results in an esti-mated 850,000 tons of fluorospar utilized in 1973 in the production of CFMs. This is to be compared with world reserves estimated to be between 73 million tons and 267 million tons, which are thus adequate for from 100 to 300 years of CFM production at the 1973 rate. An exponential growth of 10 percent per year in CFM production for the remainder of the century would consume the equivalent of 180 million tons of fluorospar. This is to be compared with the reserves of phosphate rock, by far the world's largest fluorine resource. In the United States alone, phosphate rock resources are estimated to be 60,000 mil-lion tons, which is equivalent to 3850 million tons of fluorospar, although more costly to produce. *It is clear that the projected supply of fluorine is immense and pro-vides no practical limit to CFM production.*

VIII. TOTAL BURDEN OF HALOCARBONS

The evidence that has accumulated to date indicates con-vincingly that the halocarbons that are distributed through the troposphere enter the stratosphere at rates depending only on their tropospheric concentrations. In the stratosphere, they contribute to the ozone cycle to degrees determined by their rates of stratospheric destruc-tion and the number of chlorine (or bromine) atoms that they release.

From the data discussed in Chapter 3, the greatest *present* industrial source of halogen in the stratosphere is apparently CCl_4, closely followed by the fluorocarbons F-11 and F-12, and with only smaller contributions from other halocarbons. Since the total CCl_4 atmospheric burden is now some 40 times the known annual injection rate, it is plausible to assume it is near the steady state and will not grow markedly unless emissions increase. This is true whether the steady state has been approached because of the long period of CCl_4 manufacture (Galbally, 1976) or because a strong natural source exists. However, since the present burdens of F-11 and F-12 are close to our best estimates of the total amounts that have ever been released, they are almost certainly far from a steady state, and concentrations will continue to rise if release is continued at the present rate. Accordingly, it is their effect that needs the most serious immediate attention.

The fluorocarbon F-22, CHF_2Cl, has been discussed as at least a partial replacement of other fluorocarbons as a refrigerant. It contains a C——H bond, so, as described in Appendix A, it is attacked and partly destroyed by hydroxyl radicals in the troposphere; moreover, it can release only one chlorine atom. A rough estimate is that its effect on the ozone cycle would be only about a tenth that of the same weight of F-12. Further, because of its shorter atmospheric life, there would not be long-lasting aftereffects should release be terminated. The same is true of other fluorocarbons containing hydrogen atoms and/or double bonds.

On the other hand, completely fluorinated hydrocarbons, e.g., CF_4 and C_2F_6 (and also SF_6) are almost completely inert even to solar radiation in the stratosphere. If large quantities were released and it subsequently developed that they had some deleterious environmental effect, our present knowledge indicates that the effect would persist for centuries. In short, hazards due to reactive halocarbons of short lives can perhaps be dealt with as they arise, but those from highly stable materials must be anticipated and dealt with in advance if long-term problems are to be avoided.

Furthermore, any consideration of the total halocarbon burden must be placed in context by recognizing it as but one of a variety of potential perturbations to stratospheric ozone. Already recognized as parallel sources of concern have been the SST as a possible contributor of nitric oxide and the Space Shuttle as a possible contributor

of chlorine species. Insofar as such perturbations are
operative, their effects would be in addition to those
of the CFMs, with allowance for any interactions among
them. Still other potential perturbing sources presently
under discussion include bromine-containing compounds
(used for a variety of agricultural and commercial purposes) and man's now rapidly expanding fertilizer use.
The pace of technological innovation suggests that still
other man-made substances may begin to appear in the
stratosphere to complicate further the chemistry of this
region.

Finally, in looking to the future *all major releases of
volatile halocarbons (and other pollutants) on a global
scale should be monitored via data on production and use
and, when possible, by actual monitoring of atmospheric
concentrations*. Further, as our knowledge of atmospheric
chemistry improves (e.g., through better values for hydroxyl radical concentrations and rate constants and by more
accurate assessment of inactive removal), improved calculations should be made to indicate with increasing reliability the future tropospheric concentrations to be
expected if given amounts of particular halocarbons (and
other pollutants) are released and to predict their effects on the stratosphere.

In July 1975, the Panel issued an interim report call-
ing attention to a number of important atmospheric and
laboratory studies that would aid in determining the extent to which CFMs and other halogen-containing compounds
will affect the ozone in the stratosphere. These needs
have been brought into sharper focus by the unanswered
questions that occur throughout the present report. *Most
of the investigations identified in our interim report remain important; they are reviewed and updated in Appendix
F in the light of our experience since then.*

REFERENCES

Bradbury, J. C. 1975. Illinois State Geological Survey,
    Urbana. Private communication.
Climatic Impact Committee. 1975. *Environmental Impact
    of Stratospheric Flight: Biological and Climatic
    Effects of Aircraft Emissions in the Stratosphere.*
    National Academy of Sciences, Washington, D.C.
Coakley, J. A. 1976. National Center for Atmospheric
    Research. Private communication.

Crutzen, P. J. 1974. Estimates of possible variations in total ozone due to natural causes and human activities, *Ambio* 3:201-210.

Crutzen, P. J. 1976a. National Center for Atmospheric Research and National Oceanic and Atmospheric Administration. Private communication.

Crutzen, P. J. 1976b. Upper limits in atmospheric ozone reductions following increased application of fixed nitrogen to the soil, *Geophys. Res. Lett.* 3:169-172.

Galbally, I. E. 1976. Man-made carbon tetrachloride in the atmosphere, *Science* 193:573-576.

Hahn, J. 1974. The North Atlantic Ocean as a source of atmospheric N_2O, *Tellus* 26:160-168.

Hardy, R. W. F., and U. D. Havelka. 1975. Nitrogen fixation research: A key to world food?, *Science* 188:633-643.

Hill, W. J., P. N. Sheldon, and J. J. Tiede. 1976. Analyzing worldwide ozone for trends. Paper presented at the San Francisco meeting of the American Chemical Society, August 29-September 3.

Keyser, L. 1976. Jet Propulsion Laboratory. Heterogeneous reaction of ozone with aluminum oxide. Technical memorandum 33-782.

Liu, S. C., R. J. Cicerone, T. M. Donahue, and W. L. Chameides. 1976a. Limitation of fertilizer induced ozone reduction by the long lifetime of the reservoir of fixed nitrogen, *Geophys. Res. Lett.* 3:157-160.

Liu, S. C., T. M. Donahue, R. J. Cicerone, and W. L. Chameides. 1976b. Effect of water vapor on the destruction of ozone in the stratosphere perturbed by Cl_x or NO_x pollutants, *J. Geophys. Res.* 81:311-318.

McElroy, M. B., J. W. Elkin, S. C. Wofsy, and Y. L. Young. 1976. Sources and sinks for atmospheric N_2O, *Rev. Geophys. Space Phys.* 14:143-150.

Mastenbrook, H. J. 1971. The variability of water vapor in the stratosphere, *J. Atmos. Sci.* 28:1495-1501.

Ramanathan, V. 1975. Greenhouse effect due to chlorofluorocarbons: climatic implications, *Science* 190:50-52.

Ruderman, M. A., H. M. Foley, and J. W. Chamberlain. 1976. Eleven-year variation in polar ozone and stratospheric-ion chemistry, *Science* 192:555-557.

Schulze, R. 1973. Quoted in CIAP Report of Findings. Final Report DOT-TST-75-50.

Sze, N. D., and H. Rice. 1976. Nitrogen cycle factors contributing to N_2O production from fertilizers, *Geophys. Res. Lett.* 3:343-346.

178

Whitten, R. C., W. J. Broucki, I. G. Popoff, and R. P.
    Turco. 1975. Preliminary assessment of the potential
    impact of solid-fueled rocket engines in the strato-
    sphere, *J. Atmos. Sci.* 32:613-619.
Wofsy, S. C. 1976. Harvard University. Private
    communication.
Wofsy, S. C., and M. B. McElroy. 1974. HO_x, NO_x, and
    ClO_x: their role in atmospheric photochemistry, *Can.
    J. Chem.* 52:1582-1591.

DETAILED DISCUSSION

OF REMOVAL PROCESSES

I.   REMOVAL OF HALOCARBONS AT THE SURFACE

A.   *Solubility in and Removal by the Oceans*

Although the solubilities of the halocarbons are small, both the surface area and volume of the oceans are large, and the question is whether an appreciable fraction of a halocarbon is dissolved in the oceans or may be removed thereby from the atmosphere.

The solubilities in water of several of the gaseous CFMs have been reported by Parmelee (1953), who used laboratory volumetric methods having an accuracy within ±5 percent for most measurements including those for F-12. Parmelee's paper did not report on F-11 solubility. However, solubilities of F-11 have been reported (Du Pont, 1971), based apparently on methods similar to those used by Parmelee. The measurements were made in the temperature range of 30-77°C, and extrapolation to lower temperatures would introduce error. The boiling point of F-11 is 24°C, and direct observation of the solubilities at this and lower temperatures would not have been possible using the volumetric technique of Parmelee (1953). Liss and Slater (1974) report a value for the solubility of F-11 determined by J. E. Lovelock in measurements made on board the R.R.S. *Shackleton* in 1972.

In the discussion that follows, the solubility of F-11 as measured in seawater by Lovelock will be used, i.e., 4.4×10^{18} molecules cm^{-3} atm^{-1} (at 288 K); and the solubility of F-12 will be taken to be 1.6×10^{18} molecules

cm^{-3} atm^{-1} (Parmelee, 1953). If the CFM does not undergo removal from the water, equilibrium between the CFM in the air at the ocean's surface and that in the ocean down to the depth of the thermocline is attained in the order of a few years, which is rapid relative to the total atmospheric residence time τ^0 of the CFM. The average thermocline depth in the ocean is about 50 m. For constant release at the 1973 rates, the steady-state mixing ratio of CFM in the troposphere is about 10^{-9}, a column density of $\sim 2 \times 10^{16}$ molecules cm^{-2}. The corresponding column densities in the surface water of the oceans are therefore 10^{-9} times the quoted solubilities times the thermocline depth, or 2.2×10^{13} and 8.0×10^{12} molecules cm^{-2} for F-11 and F-12, respectively. Since the oceans constitute about two thirds of the earth's surface, these numbers indicate that about 0.07 percent of the tropospheric F-11 and 0.03 percent of the F-12 could be dissolved in the oceans at equilibrium. For the other halocarbons the percentages will be proportional to their relative solubilities.

These percentages are small compared with the amounts removed annually by active (ozone reducing) processes in the stratosphere (2 and 1.1 percent yr^{-1} for F-11 and F-12, respectively). They represent a "one-shot" process not an annual rate. For solution in the oceans to be important, it would have to be followed by rapid removal ($\tau \ll 1$ yr) of the CFMs from the surface waters. For example, one removal process is mixing of the surface waters with the deep ocean. The time characterizing such exchange is about 15 yr, so it would remove about one fifteenth of the CFM in the surface water per year, or 0.004 and 0.002 percent yr^{-1} of the tropospheric F-11 and F-12 (τ's of 2.5×10^4 and 5×10^4 yr). Another possibility is hydrolysis. The hydrolysis rate for F-11 is given as $<7 \times 10^8$ molecules cm^{-3} sec^{-1} atm^{-1} in pure water. (In fact, the rates for perhalogenated compounds are immeasurably small.) From this we can calculate a removal flux of 1×10^{11} molecules cm^{-2} yr^{-1} for a tropospheric column density of 2×10^{16} molecules cm^{-2}, which gives a removal time of $>10^5$ yr. Thus, neither hydrolysis nor mixing of the surface waters with the deeper ocean is significant in comparison to active removal of F-11 and F-12 in the stratosphere.

Should other processes (such as microbial action) remove the halocarbon much more rapidly from the oceans, it is possible to place an upper limit on their overall effectiveness, which is the *maximum* rate at which halocarbons could enter the ocean from the atmosphere. As described by Broecker and Peng (1974), the model used here assumes that a thin film at the air-water interface presents the limiting resistance to gas transfer. The corresponding removal time by this process is

$$\tau \; (\text{sec}) \; \geq \; \frac{3.0 \times 10^{25} \; \text{molecules} \; \text{cm}^{-2} \; \text{atm}^{-1}}{\alpha D / Z} \qquad (A.1)$$

where

Z = film thickness $\simeq 4 \times 10^{-3}$ cm,
D = molecular diffusivity of the absorbed substance in water $\simeq 1.2 \times 10^{-5}$ cm^2 sec^{-1},
α = solubility of the absorbed gas in water, about 4.4×10^{18} molecules cm^{-3} atm^{-1} for F-11 and 1.6×10^{18} for F-12.

The value of the quantity D/Z is not very dependent on the nature of the gas. This model gives a partial removal time for F-11 of ≥ 70 yr and for F-12 of ≥ 200 yr. These are lower limits because they were derived on the assumption that the concentration of CFM below the film is zero. Note that these lower limits for the removal time by the ocean are only about twice as long as the active removal times of the respective compounds in the stratosphere.

Evidence that some oceanic process(es) may indeed remove F-11 (and CCl$_4$) at significant rates has been reported by Lovelock et al. (1973), who measured F-11 and CCl$_4$ concentrations in Atlantic Ocean surface waters and in air during the 1972 cruise of the R.R.S. Shackleton. Figure A.1 is a plot of the F-11 concentrations in water (circles) versus latitude: the measurements have been converted to equivalent equilibrium concentrations in air using 4.4×10^{18} molecules cm^{-3} atm^{-1} for the solubility of F-11. The actual measurements obtained for the F-11 concentrations in air are shown by the solid line, the best-fit third-order polynomial. If the oceans and the air were at equilibrium, and all factors affecting solubility are correct, the points would fall close to the curve. Apparently, the surface ocean waters contain less-than-equilibrium concentrations of F-11, except for the most northern measurements (40-45° N latitude). Under these circumstances, the flux of F-11 into the oceans depends on the difference between the F-11 concentration in air and in water (as equivalent concentration in air). It may be estimated with the formula of Liss and Slater (1974):

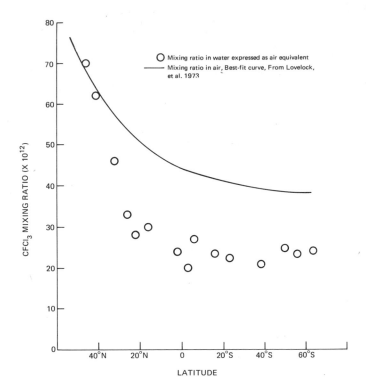

FIGURE A.1 The 1972 mixing ratios of F-11 measured in the surface waters of the ocean (as air equivalent at equilibrium) compared with that actually measured in the air, versus latitude.

$$F = k_1 \, \Delta C_1 \, n_0 \qquad\qquad (A.2)$$

where

$\Delta C_1 = s \Delta C_a RT,$

$\Delta C_1 =$ difference between equilibrium and actual mixing ratios in the liquid phase,

$\Delta C_a =$ difference between equilibrium and actual mixing ratios in air,

$s\ \ \ =$ solubility of CFM in appropriate units (the product sRT is dimensionless),

$k_1\ \ =$ D/Z from Eq. (A.1) $= 3 \times 10^{-3}$ cm sec^{-1} for F-11,

$n_0\ \ =$ molecules cm^{-3} of air at the earth's surface (2.75×10^{19}).

The flux of F-11 is found to be 6.5×10^{12} molecules cm^{-2} yr^{-1} in 1972, using $T = 288$ K, $s = 4.4 \times 10^{18}$ molecules cm^{-3} atm^{-1}, $\Delta C_a = 15 \times 10^{-12}$ as determined from Figure A.2 and a value of ΔC_l calculated to be 2.5×10^{-12}. In this manner the total global flux into the oceans is estimated as 2.4×10^{31} molecules yr^{-1}. In 1972, the average F-11 mixing ratio was about 65×10^{-12}, which yields an atmospheric burden of about 6.5×10^{33} molecules. The estimated removal time corresponding to the 1972 concentrations in the air and ocean is thus

$$\frac{6.5 \times 10^{33} \text{ molecules}}{2.4 \times 10^{31} \text{ molecules } yr^{-1}} = \sim 270 \text{ yr}$$

Within the limitations of the above analysis, it appears that there may exist an oceanic sink with a magnitude of about one fifth the active removal rate for F-11.

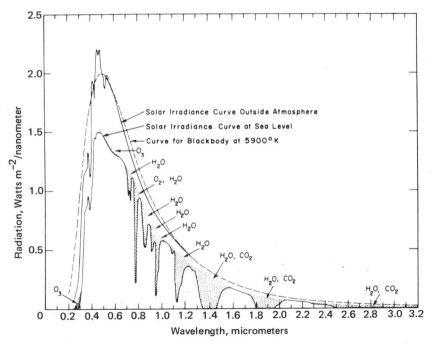

FIGURE A.2 Solar irradiance as a function of wavelength. The shaded areas show light absorbed in the atmosphere by the molecules indicated.

Since oceanic concentrations of F-12 have not been reported, no equivalent estimates of its removal can be made.
The analysis given here relies on the reported solubility
of F-11, a single set of oceanic concentrations measured
for F-11, and the validity of the thin-film model for
estimating fluxes across the air-sea interface. Uncertainties in the analysis are difficult to assess. Since
the estimated magnitude of the oceanic sink is about one
fifth that of active removal, it is possible, given reasonable uncertainties, that within the oceans the actual
amounts of F-11 destroyed could range from trivial to
nearly half of those destroyed by active removal in the
stratosphere.

Oceanic removal of CCl_4 has been suggested by Lovelock
(1975) on the basis of his unpublished measurements of its
tenfold decrease with depth at 100 m from the surface.
Measurements of F-11 exhibit a slower decrease that is
much more rapid at 14° N latitude (tenfold in 300 m) than
at 42° N (~30 percent in 200 m). This change with latitude for F-11 might reflect a temperature-dependent removal process. Improved estimates of the oceanic sink
strength would result from new laboratory measurements of
the solubilities of gaseous F-11 and F-12 in water and
more systematic measurements of CFM concentrations in the
surface ocean waters with attempts to cover large portions
of the globe and with concurrent measurements of the air
concentrations.

*In summary, solution in the oceans followed by mixing
of the surface waters with the depths or by hydrolysis
does not provide a significant sink for CFMs. Solution
in the oceans followed by its rapid removal from the surface waters by some as yet unidentified mechanism is suggested by the one set of oceanic measurements of F-11
reported so far. The limited data available indicate
that the oceanic sink strength for F-11 ($\tau \simeq 270$ yr) may
be about one fifth that of the stratospheric photolysis
sink ($\tau \simeq 50$ yr). The estimate is sufficiently uncertain
and the magnitude is sufficiently important to warrant
further investigation of the relevant phenomena.*

A paper on "The Oceans as a Sink for Chlorofluoromethanes" has just been published by Junge (1976). It covers
the same questions as considered in this section and uses
the thin-film model to obtain similar results. The value
of the parameter $k_1 = D/Z$ used by Junge was 1.4×10^{-3} cm
sec^{-1} as opposed to that of 3×10^{-3} cm sec^{-1} used above.
Newly measured values of the solubilities of F-11 and F-12
are given by Junge. While the solubility of F-12 is that

used in our assessment above, the solubility of F-11 is
about 70 percent of that used above. Junge's estimated
minimum lifetime for F-11 of $\gtrsim 200$ yr differs from our estimate of $\gtrsim 70$ yr by the combined factors of the values of
D/Z and solubility. In estimating a value of the actual
lifetime for F-11 removal, Junge, also using Lovelock's
ocean concentrations, adopted a somewhat different approach.
The lifetime estimated by him (800 yr) is similarly larger
than our estimate of 270 yr.

B. Absorption and Microbial Action in Soil and Vegetation

Studies of the possible uptake of halocarbons by representative soils (including resident microbes) and vegetation
were performed by a group at the University of California,
Riverside, under the direction of Taylor (1975). Within
instrumental error (± 10 percent) no loss of halocarbons
nor fluorine uptake was detected during exposures of up to
20 days (Hester, 1975). It is possible to deduce some
limits on the possible effects of microbial removal from
this admittedly crude experiment, using the following assumptions. The flux to the surface ϕ can be described as

$$\phi = wc \text{ (molecules cm}^{-2} \text{ day}^{-1}) \qquad (A.3)$$

where c is the halocarbon concentration and w is the deposition velocity in cm day^{-1}. The value of w has been
found to be independent of concentration over a considerable range for a number of substances (SO_2, CO, H_2, and
aerosol) and in these cases amounts to about 9×10^3 to
9×10^4 cm day^{-1}. If we assume that the uncertainty of
10 percent in the experiment represents the actual removal,
we can obtain a value of

$$w = 0.1z/20 \text{ cm day}^{-1} \qquad (A.4)$$

where z is the height of the air above the soil in the
experiment.
   The atmospheric residence time for such a process will
be approximately given by

$$\tau = H/w \qquad (A.5)$$

where H is the scale height = 8.4 km. It is reasonable to
use this treatment, which assumes the microbial removal
rate to be first order, since vertical mixing in the

troposphere will be rapid relative to the uptake time. Upon substituting Eq. (A.4) into Eq. (A.5), we obtain

$$\tau \simeq (20/0.1)(H/z) \qquad \text{(A.6)}$$

In the experiment z was about 10 cm, and therefore

$$\tau \gtrsim 5 \times 10^4 \text{ yr} \qquad \text{(A.7)}$$

This is a very strong indication that *removal by soils and vegetation is a negligible sink for halocarbons.*

C. *Entrapment in Ice*

In January 1975, the team of Rasmussen, Allwine, and Zoller obtained air and snow samples from the geographic South Pole, which were analyzed for CFMs. The surface air concentration of F-11 was 90 pptv (Rasmussen *et al.*, 1975), and the concentration of F-11 in gas released by melting the snow collected at a depth of 60 cm (single sample) was found to be 1.6 ppbv, giving an enrichment factor of about 20. They found no enrichment for CCl_4 but enrichments ranging from 20 to 30 for other halocarbons. They suggested that CFMs and other condensable trace gases may become entrapped in clathrate cages as H_2O sublimes onto the surface ice and snow, i.e., form CFM hydrate crystals in which hydrogen-bonded lattices represent sites suitable for occupancy by CFM molecules.

Hydrates of the CFMs are indeed known to exist, and thermodynamic data have been reported (Wittstruck *et al.*, 1961). However, the equilibrium vapor pressures of the CFMs for these solids are very much higher than their partial pressures in the atmosphere. For example, at -65°C the equilibrium pressure is about 10^{-4} atm for the F-11 hydrate and about 10^{-3} for the F-12 hydrate. Since these pressures are at least 10^8 times larger than the partial pressures of the respective CFMs in the atmosphere, hydrate (or clathrate) formation is not possible.

Other mechanisms for removal of CFMs by ice are unlikely. Adsorption can be ruled out as an important mechanism in the same manner as is done for adsorption onto urban aerosols (Section II.D.3), because of the low capacity of nonporous surfaces of polar substances for nonpolar molecules such as CFMs. Similarly, the possible importance of solid solutions can be ruled out by noting that, unless the behavior of CFMs in ice is anomalous, their solubility in ice is much lower than that in liquid water. The surface layers of the oceans can dissolve only a negligible fraction ($\lesssim$0.05 percent) of the CFMs (Section I.A), and the global volume of ice available for absorption of CFMs is much smaller than the volume of the ocean surface layers. However, even if we accept the observations as representative, it can be shown that ice caps and glaciers cannot provide a significant sink for CFMs.

Rasmussen and Robinson (1975) have performed calculations of the burden in ice of F-11 based on the polar observations. They estimate that the ice contains no more than approximately 2×10^{-6} g m^{-2} of F-11, down to a depth of 15 cm, which represents approximately a year's accumulation of snow. The area of permanent ice cover on the earth is about 16×10^{12} m^2. If it is assumed that the burden per unit area estimated by Rasmussen and Robinson is representative of the permanent ice of the world, then the maximum total global burden of F-11 bound in the top 15 cm of ice (snow) is calculated to be 2×10^{-6} g m^{-2} $\times$ 16×10^{12} m^2 = 3×10^7 g (per year). With the atmospheric burden of F-11 of 2.9×10^{12} g given in Table 3.3, the estimated shortest partial removal time by trapping in the ice caps and glaciers is approximately 10^5 yr. This estimate must be considered as preliminary since it is based essentially on a single sample. However, it appears that *glaciers and polar ice caps are not likely to remove significant amounts of halocarbons from the atmosphere.*

II. REMOVAL PROCESSES IN THE TROPOSPHERE

A. *Photolytic Processes*

Figures A.2 and A.3 show that in the troposphere, irradiation by the sun is effectively confined to wavelengths longer than 280 nm by the ultraviolet absorption spectra of oxygen and ozone. Although light with wavelengths less than 400 nm has sufficient energy to dissociate the C-Cl bonds in chlorofluoromethanes and ethanes, for which the dissociation energy is 70-75 kcal mole^{-1} (Kaufman and Reed, 1963; Gurvich *et al.*, 1974), the continuous absorption spectra corresponding to the $n \rightarrow \sigma^*$ transition, which results in C—Cl bond rupture, lie at much shorter wavelengths (around 200 nm); in addition, no absorption in the ultraviolet between 280 and 400 nm has been reported for the CFMs or for CH_3Cl or CCl_4 (see, e.g., Figure A.4). The

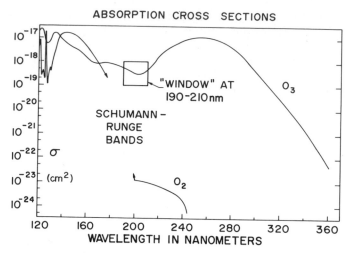

FIGURE A.3 Absorption cross sections for oxygen and ozone between 120 and 360 nm.

observed rate of decrease of absorption coefficient with wavelength at the long-wavelength edge of these bands is at least a factor of 10 for each 10-nm increase in wavelength. Therefore, the cross section at 280 nm is, at most, 10^{-26} cm^2, corresponding to a partial removal time of at least 5×10^3 yr.

The presence of a forbidden transition to a lower-lying excited state could modify such an extrapolation. The only predicted state would be the triplet (n, σ^*) state, which would lie directly below the repulsive singlet (n, σ^*) state responsible for absorption in the solar window. No such state has been detected for any halomethane, not even for methyl iodide (Calvert and Pitts, 1966), for which the presence of the heavy iodine atom would enhance the intensity of the forbidden transition from a ground singlet to an excited triplet state by several orders of magnitude as compared with chloromethanes and fluoromethanes. For the chlorofluoromethanes, the displacement toward longer wavelengths of the maximum of this forbidden band relative to the observed allowed transition should be less than 20 nm. Allowing for the lower intensity of the forbidden band, its contribution at longer wavelengths will always be much less than the extrapolated contribution of the allowed transition.

It is therefore most unlikely that the photolysis of chlorofluoromethane in the troposphere, where solar

FIGURE A.4 Absorption cross sections for halo-gen compounds in the solar window. F-11 $(CFCl_3)$, F-12 (CF_2Cl_2), CCl_4 from Rowland and Molina (1975); F-21 $(CHFCl_2)$, F-22 (CHF_2Cl), F-113 $(CFCl_2 \cdot CF_2Cl)$, Cl_2CO, CFClO from Row-land *et al.* (1975); HCl from Watson (1974); CH_3Cl from Gedanken and Rowe (1975).

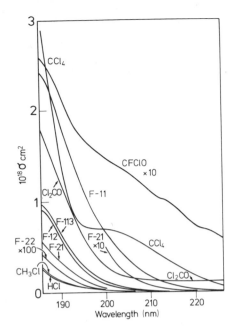

irradiation is restricted to wavelengths above 280 nm, contributes significantly to their removal. Laboratory verification of this conclusion would require very ac-curate measurements on the highly purified liquid using an absorption path of at least 1 m.

Another potential removal process is photosensitized decomposition, in which another molecule, S, absorbs ultraviolet light and transfers sufficient energy (~ 70 kcal mol^{-1}) in a collision to dissociate the CFM molecule. Because the CFM mixing ratio is low, the main competitive process will be quenching of S by the bulk constituents O_2 (primarily) and N_2. If all these energy transfer process-es have comparable collisional efficiencies, the correspond-ing lifetime for the CFM is given by $1/J_S f_S$, where J_S is the photoabsorption rate constant of the sensitizer and f_S is the mixing ratio. For a strongly absorbing sensi-tizer with $J_S = 10^{-3}$ sec^{-1}, f_S would have to exceed 3×10^{-7} to give a tropospheric lifetime for the CFM of less than 100 yr. In the case of weaker absorbers or species that are rapidly decomposed by light (e.g., NO_2 at wave-lengths below 396 nm), the mixing ratio of the sensitizers would have to be correspondingly larger. No sensitizers with the right combination of light absorption and mixing ratio have been detected nor could they be present with so high a mixing ratio and J and yet remain undetected in the troposphere. We conclude that *photolytic processes in the troposphere do not constitute a significant removal process for CFMs*.

B. Chemical Reaction with Neutral Molecules

In this section, tropospheric halocarbon removal processes by gas phase reactions with neutral species are discussed. First, the general principles of these interactions for the different classes of halocarbons are established in terms of their chemical structure and reactivity; second, the results for the important chlorofluorocarbons F-11 and F-12 are summarized.

The principal reactive species of the normal troposphere, in decreasing concentrations, are O_3 (10^{12}), HO_2 (4×10^8), HO (4×10^6), $O(^3P)$ (10^3), and $O(^1D)$ ($<10^{-2}$). The numbers given in parentheses are typical values for the molecular concentration (molecules/cm^3) of the particular species at ground level at noon and at midlatitude as obtained from one-dimensional model calculations described in Chapters 5 and 7. The values for O_3 and HO are supported by the experimental measurements summarized in Appendix C. This list of species is incomplete and omits several, such as NO, NO_2, and $O_2(^1\Delta)$, mainly because their reactions with halocarbons are slow.

Rates of most chemical removal processes in which an atmospheric species X reacts with a halocarbon have a positive temperature dependence (energy of activation). In this event, the reaction rate decreases with increasing altitude in the troposphere. For a mean temperature decrease (lapse rate) of 6.5 deg per km typical of the midlatitude troposphere this effect may be sizable, and information on the temperature dependence of the rate constant k_x is therefore normally required. The upper limit of k_x is approximately 10^{-10} cm^3 sec^{-1} and corresponds to reaction at every molecular collision. The most reactive atmospheric species, $O(^1D)$, may be dismissed from consideration here because of its small tropospheric concentration, even though its k_x's are very large. Because of this, the local removal time t_c for reaction of halocarbons with $O(^1D)$ ranges from longer than 30,000 yr near ground level to about 3000 yr near the tropopause, far longer than the total residence time τ^0.

For the purpose of discussing the nature and speed of reactions that may remove halocarbons, we subdivide the latter into saturated and unsaturated (double-bonded) organic compounds and further subdivide the saturated molecules into fully halogenated and partially halogenated, i.e., molecules that contain carbon-hydrogen bonds.

Fully halogenated, saturated halocarbons (e.g., F-11, F-12, and CCl_4) are unreactive toward all the above atmospheric species, except for $O(^1D)$, whose concentration is too small to make it a competitive removal process. $O(^3P)$ does react slowly with CCl_4 giving t_C values in the range of 10^5 to 10^6 yr, and the CFMs react even more slowly if at all. There are no exothermic reaction channels for direct attack by O_3 and HO_2. The possibility of reactions with HO is discussed below.

Saturated, partially halogenated molecules (with hydrogen atoms) are not attacked by O_3 or HO_2, but they are susceptible to hydrogen abstraction reactions by O or HO. Of these, the O atom reactions have larger activation energies than the HO reactions, and since O atom densities are several orders of magnitude smaller than those of HO in the lower troposphere we need only consider HO as a potential tropospheric reactant. For the halogen-substituted methanes, the following qualitative trends are observed: replacement of H by Cl or Br increases the rate constant of reaction with HO, and replacement by F increases it less or decreases it, particularly when the molecule already contains one or more Cl's.

Table A.1 lists recent experimental results (k) of halocarbon + HO reaction rate studies by several investigators. In it, the t_C values based on a diurnally averaged HO density of 2×10^6 cm^{-3} range from about 6 weeks to 4 yr for the H-containing halomethanes. These values are inversely proportional to the HO density and indicate the need for the experimental measurement of this species in the troposphere. Another set of calculated t_C's is also presented, based on a more elaborate, recent 2-D model of HO-halocarbon reactions and on tropospheric HO-concentration measurements. The two sets are in very good agreement. The values of t_C are based on rate constants at 298 K and are therefore lower bounds, because the HO reactions have activation energies in the 2 to 3 kcal/mole range, which result in a threefold to sixfold decrease of the rate constants between 298 and 220 K. The kinetic data presented in Table A.1 have accuracies in the range of ± 20 to 30 percent. Temperature-dependence data are somewhat preliminary, but activation energies fall within a narrow range. For example, the activation energies for the reactions HO + CH_3Cl, CH_2Cl_2, $CHCl_3$, or CHF_2Cl (F-22),

TABLE A.1 Rate Constants for Reactions of Halocarbons with HO and Corresponding Tropospheric Local Removal Times, t_c, at 298 K

	Saturated Fully Halogenated			Saturated H-Containing							Unsaturated		Ref.
	$CFCl_3$ (F-11)	CF_2Cl_2 (F-12)	CCl_4	$CHFCl_2$ (F-21)	CHF_2Cl (F-22)	CH_3Cl	CH_2Cl_2	$CHCl_3$	CH_3CCl_3	CH_3Br	C_2Cl_4	C_2HCl_3	
k (10^{-15} cm^3 sec^{-1})	<0.5	<0.4	<4	26	3.4	36	155	101	15	35	180	2240	a
	<1	<1	<1	27	4.8	44	145	114	16	37			b
				32	4.8	45,43	116,150		22		150	2300	c
				30	4.3						160	2400	d
t_C, yr [HO] = 2 × 10^6 cm^{-3}	>30	>40	>15	0.5	4	0.4	0.12	0.15	1	0.4	0.1	0.007	
	>200^f	>200^f											
t_C, yr[e]				0.7	4.6	0.4	0.2	0.2	1.4	0.5	0.2		

[a]Howard and Evenson (1976a, 1976b).

[b]Atkinson et al. (1975); Perry et al. (1976).

[c]Davis and Watson (1975).

[d]Chang et al. (1976).

[e]Based on a 2-D tropospheric model including measured HO concentrations and taking account of temperature dependence (Davis et al., 1976).

[f]Based on an upper limit of 1 × 10^{-15} cm^3 sec^{-1} for the rate constant measured at 480 K (Chang and Kaufman, 1976) and on a lower limit of 1 × 10^{-13} for the temperature-independent factor in the Arrhenius expression.

respectively, are 2.1, 2.2, 2.3, and 3.2 kcal/mole. Such
an activation energy will have the effect of increasing
the tropospheric residence time, τ_T, to values somewhat
larger than those given in Table A.2 for t_c, but τ_T will
still be considerably smaller than the removal time by
stratospheric photolysis, τ_S.

The unsaturated (double-bonded) molecules may undergo
addition reactions with O_3, O, and HO and will have short
chemical lifetimes mainly due to reactions with O_3 and
HO. For HO reactions with ethylene and chloroethylene,
rate constants of 2×10^{-12} to 6×10^{-12} cm^3 sec^{-1} (Howard
and Evenson, 1976b) lead to very short t_c's and indicate
rapid tropospheric removal. The fully chlorinated molecule, C_2Cl_4, seems to react at least an order of magnitude
more slowly. Rate data for C_2HCl_3 indicate that it reacts
with HO about as fast as C_2H_4. A similar decrease in reactivity seems to hold for ozone reactions, i.e., C_2Cl_4
reacts much more slowly with O_3 than does C_2H_4, leading
to a t_c in the range of 30 to 3000 yr, very much longer
than those of the HO reactions. As in the case of H abstraction, the O atom addition reactions are slower than
those of HO, and the small O atom densities make their
contribution to tropospheric removal of haloethylenes a
negligible one. The reactions of HO_2 with olefins to form
alkyl radicals plus O_2 are sufficiently endothermic to be
unimportant. Addition reactions to form peroxy radicals
are possible and would add further to the already rapid
removal by HO.

The central role played by the HO radical in the various halocarbon removal processes points up the great importance of knowing its reaction rate parameters and its
tropospheric concentration profile. The former are increasingly well known, at least for most of the initial
reactions. The subsequent removal of chlorine from a halocarbon radical such as CF_2Cl via oxidation and formation
of CF_2O is probably fast. Addition of HO to unsaturated
halocarbons may be followed by loss of one chlorine atom
with rearrangement to an acid chloride:

$$C_2Cl_4 + HO \rightarrow CHCl_2COCl + Cl \qquad (A.8)$$

which could photolyze with loss of CO and formation of
another halocarbon molecule

$$CHCl_2COCl + h\nu \rightarrow CO + CHCl_3 \qquad (A.9)$$

The further chemical stability of $CHCl_3$ must then be
considered.

The measurement of the tropospheric HO concentration is in a more unsatisfactory state (Chapter 6 and Appendix C). Although semiquantitative measurements in the ambient atmosphere have been carried out successfully by laser fluorescence in the near ultraviolet (Wang et al., 1975; Davis, 1975), they are enormously difficult, because they require the detection of one HO per 3×10^{13} air molecules, i.e., if 10 percent accuracy is desired, determination of a mixing ratio to about 3×10^{-15}. It is not yet clear whether such measurements can be carried out over the required range of ground locations and altitudes.

Modeling calculations have been carried out for HO as described in Chapter 7 and elsewhere (Levy, 1973; Crutzen, 1974) and appear to give reasonable results. The principal source term, the reaction of H_2O with $O(^1D)$ from O_3 photolysis is well understood, but some of the suggested HO removal steps are either kinetically uncertain (reaction with HO_2) or may be complicated by subsequent regeneration or even net production of HO_x. For these reasons, the local chemical removal times listed in Table A.1 must only be considered order-of-magnitude estimates, primarily useful when they are very long, indicating no halocarbon removal, or very short, indicating fast removal even in the face of large uncertainties.

The tropospheric reactions of the chlorofluorocarbons F-11 and F-12 may be summarized as follows: For $O(^1D)$, rapid reaction is known to take place, but because of its extremely small concentration τ_T is longer than 10,000 yr. With all other reactants, chemical reactions are either very slow or do not occur at all; therefore, τ_T for them is even longer. Thus, gas-phase reactions with neutral species do not, according to present knowledge, provide a significant sink for these CFMs. It is unlikely that future studies will change these conclusions. For example, in order for tropospheric reaction of HO with F-11 to be competitive with "active" photolytic removal in the stratosphere, the rate constant k_x would need to be $\geq 1.5 \times 10^{-16}$ over the *entire* troposphere ($\tau_T \lesssim 100$ yr). The upper limit for this constant (Table A.1) is 5×10^{-16} cm^3 sec^{-1} at room temperature, and this might conceivably provide an "inactive" removal process. However, this upper limit was set by the detection limits in the laboratory studies of the reaction, and there is no reason whatever to believe that k_x is as large as 5×10^{-16}. This conclusion is further supported by recent work (Table A.1), which shows the reaction to be undetectably slow at 480 K. The only conceivable mechanisms for such a reaction would be direct substitution of Cl by HO. Reactions of this type are known to occur in the chemistry of translationally hot atoms following nuclear recoil processes but are very rare in thermal systems, since they have higher activation energies than competing processes. Even if the rate constant were to be as high as 5×10^{-16} cm^3 sec^{-1} for this reaction at 298 K, the activation energy would likely be of the order of 4.5 kcal/mole. This means that the rate constant would drop rapidly with altitude, and this variation would increase the value of τ_T to more than 70 yr. However, the true value of τ_T is likely to be several orders of magnitude larger than this.

In summary, this section has shown that all unsaturated halocarbons and partially halogenated saturated hydrocarbons have short tropospheric lifetimes (0.1 to 5 yr) due to homogeneous chemical reactions. However, *completely halogenated halocarbons, such as the CFMs and CCl$_4$, are by all atmospheric and laboratory evidence chemically inert in the troposphere*, i.e., have lifetimes for chemical removal that are much longer than their total atmospheric residence times. Therefore, in the following two sections we will consider other possible tropospheric removal processes for the completely halogenated saturated halocarbons only.

C. Ionic Processes

1. Loss by Direct Ionization Since CFMs have large cross sections for dissociative electron capture, cosmic-ray bombardment has been suggested as a removal process. However, molecular oxygen also has an appreciable electron capture cross section and will compete with the CFM for the bombarding electron. The local loss rate constant L (cf. Chapter 4) for this process can be calculated from the expression

$$L = \frac{k_{CFM}}{k_{O_2} [O_2]} I \qquad (A.10)$$

where k_{CFM} and k_{O_2} are the rate constants for electron capture by the CFM and by O$_2$, respectively, and I is the total ionization rate. Values of I in the troposphere are of the order of about 10 ions cm^{-3} sec^{-1} at ground level (partly because of radioactive sources), decrease to about 3 cm^{-3} sec^{-1} at 2 km, and then increase to a peak value of about 25 cm^{-3} sec^{-1} at the tropopause (Cole and Pierce, 1965). F-11 has the largest value of $k = 9.5 \times 10^{-8}$ cm^3 sec^{-1}. Electron capture by O_2 is a three-body process at low pressure but is probably at its second-order limit in the troposphere for which the lowest reported value of k_{O_2} is 9.5×10^{-11} cm^3 sec^{-1} (Phelps, 1969). The smallest value of $t_c = 1/L$ for this process in the troposphere is then 10^6 yr. Therefore, τ_T is at least this large and *we can dismiss direct ionization from further consideration* as a significant sink for CFMs.

2. *Loss by Ion-Molecule Reactions* Although cosmic rays are ineffective in removing halocarbons directly, they do produce ion pairs of other species, which, conceivably, might react with halocarbons. The primary positive ions formed are N_2^+ and O_2^+. These are rapidly converted to oxonium cluster ions $H_3O^+(H_2O)_n$ by reaction paths that are well established in the laboratory and verified by observations in the D region of the atmosphere (Ferguson, 1971). All of these reactions occur with fast rates close to their classical limits and involve O_2 and H_2O, which are present in the troposphere in amounts so much greater than CFMs that the latter cannot compete with the conversion to oxonium cluster ions. If NH_3 is present in concentrations greater than about 10^6 cm^{-3} (a condition that may apply in the lower troposphere), the oxonium ion may be converted to ammonium ion clusters

$$H_3O^+(H_2O)_n + NH_3 \rightarrow NH_4^+(H_2O)_{n-m} + (m+1)H_2O \quad (A.11)$$

We need, therefore, consider only possible reactions of CFMs with H_3O^+ and NH_4^+ cluster ions.

Atmospheric reactions of chlorine compounds with ions have been studied by Fehsenfeld et al. (1976). The reactions of F-11 and F-12 with NH_4^+ or its clusters were found to be too slow to be observed in the laboratory, which is to be expected, since they are almost certainly endothermic. No reaction was observed in the laboratory between oxonium ions and F-12, and these reactions are also likely to be highly endothermic. The reaction of H_3O^+ with F-11 does occur with a rate constant of 4×10^{-10} cm^3 sec^{-1}. But the rate constant decreases rapidly with increasing hydration of the ion. For $n = 2$, k was found to be no greater than 10^{-12} cm^3 sec^{-1}, the lower limit measurable in the experiment. For higher values of n, the reaction becomes endothermic. Since the predominant ions in the troposphere have values of n from 5 to 7, the corresponding (average) rate constant will almost certainly be less than 10^{-13} cm^3 sec^{-1}. The ion densities in the troposphere are given in the CIAP report and range from about 2×10^3 cm^{-3} near ground level to a peak concentration of about 10^4 at 10 km. If it is assumed that all the positive ions are of the $H_3O^+(H_2O)_n$ type and react with a rate constant of 10^{-13} cm^{-3} sec^{-1}, then $\tau_T = 100$ yr. However, it is much more likely that the rate constants for the dominant ions with F-11 are at least an order of magnitude smaller than this, giving $\tau_T \geq 1000$ yr. Therefore, it is safe to conclude that *positive ions do not provide a significant tropospheric sink for CFMs.*

The primary negative ion formed in the troposphere is O_2^-, which will rapidly hydrate to form $O_2^-(H_2O)_n$. The value of n in the troposphere can be calculated from thermodynamic data to be greater than 5. The unhydrated O_2^- ions react rapidly with F-11 and F-12 by charge transfer and by the reaction

$$O_2^- + CFCl_3 \rightarrow Cl^- + CFCl_2 + O_2 \qquad (A.12)$$

However, both channels become endothermic for hydrated O_2^-. For example, the reaction of F-11 with $O_2^-(H_2O)_2$ is 17.8 kcal $mole^{-1}$ endothermic for charge transfer and 2.5 kcal $mole^{-1}$ endothermic for Reaction (A.12) if the ion product is $Cl^-(H_2O)_2$. The value for k would then be less than 10^{-14} cm^3 sec^{-1} in agreement with the observed upper limit of 10^{-13} cm^3 sec^{-1} (Fehsenfeld *et al.*, 1976). For $n = 5$, k would be orders of magnitude smaller than this.

Reaction paths for $O_2^-(H_2O)_n$ ions with CO_2 and NO_2 exist, which produce CO_3^-, NO_2^-, and NO_3^- and which, in turn, will be rapidly hydrated. However, no exothermic channels have been found for reaction of any of these ions with CFMs. Therefore, *negative ions are ineffective as a tropospheric sink for CFMs.*

D. *Heterogeneous Processes*

The possibility of trace gas removal from the troposphere by physical and chemical interaction with aerosols will be examined next.

1. *Absorption in Cloud Droplets and Rain-out* The maximum rate of removal would occur if all water droplets in the atmosphere were saturated with F-11 at the steady-state

partial pressure of $\sim 1 \times 10^{-9}$ atm. The average annual global precipitation rate is 70 cm yr^{-1}. If one takes the solubility of F-11 in pure water at 20°C and 1 atm to be 4.4×10^{18} molecules cm^{-3}, the amount dissolved would be $(1 \times 10^{-9}$ atm$)(4.4 \times 10^{18}$ molecules cm^{-3} $atm^{-1})(70$ cm $yr^{-1}) = 3.1 \times 10^{11}$ molecules cm^{-2} yr^{-1}. Hydrolysis would result in an insignificant increase in this removal flux (see discussion in Section I.A). The column density of F-11 at steady state would be about 2×10^{16} molecules cm^{-2}. Therefore, the removal time would be $(2/3.1) \times 10^5$ yr or $\gtrsim 6 \times 10^4$ yr, and we conclude that *rain-out provides a negligible removal process for CFMs*.

Much of the tropospheric aerosol is hygroscopic and, as pointed out by Junge (1963), most of its mass is in cloud droplets and would not be expected to create any differences in the dissolution rate of the CFM discussed above. Hydrolysis is faster in alkaline solution, but most of the aerosols are acidic. Therefore, aerosols are not expected to increase the rate of solution or hydrolysis of CFMs over the values given above for cloud droplets or rain.

2. Chemical Reaction with Aerosols Aside from hydrolysis reactions, CFMs do not react in aqueous or polar media with the inorganic substances commonly associated with aerosols (e.g., NH_4^+, $SO_4^=$, H_3O^+, Cl^-, Ca^{++}, NO_3^-).

Organic molecules associated with tropospheric aerosols may be expected to contain a significant fraction of oxygenated compounds such as carboxylic acids, aldehydes, alcohols, and nitrates. Such compounds have been detected in urban aerosols (Scheutzle *et al.*, 1975) and in smog chamber studies (Schwartz, 1974). Organic molecules may constitute one third of the aerosol mass, and their composition is expected to be strongly dependent on local influences such as pollution emissions and the presence of vegetation. There is no evidence that CFMs react with such organic substances under atmospheric conditions.

Adsorption followed by chemical reaction with other reactive species may be important if the adsorbed molecules are thereby rendered more reactive than they are in the gas phase. Such adsorption of molecules is a phenomenon associated with porous surfaces and substances that can form weak physical associations. Since most tropospheric particles have the properties of hygroscopic acidic salts with varying amounts of associated water, it is probable

that nonpolar CFM molecules would not be adsorbed at all--
and even if they were, the adsorption would be weak and
reversible, rather than leading to chemical reaction and
true removal.

The only known reaction of CFMs with a tropospheric
constituent is that reported by Siegemund (1973) in which
$CFCl_3$ reacts with SO_3 in the presence of concentrated sulfuric acid to produce COFCl at room temperature, but the
reaction rates were not measured. Even if it were assumed
that a CFM can react with SO_3 in the gas phase, there is
so little SO_3 present that even the fastest possible reaction rate would be insufficient to give a partial removal
time of less than 10^{10} yr.

3. *Enhanced Removal in Urban Areas* Most of the CFMs are
released initially in densely populated urban areas that
have local atmospheres heavily laden with industrial particulates, which should enhance the heterogeneous removal
in these areas. However, urban areas are such a small
fraction of the troposphere that even with enhancement
factors of several orders of magnitude, their contribution
to the total global removal rate is not significant. The
experiments of Hester *et al.* (1975) were intended to simu-
late a smoggy atmosphere with sulfuric acid aerosol present. Destruction of F-11 and F-12 in this atmosphere was
not detected within experimental error. The sensitivity,
however, was not sufficient to rule out a small removal
flux.

The pollutants in urban smog are believed to contain
some substances with considerable chemical reactivity [e.g.,
peroxyacetyl nitrate (PAN), O_3, HO, $O(^3P)$, $O(^1D)$, organic
oxy and peroxy free radicals]. Of these only $O(^1D)$ is
known to react with CFMs. However, as was shown in Section II.B, $O(^1D)$ is present in too small concentrations
in the troposphere to constitute a significant sink. Even
with the increased O_3 concentrations in urban smog atmospheres, the lower $O(^1D)$ yield from O_3 photolysis in the
troposphere, the rapid quenching of $O(^1D)$ by N_2 and O_2,
and the low fraction of the earth's surface occupied by
urban regions with high O_3 densities lead to the conclusion that these reactions will be unimportant. To be significant a local sink must, of course, be much stronger
than one that operates over the entire globe.

*In summary, then, the known heterogeneous reactions of
CFMs are unlikely to produce an effective removal flux.*

E. Removal by Lightning

1. Thermal Decomposition In the core of lightning, arc
temperatures of more than 5000 K are reached and all CFMs
will, of course, be broken down into atomic or diatomic
fragments, which are removed from the troposphere. Over
somewhat larger volumes and for average temperatures down
to about 1500 K, the thermal decomposition of such halomethanes may still be rapid, but any longer-range (or
larger-volume) thermal effects must be negligible because
of the very large activation energy of the decomposition
reaction. It has been estimated that the global source
of NO due to lightning is of the order of 10^{35} molecules
per year or 10^{-9} of the total atmosphere. Allowing for
at least one order of magnitude for the lower efficiency
of NO production (~10 percent NO yield) compared with the
total CFM destruction, we arrive at a rough estimate of 10^{-8}
per year or a lifetime of 10^8 yr for CFM removal by lightning in a thermal process.

This is six orders of magnitude longer than the range
of values that are of interest here and shows that even
with much lower relative NO yields and much larger post-
lightning volumes of heated gas this potential removal
process is negligible. Any proposed electron/ion processes of a nonthermal nature arising from lightning are
limited by the considerations of Sections II.C.1 and II.C.2
above and may also be dismissed.

2. Photolysis A rough estimate may also be made of an
upper limit to tropospheric photolysis of CFMs due to the
emission of radiation from lightning strokes in the atmospheric window between 185 and 225 nm. The average energy
of a lightning stroke is about 5×10^8 J and the average
global frequency is 100 per sec. The light emission in
the ultraviolet region of the spectrum consists at least
partly of discrete atomic lines and molecular bands, but
it will here be approximated by blackbody emission at
20,000 K. This represents a large overestimate of its
total intensity. The total radiant energy is extrapolated
from field observations in the visible and near infrared
regions to be about 3 percent of total energy, with 0.9
percent of the radiant energy emitted in the window near
200 nm. On that basis, about 10^{33} photons are emitted per
year in that wavelength range. Considering only the continuous absorption by O_2 in the Herzberg continuum, i.e.,

neglecting the stronger, discrete absorption in the Schumann-Runge bands, and averaging the absorption cross sections for both the CFMs and for O_2, we obtain fractions of 2×10^{-5} (F-11) and 1×10^{-5} (F-12) of the emitted light that is absorbed by the CFMs. These fractions use the present tropospheric mixing ratios of 1.2×10^{-10} (F-11) and about 2×10^{-10} (F-12). This absorbed light dissociates the CFMs and amounts to a global destruction rate of about 3 tons per year for F-11 and F-12 compared with the accumulated atmospheric burden of about 7 megatons. *Photolysis by lightning therefore corresponds at best to removal times of order 10^6 yr and may be disregarded.*

F. *Thermal Decomposition in Combustion Processes*

Although temperatures in most combustion processes are not nearly so high as those in lightning, CFMs will be destroyed in them, and the products will be removed subsequently from the troposphere. The amount of such removal may be estimated from the fraction of the earth's atmosphere that passes through high-temperature combustion processes annually.

Strehlow (1976) has calculated the latter in two different ways, one based on the total energy from combustion worldwide and the other on the rate of accumulation of CO_2 in the atmosphere. Using as a starting point an estimate by Perkins (1974) that world energy input was 0.15 Q (10^{18} BTUs) per year in 1970-1971, Strehlow estimates the total energy produced by high-temperature combustion to be 0.25 Q per year (an upper bound) at present. In turn, such combustion would require about 1.6×10^{-5} of the atmospheric O_2 per year. Similarly, the 0.7 ppm annual increase of CO_2 in the atmosphere (Perkins, 1974) reflects an annual production rate of about 2.1 ppm per year, which consumes 1.1×10^{-5} of the atmospheric O_2.

The fraction of CFM destroyed may be somewhat greater than these two figures, because of the concentration of CFM release in urban areas, where much of the combustion is also localized. On the other hand, it seems likely that some of the CFM passing through combustion processes will not be decomposed. In any case, *the removal time of CFMs by combustion will be at least 10^4 yr, which is negligible compared with the stratospheric removal.*

III.   STRATOSPHERIC PROCESSES AND OZONE DESTRUCTION

Stratospheric removal of the halocarbons will generally be active with respect to ozone destruction. The effectiveness with which decomposition of the halocarbon leads to ozone destruction under steady-state conditions depends directly on the ratio of the amount of chlorine in the reactive products (ClO_x) to that in unreactive forms (e.g., HCl, $ClONO_2$, and unreacted halocarbon). The processes that affect this partitioning will be considered in detail. We also give consideration to any process that might lead to inactive removal of the halocarbon. To be inactive, such a process must lead to a product that is not converted to the active ClO_x form in a time short compared with its residence time in the stratosphere.

A.   Photodissociation of Halocarbons

The amount of ultraviolet light in the region of 280 to 320 nm increases with increasing altitude. Above 20-km altitude, light is also transmitted in the so-called "solar window" between 185 and 225 nm, which is bounded on the short- and long-wavelength sides by strong absorption by O_2 and O_3, respectively, as shown in Figures A.2 and A.3. The complex and highly structured absorption by the Schumann-Runge bands of O_2 in this region make the light fluxes in this solar window hard to calculate (Kockarts, 1971), and the uncertainty may be as great as a factor of 2.

Compounds containing C—Cl and C—Br bonds absorb sunlight in the region of the solar window but not above 280 nm. Absorption spectra of $CFCl_3$ (F-11), CF_2Cl_2 (F-12), and other halogenated compounds at laboratory temperatures are shown in Figure A.4. In halocarbons, the C-Cl and C-Br linkages are weaker than the C-H and C-F bonds (Kerr, 1966). The energy of the absorbed photon in the solar window is 125 to 150 kcal/mole, and the C-Cl bond energies in chlorofluoromethane and halogenated methanes are about half of this (Kaufman and Reed, 1963; Gurvich et al., 1974). As the absorption spectra are continuous and no fluorescence is observed, a quantum yield of unity for the dissociation of the parent molecule is expected; this is consistent with the interpretation of the absorption spectra as $n \rightarrow \sigma*$ transitions. The photochemistry of halogenated hydrocarbons has been reviewed by Majer and Simons (1964), and the photochemical oxidation of the

perhalogenated compounds by Heicklen (1969). The clear
conclusion from the investigations reviewed is that, in
addition to the chlorine or bromine atoms liberated in
the initial photochemical act, the action of atmospheric
O_2, O_3, and O on the halogenated radicals produced rapidly liberates the remaining halogen atoms as ClX or BrX
(e.g., Cl, ClO, or HCl). Some recent data and stratospheric calculations on chlorocarbons are given by Crutzen
and Isaksen (1976). The following discussion is limited
to the chlorofluoromethanes.

Milstein and Rowland (1975) have photolyzed CF_2Cl_2 at
184.9 nm in the presence of oxygen and shown that the
quantum yield is 1.1 ± 0.1 and that both chlorine atoms
would become available to catalyze ozone destruction.
Rebbert and Ausloos (1975) have shown that the primary
processes at 213.9 nm are predominantly

$$CF_2Cl_2 + h\nu \rightarrow CF_2Cl + Cl \qquad (A.13)$$

$$CFCl_3 + h\nu \rightarrow CFCl_2 + Cl \qquad (A.14)$$

the quantum yields for chlorine atom formation again being
1.1 ± 0.1. With increasing photon energy there is a greater probability of a second chlorine atom being eliminated
in the primary photochemical process; the quantum yield
for this process reaches approximately 0.3 at 185 nm.
Similar behavior has recently been reported in the photolysis of carbon tetrachloride over a somewhat greater wavelength range (Davis *et al.*, 1975b).

The total number of chlorine atoms released from a
fluorocarbon is governed by the fate of the free radicals
formed in the primary photochemical act. By analogy with
the behavior of CF_3 and CCl_3 radicals (Heicklen, 1969), the
dominant reactions in the presence of oxygen are expected
to be

$$CF_2Cl + O_2 \rightarrow F_2CO + ClO \qquad (A.15)$$

$$CFCl_2 + O_2 \rightarrow FClCO + ClO \qquad (A.16)$$

Any CFCl formed at shorter wavelengths will be oxidized to
yield ClX species. Thus, the formation of two ClX species
in the photolysis of both CF_2Cl_2 and $CFCl_3$ is effectively
instantaneous. Release of the third chlorine atom from
$CFCl_3$ depends on the reactivity of CFClO, about which
little is known. However, the absorption spectrum of
CFClO has recently been determined (see Figure A.5) by

Rowland *et al.* (1975). The local removal time for its photodissociation to yield chlorine atoms is about 70 days at 30 km and 7 days at 40 km. These removal times are similar to those for $CFCl_3$ from which it is formed.

Rowland and Molina (1975) calculate the local photodissociation removal times ($t_C = 1/L$) of $CFCl_3$ and CF_2Cl_2 (F-11 and F-12) to be 6.6 and 66 yr at 20 km, 1.3 and 11 months at 30 km, and 4.7 and 40 days at 40 km altitude, which lead to atmospheric residence times of about 50 and 100 yr, respectively. A change in the assumed values of the solar flux would result in a proportional change in the residence time. Rebbert and Ausloos (1975) have recently shown that the absorption cross section of CF_2Cl_2 is strongly temperature-dependent in its long wavelength tail, its value of 213.9 nm falling from 0.27×10^{-20} cm^2 at 295 K to 0.15×10^{-20} and 0.08×10^{-20} cm^2 at 249 K and 234 K, respectively. Such behavior, due to absorption by vibrationally excited molecules near the long wavelength limit, is expected with all halogenated hydrocarbons. The maximum effect would be observed with CF_2Cl_2 because the onset of its absorption coincides with the solar window. For a temperature of 195 K the photochemical local removal time of CF_2Cl_2 at 30 km is increased by a factor of 1.5 from its value at 298 K; however, this only increases the calculated atmospheric residence time by 10 percent because of the steep increase in its photolysis rate with altitude (Rowland, 1975). The overall effect will be negligible for $CFCl_3$, FClCO, and other species where the maximum absorption is closer to the solar window.

This effect may need to be considered in the stratospheric photolysis of such substances as $CHFCl_2$ and CHF_2Cl (F-21 and F-22), where the tail of their light absorption coincides with the solar window and their strongest absorption lies at shorter wavelengths. Such compounds would be photolyzed at higher altitudes (higher temperatures), where the destruction of CFMs by $O(^1D)$ is also significant. The products of their decomposition still have to pass down through the stratospheric ozone.

B. *Photodissociation of ClO and HCl*

The photolysis of ClO reduces the efficiency of the O and O_3 removal cycle

$$O + ClO \rightarrow Cl + O_2 \qquad (A.17)$$

$$Cl + O_3 \rightarrow ClO + O_2 \qquad\qquad (A.18)$$

by substituting the cycle

$$Cl + O_3 \rightarrow ClO + O_2 \qquad\qquad (A.18)$$

$$ClO + h\nu \rightarrow O + Cl \qquad\qquad (A.19)$$

$$O + O_2 + M \rightarrow O_3 + M \qquad\qquad (A.20)$$

which does not remove O and O_3. Therefore, we need to know the extent to which photolysis of ClO competes with its role in the destruction of ozone.

ClO shows banded absorption followed by a continuum in the region from 340 to 230 nm. All of these bands are diffuse, and no fluorescence has been detected from them. Light absorption in both the bands and continuum would therefore lead to dissociation. However, ClO is largely protected from photolysis by the strong Hartley band absorption of ozone. At altitudes of 30 to 40 km, where the decomposition of ozone by ClO_x from F-11 and F-12 is most important, photolysis of ClO can occur down to 290 nm. The absorption coefficients of ClO given by Basco and Dogra (1971) and the Franck-Condon factors and absorption coefficients recently calculated by Coxon (1976) lead to a value of $J_{ClO} \sim 4 \times 10^{-4}$ sec^{-1} at 40 km for the process

$$ClO + h\nu \rightarrow Cl + O \qquad\qquad (A.19)$$

The next band system of ClO lies beyond the solar window, and at 175 nm the absorption cross section of ClO has fallen to about 5×10^{-19} cm^2 (Basco and Morse, 1973). With the assumption that it rises uniformly to $\sigma = 4 \times 10^{-18}$ cm^2 at 258 nm, radiation in the solar window would only contribute about 10^{-5} sec^{-1} to J_{ClO} at altitudes of 30 to 40 km. Using the known rate constants (Appendix D) for the reactions of ClO with NO ($k = 2.7 \times 10^{-11}$ cm^3 sec^{-1}) and ClO with O ($k = 4.5 \times 10^{-11}$ cm^3 sec^{-1}) as well as the concentrations of NO (5×10^8 molecules/cm^3) and O (5×10^8 molecules/cm^3) at 35 km, we calculate the rate of removal of ClO by chemical reaction to be 3.6×10^{-2} sec^{-1} at 35 km. The average rate of removal of ClO by photolysis ($J_{ClO} \sim 4 \times 10^{-4}$ sec^{-1}) given above is, thus, about a hundred times slower than the removal of ClO by O and by NO in the important altitude range of 30 to 40 km. The photochemical decomposition of ClO therefore has a negligible effect on its removal of O and O_3.

Photolysis of HCl to produce Cl would increase the ratio of chlorine in the ClO_x form and thereby increase the ozone destruction. However, HCl does not absorb strongly in the solar window, and its photolysis to yield Cl contributes about 1 percent at 40 km and 3 percent at 50 km to its reconversion back to an active form, as compared with the dominant reaction (HO + HCl), which also yields Cl. Nonetheless, the photolysis of HCl (and ClO) is included in the reaction scheme employed for our 1-D calculations along with a number of other relatively minor photochemical and chemical reactions (Appendix D).

C. *Halocarbon Reactions with Neutral Molecules*

Many of the tropospheric chemical reactions discussed in Section II.B apply to the stratosphere as well. However, the concentrations of some of the reactive atmospheric species are different by several orders of magnitude. The model calculations of Chapter 7 show that ground-state O atom (^3P) concentrations, for example, range from about 10^5 to 10^{10} cm^{-3} as one ascends through the stratosphere and that $O(^1D)$ concentrations rise from about 0.1 to 500 cm^{-3}. HO concentrations decrease by about a factor of 5 from ground level to 20 km altitude but then rise to a peak value of 1×10^7 to 2×10^7 cm^{-3} just below the stratopause. The HO_2 concentration is calculated to decrease through the troposphere and be somewhat below 10^8 cm^{-3} throughout the stratosphere. The effects of these changes on the atmospheric residence times of halocarbons are readily calculated by the method outlined earlier, but they are small relative to the principal removal process, photolysis. This is especially true for the CFMs. But, as noted in Chapter 4, halocarbon breakdown in the stratosphere, whether by photolysis or chemical reaction will probably constitute active removal, producing ClO_x and leading to O_3 destruction. *No inactive removal process has been found for reactions of halocarbons with neutral molecules in the stratosphere.*

It should be noted that for F-11, where reaction with $O(^1D)$ competes with stratospheric photolysis, the latter is about two orders of magnitude more efficient than $O(^1D)$ reaction in the 30- to 40-km region; for F-12, photolysis is only one order of magnitude more efficient than $O(^1D)$ reaction, and the latter must be taken into account in calculations of atmospheric residence times. This was done for the value of τ given in Table 4.2 (90 yr).

D. *Reactions of ClO$_x$*

As mentioned in the introduction to Chapter 4, the effectiveness of the chlorine-catalyzed destruction of ozone depends on the fraction of chlorine that is present as ClO$_x$. Once Cl atoms are generated in the stratosphere by photolysis or reaction of halocarbons, they take part in the catalytic chain decomposition of ozone by a sequence of reaction steps. The primary cycle consists of the two reactions

$$Cl + O_3 \rightarrow ClO + O_2 \tag{1}$$

$$ClO + O \rightarrow Cl + O_2 \tag{2}$$

which are equivalent to the corresponding NO$_x$ reactions

$$NO + O_3 \rightarrow NO_2 + O \tag{1'}$$

$$NO_2 + O \rightarrow NO + O_2 \tag{2'}$$

Further reactions of ClO,

$$ClO + NO \rightarrow Cl + NO_2 \tag{3}$$

and

$$ClO + h\nu \rightarrow Cl + O \tag{4}$$

may partially reduce the catalytic action, because Reaction (3) is followed by photolysis of NO$_2$, which regenerates "odd oxygen" (O and O$_3$), while Reaction (4) regenerates it directly as discussed in Section III.B.

Cl atoms may be transformed into inactive HCl by the following processes:

$$Cl + CH_4 \rightarrow HCl + CH_3 \tag{5}$$

$$Cl + H_2 \rightarrow HCl + H \tag{6}$$

$$Cl + HO_2 \rightarrow HCl + O_2 \tag{7}$$

$$Cl + H_2O_2 \rightarrow HCl + HO_2 \tag{8}$$

Other, less important reactions producing HCl are included in many models (see, e.g., Appendix D). The HCl is transported downward into the troposphere and is removed by rain-out; however, active Cl is regenerated by Reaction (9) of HCl with HO

$$HCl + HO \rightarrow Cl + H_2O \tag{9}$$

The above reactions clearly represent a small fraction of all possible steps involving Cl species. They were thought to include the most important chemical and photochemical processes of stratospheric importance until recent laboratory measurements (Rowland et al., 1976) of the absorption cross section of chlorine nitrate, $ClONO_2$, in the ultraviolet wavelength region showed this compound to be much more weakly absorbing than had been anticipated, so that it had to be added to the main scheme. Since $ClONO_2$ apparently does not react with ozone and only slowly with atomic oxygen (see below), it represents an inactive reservoir of Cl as well as of NO_x. It is formed and removed by the reactions

$$ClO + NO_2 + M \rightarrow ClONO_2 + M \tag{10}$$

$$ClONO_2 + h\nu \rightarrow ClO + NO_2 \text{ or other products} \tag{11}$$

$$ClONO_2 + O \rightarrow ClO + NO_3 \text{ or ClONO} + O_2 \tag{12}$$

Reactions (10), (11), and (12) are a minimal set of such processes. Even so, the photolysis products of Reaction (11) and the reaction products of (12) are still unknown, although the rate constants are being determined. If Reactions (11) and (12) regenerate ClO, they return active ClO_x species to the catalytic cycle. However, if ClONO is a major product of either process, the photolysis and further reactions of that compound must also be considered. In the following discussion, the scheme (1) to (9) will first be examined, i.e., the effect of $ClONO_2$ will be left out, and that perturbation will be introduced later.

Reactions (1) to (9) and our present knowledge of their rate parameters lead to the following qualitative conclusions: Since Reactions (1) and (2) are both very fast, the primary, catalytic destruction of "odd oxygen" by this chain process is clearly fast; under steady-state conditions its overall rate equals twice the rate of Reaction (2) and is thereby proportional to the concentration of ClO, i.e., to the fraction $[ClO]/[ClX]$, where $[ClX]$ is the sum of the concentrations of all those Cl-containing species (Cl, ClO, HCl, $ClONO_2$, etc.) interconnected by fast reactions. The formation of HCl in Reactions (5) to (8) is not really chain-breaking because "active" ClO_x is regenerated in Reaction (9). A similar remark applies to $ClONO_2$ in its formation and redissociation, Reaction (10), and Reactions (11) and (12).

In those regions of the stratosphere where chemical reactions are much faster than transport, above about 30 km at low and midlatitudes, simple steady-state analysis of the chemical scheme can provide a good estimate of the local reduction of "odd oxygen" by chlorine catalysis. Such calculations show, for example, that ClO represents one half to one third of all ClX at 30 to 40 km altitude but is reduced to about one tenth in the lower stratosphere if transport is neglected.

Even in this oversimplified form, a large amount of information is required to assess the effect of the chlorine cycle. Accurate rate constants are needed for all important reactions. This requirement is modified slightly by the fact that only the sum of certain rates need to be known accurately, such as those that form HCl, Reactions (5) to (8), for example. Larger uncertainties are allowed for the slower of several parallel steps. Concentrations of minor atmospheric species such as O_3, O, NO, CH_4, and HO must also be known, and these species take part in a host of other reactions and catalytic cycles. Wherever possible, direct measurements of any of the minor species should, of course, be available to verify model calculations.

It is instructive to carry out an oversimplified steady-state calculation of the ClO_x-O_3 catalysis, without $ClONO_2$, for the midstratosphere in order to show its magnitude as well as the partitioning of ClX. Here, we neglect Reactions (6), (7), and (8) and assume an altitude of 35 km and a temperature of 237 K. The lifetimes, t_i, of ClO_x species for all the reactions considered here are very short. These lifetimes are defined as the inverse of pseudo-first-order rate constants, for example, for Cl in Reaction (1), $t_1 = 1/k_1[O_3]$ is about 1/20 sec, t_2 for ClO in Reaction (2) is about 70 sec, t_3 is 80 sec, t_5 is 200 sec, and t_9 is 3×10^5 sec. We see that t_1 is much shorter than t_2 even though k_1 and k_2 are both large. This is because $[O_3]$ is much larger than $[O]$, their ratio being controlled by the very fast recombination of O with O_2 and the somewhat slower photolysis of O_3. Consequently, $[Cl]$ is much smaller than $[ClO]$. The fraction $[ClO]/[ClX]$ is then approximately equal to $t_5 t_{23}/(t_5 t_{23} + t_1 t_9)$, where t_{23} is the inverse of the total rate constant for ClO removal, i.e., $1/t_{23} \equiv (1/t_2) + (1/t_3)$. The ClO fraction is approximately 1/3. At lower altitude, say 25 km, it decreases to about 0.1, mainly because [O] is lower, whereas [CH_4] is higher, so that the formation of HCl is favored. Table A.2 lists currently available values of ClX rate constants (and sources), many of them as yet unpublished or in process of publication.

With the addition of $ClONO_2$ to the scheme, assuming that Reactions (10), (11), and (12) determine the ratio [$ClONO_2$]/[ClO] and that Reactions (11) and (12) regenerate active ClO_x, one calculates a decrease in [ClO]/ [ClX] of only about 10 percent. The effect of $ClONO_2$ in the midstratosphere is thus quite small. But when it is considered that Reaction (10) also brings about the temporary removal of one NO_x for each ClO_x and that this three-body reaction becomes faster at lower altitudes both because of its pressure--and its temperature dependence--the perturbation due to $ClONO_2$ formation becomes much larger in the lower stratosphere. In that region, transport is no longer slow compared with chemical reaction and simple steady-state calculations are no longer useful.

The magnitude of all catalytic chain cycles is strongly dependent on the hydroxyl concentration. For example, HO regenerates Cl from HCl, it removes NO_2 in the lower stratosphere by formation of HNO_3, and it reacts directly with O_3 in the first step of a catalytic cycle that recombines "odd oxygen" as do Reactions (1), (2), and (1'), (2'). The reaction

$$HO + HO_2 \rightarrow H_2O + O_2 \qquad (13)$$

is the principal removal process for HO_x, and its rate constant is both the most important and the least well known among all those listed in Table A.2. Experimental and theoretical arguments may be advanced for and against the high value of $k_{13} \sim 1$ to 2×10^{-10} cm^3 sec^{-1}. The case against this value seems on firm ground, and on consideration of a wide range of evidence we recommend 2×10^{-11} cm^3 sec^{-1}, albeit with large error bars. Details are given near the end of this section.

In order to assess the magnitude of the catalytic destruction of "odd oxygen" by ClO_x locally in the midstratosphere and at midlatitudes, it is instructive to compare its contribution, which equals $2k_1$[ClO][O] under steady-state conditions, with its NO_x counterpart, $2k_2$·[NO_2][O] and with pure oxygen (Chapman) rate $2k_{14}$[O_3][O] of the reaction

$$O + O_3 \rightarrow 2O_2 \qquad\qquad (14)$$

If we assume a present tropospheric mixing ratio of 1×10^{-10} for F-11 and 2×10^{-10} for F-12 and also assume complete liberation of all Cl above about 25 km, the total corresponding mixing ratio of active Cl species would be 7×10^{-10}, and [ClO] at 35 km would then be about 3×10^7 cm^{-3}. This assumes steady-state conditions for present CFM concentrations and is therefore in error. It over-estimates Cl species by a factor of about 1.5 as shown by 1-D calculations, and a value of 2×10^7 cm^{-3} will here be assumed for the present ClO concentration at 35 km arising from the photolysis of F-11 and F-12. The cata-lytic ClO_x rate constant, $2k_2[ClO]$, is then approximate-ly 1.8×10^{-3} sec^{-1} (setting $k_2 = 4.5 \times 10^{-11}$ cm^3 sec^{-1}). The corresponding NO_x rate constant, $2k_{2'}[NO_2]$ equals 2.5×10^{-2} sec^{-1}, and the pure oxygen rate constant, $2k_{12}[O_3]$, equals 4×10^{-3} sec^{-1}. The current local ozone destruction rates at 35 km due to ClO_x, NO_x, and O_x thus stand in the approximate ratios 1:14:2. These estimates have used the atmospheric concentrations presented in Chapter 7, which are the results of state-of-the-art one-dimensional model calculations and of the best present estimates of the rate coefficients.

There is an appreciable understanding of the role of NO_x and O_x chemistry in ozone destruction. There is now general agreement that NO_x is responsible for about two thirds of the ozone destruction in the total stratosphere and O_x for about 15 percent. If it were assumed that the behavior at 35 km were typical of all altitudes, then the highly simplified calculation given above would predict that CFMs presently represent a sink corresponding to about 7 percent of all the ozone removal processes. The behavior at 35 km is, however, far from typical and over-estimates the contribution of ClO_x to ozone destruction for a number of reasons. For one, the ClO_x/ClX ratio is near its maximum at 35 km and decreases at 25 km, where the ozone density is at its maximum. Furthermore, there is a partial healing effect due to the partial overlap of the uv absorption by O_2 and O_3. A fraction of the light that is not absorbed at a given altitude because of re-duced ozone concentration will dissociate oxygen and pro-duce more ozone below that altitude. Two remarks are in order concerning this "partial healing" effect: (1) it is necessarily incomplete, i.e., at best it regenerates

TABLE A.2 Recent Laboratory Results of Rate Constant Measurements

Reaction	Temperature Dependence $k(T)$, cm^3 sec^{-1}	$k(T = 230)$	Reference
$Cl + O_3 \rightarrow ClO + O_2$ (1)	3.08×10^{-11} exp$(-290/T)$	0.87×10^{-11}	Watson et al., 1976
	2.72×10^{-11} exp$(-298/T)$	0.74×10^{-11}	Kurylo and Braun, 1976
	5.18×10^{-11} exp$(-418/T)$	0.84×10^{-11}	Clyne and Nip, 1976b
	2.17×10^{-11} exp$(-171/T)$	1.03×10^{-11}	Zahniser et al., 1976
$ClO + O \rightarrow Cl + O_2$ (2)	1.07×10^{-10} exp$(-224/T)$	4.0×10^{-11}	Clyne and Nip, 1976a
	3.38×10^{-11} exp$(+75/T)$	4.7×10^{-11}	Zahniser and Kaufman, 1976
$ClO + NO \rightarrow Cl + NO_2$ (3)	$(1.7 \times 10^{-11}$ at 298 K$)$		Clyne and Watson, 1974
	1.1×10^{-11} exp$(+200/T)$	2.7×10^{-11}	Zahniser and Kaufman, 1976
$Cl + CH_4 \rightarrow HCl + CH_3$ (5)	0.74×10^{-11} exp$(-1226/T)$	3.6×10^{-14}	Watson et al., 1976
	3.76×10^{-11} exp$(-1701/T)$	2.3×10^{-14}	DeMore and Leu, 1976
	0.84×10^{-11} exp$(-1328/T)$	2.6×10^{-14}	Zahniser and Kaufman, 1976
$HO + HCl \rightarrow H_2O + Cl$ (9)	4.1×10^{-12} exp$(-530/T)$	4.1×10^{-13}	Smith and Zellner, 1974
	2.0×10^{-12} exp$(-313/T)$	5.1×10^{-13}	Zahniser et al., 1974
$ClO + NO_2 + N_2 \rightarrow ClONO_2 + N_2$ (10)	1.5×10^{-31} cm^6 sec^{-1} at 297 K	(3.2×10^{-31})	Birks, 1976
	1.5×10^{-31} cm^6 sec^{-1} at 301 K	(3.2×10^{-31})	Zahniser and Kaufman, 1976
	1.8×10^{-31} cm^6 sec^{-1} at 298 K		DeMore and Leu, 1976
$ClONO_2 + O \rightarrow$ products (12)	2.1×10^{-13} at 245 K		Davis, 1976
$HO + HO_2 \rightarrow H_2O + O_2$ (13)	2×10^{-10} at 298 K		Hochanadel et al., 1972
	1×10^{-10} at 298 K		DeMore and Tschuikow-Roux, 1974
	$<3 \times 10^{-11}$ at 298 K		Hack et al., 1975
	$\sim3 \times 10^{-11}$ at ~1500 K		Dixon-Lewis et al., 1975

only a fraction of the removed ozone; (2) it is included in the present one-dimensional model calculations.

It is important, however, to point out that simple steady-state calculations such as the above should reflect local behavior reasonably accurately, since they are independent of the vagaries of atmospheric transport. Assuming, for example, that continued release of chlorofluoromethanes at the 1973 production rate may lead to steady-state mixing ratios of about 5×10^{-9} for ClX, the ClO_x destruction rate of "odd oxygen" at 35 km altitude would be nearly 11 times larger than that given above. Its contribution to ozone destruction would then be almost as large as that of NO_x, and this would clearly result in a substantial reduction of local ozone concentration. Even if the total ozone column density were to change only slightly, because of regeneration and interaction with NO_x chemistry at lower altitudes, there seems to be no way of avoiding large changes in the altitude profile of ozone concentration.

The simplified steady-state calculations given above of ClO_x-O_3 catalysis are for the effects of Cl from F-11 and F-12. They will be in addition to those of Cl from other sources such as CCl_4, CH_3Cl, and volcanic HCl.

The quantitative evaluation of ClO_x catalysis depends, of course, on the accuracy with which various rate parameters are known. A brief discussion of this subject is therefore in order. The National Bureau of Standards (1975) has published data on which we have relied for most of our values (tabulated in Appendix D), especially for the unperturbed atmosphere. The NBS evaluation includes references to the voluminous literature, prior to 1975, which we have not duplicated. Instead, we have focused on aspects of the ClO_x chemistry not included in the NBS report and on results obtained since then for the most important, less well-established rate constants.

Table A.2 presents such information for the rate constants of eight of the reactions listed in this section. It leaves out the two photochemical processes, Reactions (4) and (11), which are discussed elsewhere (Section III.B and earlier in this section); Reactions (6), (7), and (8), which are less important than (5); and Reaction (14), which does not involve ClO_x and has been amply reviewed. Table A.2 is not intended to be an exhaustive review, but it includes recent laboratory data known to the Panel. A good fraction of the most recent work is as yet unpublished and does not, therefore, carry as much authority as a refereed journal article. However, such results come from well-known research groups and were included only if the authors deemed the work sufficiently detailed to warrant dissemination.

Atom- or radical-molecule reactions are now studied over wide ranges of temperature by a variety of experimental techniques, including mainly flash photolysis or discharge-flow. The best experimental results may show reproducibility of ±5 percent and accuracy of ±10 to 20 percent within one study, as well as agreement to ±20 to 30 percent between different studies employing substantially different techniques. These uncertainty limits are single standard deviations. It can be seen in Table A.2 that such agreement has been reached between two to four research groups for Reactions (1), (2), and (9) and has probably been reached for Reactions (3), (5), and (10), although the latter reaction has only been investigated for a very short time in three laboratories. The rate constant of the (HO + HO$_2$) Reaction (13), however, is very uncertain, and more accurate determinations are needed. The experimental difficulties of such a measurement are, of course, vastly greater than those for a radical-molecule reaction, where only one type of unstable species needs to be produced and where the reaction can often be studied with a large excess of stable, molecular reactant, i.e., as the pseudo-first-order decay of the radical. In order to measure k_{13}, on the other hand, both HO and HO$_2$ must be produced at known concentrations, their rate of change measured, and the effects of other reactive species such as H and O, which are tied to HO$_x$ chemistry by fast reactions, must also be monitored.

A brief discussion of Reactions (5), (10), (12), and (13) seems warranted. Although the three most recent Arrhenius expressions for k_5 (Cl + CH$_4$) are still in some disagreement, the values at 230 K, a typical temperature in the lower stratosphere, are within 30 percent of the mean of 2.8×10^{-14} cm^3 sec^{-1}. The uncertainty of k_5 is still somewhat larger than that of k_1 or k_9, but it is now in a range where further improvement will not have a major effect. It should also be noted that all three latest studies (see Table A.2) have found k_5 to be somewhat smaller near 230 K than the values reported in early 1975 (Davis *et al.*, 1975a).

For Reaction (10), the formation of ClONO$_2$, gratifying agreement exists between three studies, all in early stages but seemingly sound. A comparison with the well-studied recombination

$$HO + NO_2 + M \rightarrow HNO_3 + M \qquad (15)$$

which is similar in every respect except that the bond formed is nearly twice as strong as that in $ClONO_2$ shows the expected behavior: k_{10} is about 15 times smaller than k_{15}, and both its M effect (N_2 about twice as effective as He) and the temperature dependence (k_{10} proportional to $T^{-3.1}$, $k_{15} \propto T^{-2.6}$) are quite similar.

For Reaction (12), of $ClONO_2$ with O, preliminary results from only one laboratory are available. Several other measurements are just getting under way. The importance of this reaction is overshadowed by the photolysis rate constant of $ClONO_2$, J_{11}.

For k_{13} (HO + HO_2), the least well-known and perhaps most important rate constant, the uncertainties are large. Two indirect measurements listed in Table A.2 give very high values. Of these, the study by DeMore and Tschuikow-Roux (1974) is the more believable. It leads to a k_{13} of 1×10^{-10} cm^3 sec^{-1} at 298 K when its mechanism is assumed valid and best available values of other rate constants are used in the proposed scheme. Yet, that study suffers from certain experimental shortcomings in addition to the possible incompleteness of the assumed reaction mechanism. One recent, fragmentary discharge-flow study (Hack et al., 1975) sets an upper limit for k_{13} at about 3×10^{-11} cm^3 sec^{-1}, in strong disagreement with the other two experiments. A study of H_2-O_2 flames has led Dixon-Lewis et al. (1975) to assign a range of $(3 \pm 1.5) \times 10^{-11}$ to k_{13} at 1300 to 1600 K. Moreover, the first successful measurements of HO in the stratosphere by Anderson (1976) also support a low value of k_{13}. Above all other considerations, the high value of k_{13} is greatly in excess of reasonable bounds arising from transition state or thermochemical theory as well as of the rate constants or pre-exponential factors for other reactions of these or similar radical species. The present recommendation is therefore to use a lower value of 2×10^{-11} cm^3 sec^{-1} but to consider it uncertain to $\pm$ a factor of 3. Some of the earlier model calculations have used the largest k_{13}, 2×10^{-10} cm^3 sec^{-1}, which now seems improbable. The effect of a smaller k_{13} is, in qualitative terms, to decrease the catalytic destruction of O_3 by NO_x (because larger [HO] removes more NO_2 to form HNO_3) but to increase that by ClO_x (because larger [HO] releases more Cl from HCl). It must always be borne in mind that the reactions discussed here form a small subset of all of those that must be considered and that much larger sets are used in model calculations. In that sense, should it turn out that k_{13} is indeed small, several other processes such as Reaction (7), $HO_2 + HO_2 \rightarrow H_2O_2 + O_2$, or $HO + H_2O_2 \rightarrow H_2O + HO_2$ may take on added importance.

For the related reaction $Cl + HO_2 \rightarrow HCl + O_2$, Leu and DeMore (1976) have recently determined a rate coefficient of $(3 + 4.5, -1.8) \times 10^{-11}$ cm^3 sec^{-1} at 295 K close to the value for k_{11} adopted here. They obtained a rate coefficient of $(6.8 \pm 3.4) \times 10^{-15}$ cm^3 sec^{-1} also at 295 K for the reaction $Cl + HNO_3$. This is another potential HCl conversion reaction, but the rate coefficient is too low for it to contribute significantly in the stratosphere.

E. Ozone Destruction by Bromine and Fluorine

Potential catalytic ozone removal cycles due to other halogen species can also be discussed briefly. For bromine, the reactions equivalent to (1), (2), and (3) in the previous section appear to be fast. However, those equivalent to (5) and (6) are endothermic and could therefore not take place at stratospheric temperatures. The reaction with H_2O_2 is probably slightly endothermic and thereby slower than the chlorine reaction. This leaves only the $Br + HO_2$ reaction to produce HBr and thereby increases the catalytic effect on O_3. The equivalent of Reaction (9), $HO + HBr \rightarrow H_2O + Br$, is a rapid reaction, and the regeneration of Br from HBr may be further enhanced by the slightly exothermic and probably rapid reverse of the H_2O_2 Reaction (8), i.e., $HO_2 + HBr \rightarrow H_2O_2 + Br$. It appears, therefore, that the BrO_x catalysis scheme is more efficient than its ClO_x analog, although at this time the source strength of bromine-containing hydrocarbons is not significant (see Chapter 3). Our knowledge of the relevant rate constants is less extensive for the BrO_x system than for ClO_x, and further experimental studies are needed.

The possibility of an analogous FO_x chain can be ruled out on thermochemical grounds. The basic ozone reactions are feasible, but the steps that form HF are much faster than their Cl analogs, and the H-F bond is so strong that HF becomes a permanent product, which is slowly transported to the troposphere and removed by rain-out.

F.   Other Stratospheric Reservoirs for Chlorine

The recent discovery that $ClONO_2$, a rather unusual and
fairly unstable compound, may play an important role in
the stratospheric ClX budget should not be construed to
mean that an endless succession of ClO_x-NO_y species may
have to be considered in great detail.   The reason is that
many of these compounds do have strong absorption bands
or continuous absorption in the visible or near ultraviolet wavelength range leading to photolysis and to the
regeneration of active ClO_x species at high rates.   The
effective first-order rate constant, J_i (sec^{-1}) for such
a process is given by the product of the solar flux at
that altitude, the absorption cross section of the compound, and the quantum yield (fraction of absorption
events leading to dissociation) integrated over all
wavelengths.   For a J_i of 10^{-3} sec^{-1}, for example, the
compound i would be reduced to $1/e$ of its initial concentration (assuming no other reactions and an average fixed
solar elevation) in 10^3 sec, i.e., 17 min.
      Considering first the species Cl_2, $OClO$, and its asym-
metrical isomer $ClOO$, we find that they all have strong
absorption spectra in the ultraviolet region and are
readily photolyzed. Cl_2 has its maximum absorption at
330 nm and is dissociated into atoms by sunlight with a
lifetime of the order of 10 min even down to ground level.
$OClO$ shows very strong banded absorption in the range
300 to 400 nm.   The spectrum is predissociated below 375
nm (Herzberg, 1966), and there is clear evidence that the
photodissociation products are ClO + O (Lipscomb et al.,
1956).   Clyne and Coxon (1968) give an absorption cross
section of $\sigma = 12 \times 10^{-18}$ cm^2 for the strongest band at
351 nm, and an estimate of the intensities and widths of
the bands in published photographic spectra gives averaged
values of J in the range 10^{-3} to 10^{-2} sec^{-1}, i.e., a lifetime of a few minutes for the process $OClO$ + $h\nu$ = O + ClO
at all altitudes down to ground level.
      The isomeric species $ClOO$ absorbs strongly in the ultra-
violet region between 220 and 280 nm, and the continuous
absorption spectrum published by Johnston et al. (1969)
closely resembles the Hartley bands of ozone both in contour and in absolute intensity.   However, the recent kinetic data of Clyne et al. (1975) suggest a lower heat of
formation of $ClOO$, which would increase its calculated
concentrations in the laboratory experiments and thereby
reduce the published cross sections by a factor of about
4.   This uncertainty and the strong altitude dependence

of its light absorption introduced by the overlapping ozone absorption make it difficult to estimate the rate of photochemical decomposition of ClOO. It is, however, clearly going to yield either Cl or ClO as will the reactions of ClOO with atoms and free radicals. The thermochemistry and kinetics of ClOO are of great importance, since, at stratospheric pressures, it is in equilibrium with Cl and O_2. If the [ClOO]/[Cl] ratio approaches or exceeds unity at stratospheric temperatures, the reaction of ClOO with O atoms and O_3 will have to be studied.

Three-body recombination reactions of Cl with NO and NO_2 are known to take place in the laboratory, but their effective two-body rate constants in the stratosphere are small, the product species ClNO and $ClNO_2$ are likely to react rapidly with O, and they are also photolyzed with regeneration of ClO_x molecules. The chemistry of $ClNO_2$ and its isomer ClONO needs further study, not so much because of its formation by Cl + NO_2 but because it may be a photolysis or reaction product of $ClONO_2$, which is, in turn, formed from the much more abundant ClO radical.

There are insufficient data on the absorption spectrum of nitryl chloride ($ClNO_2$) to estimate its rate of photolysis. It is agreed that purified nitryl chloride does not absorb in the visible (Yost and Russell, 1946; Collis et al., 1958), but unstructured absorption has been reported in the near ultraviolet region with $\sigma = 5 \times 10^{-20}$ cm^2 at 340 nm (White, 1970). This indicates that J is in the range 10^{-4} to 10^{-3} sec^{-1} down to ground level, presumably for the process $ClNO_2 + h\nu = NO_2 + Cl$.

A recent suggestion has been made (Simonaitis and Heicklen, 1975) that two consecutive recombination reactions

$$Cl + O_2 + O_3 \rightarrow ClO_3 + O_2 \qquad (A.21)$$

followed by

$$ClO_3 + HO + M \rightarrow HClO_4 + M \qquad (A.22)$$

may take place in parallel with HCl formation and further reduce the ClO_x catalysis by rapid formation, downward transport, and rain-out of $HClO_4$. This must be evaluated carefully but seems unlikely for a number of reasons. (1) The proposed rate constant for the ClO_3 formation step (3×10^{-30} to 6×10^{-30} cm^6 sec^{-1}) appears to be much too large for a reaction that forms such a weakly bound molecule. (2) Any unsymmetrical form of ClO_3, representing the structure that should result from the direct Cl + O_3 recombination, will rapidly decompose into ClO + O_2, the normal products of the Cl + O_3 reaction. (3) The only reasonable manner in which symmetrical ClO_3 may be formed is by way of the ClOO + O_3 reaction, but recent high pressure photolysis studies (Lin and DeMore, 1975) suggest a very small rate constant for that reaction, too small for the required stratospheric ClO_3 formation rate by 5 to 6 orders of magnitude. (4) The proposed rate constant of 2×10^{-11} cm^{-3} sec^{-1} for the formation of $HClO_4$ is too high for the midstratosphere, i.e., a recombination reaction that involves a total of six atoms is not likely to be at its high pressure limit at 2 to 10 Torr pressure. Yet, even if these two extreme kinetic parameters considered in the above discussion were to be accepted, the overall diminution of the catalytic ozone removal as a consequence of this scheme is less than a factor of 2. Further experimental work both on the mechanism of Cl_2-O_3 oxidation and on some of the proposed rate processes is needed, but the hopes for substantial relief from the generally accepted catalytic ClO_x scheme are slight.

It has also been suggested that HOCl may be formed by the reaction HO_2 + ClO $\rightarrow$ HOCl + O_2 and may thereby remove some active ClO_x. Hypochlorous acid vapor (HOCl) shows strong continuous absorption in the ultraviolet region with maxima at 320 nm and near 200 nm (Ferguson et al., 1936), which cannot plausibly be attributed to impurities. The absorption data yield a 24-hour averaged value of J that is $\sim 2 \times 10^{-3}$ sec^{-1} down to the lower stratosphere. Molecular orbital considerations strongly support the view that the dissociation process is HOCl + $h\nu$ $\rightarrow$ HO + Cl.

Some uncertainty remains regarding the speed of the reaction ClO + O_3 $\rightarrow$ ClO_2 + O_2. Its rate constant may be as large as 1.5×10^{-15} cm^3 sec^{-1} (Birks, 1976), or it may be much smaller (DeMore and Lin, 1975). In either case it is of little consequence, since, if the product is ClOO, the Cl + O_2 equilibrium cannot be substantially affected, because its association and dissociation rates are several orders of magnitude faster than the ClO + O_3 rate.

Lastly, the formation, photolysis, and reaction of $ClONO_2$ require further discussion. The recently determined ClO—NO_2 bond strength of about 25 kcal (Knauth et al., 1974) is very similar to that of O—O_2 in ozone. At stratospheric temperatures, the molecule is therefore stable even though its rate constants for formation and dissociation are much larger than those of O_3. Its first-order photodissociation rate constant, J_{11}, has been reported to be about 5×10^{-5} sec^{-1} at 35 km using the approximation of a fixed solar elevation and 24-hour average (Rowland *et al.*, 1976). This surprisingly small value, i.e., t_{11} of about 6 hours, precipitated the recent re-evaluation of ClO$_x$ catalysis, but, as stated above, near the end of Section III.D, further measurements of the absorption coefficient above 300 nm have not yet been carried out, and J_{11} must therefore be considered a lower bound. Laboratory study is also needed on its possible reaction with HO. Reaction with stratospheric aerosol particles is also possible but would be an effective sink only if it occurs with high collision frequency.

G. Removal of HCl from the Stratosphere

Removal of chlorine-containing products of halocarbon decomposition from the stratosphere is achieved almost exclusively by transport down into the troposphere. Since the predominant form of such products in the lower stratosphere is HCl, it is only necessary to consider the removal process for this compound, which is rain-out. The average lifetime of HCl in the upper troposphere is expected to be longer than that of HCl near the surface or generally distributed throughout the lower troposphere. Assuming that HCl is about as efficiently removed by precipitation as are aerosols, we estimate the lifetime of HCl entering the troposphere from above to be about 30 to 40 days. This lifetime should be compared with a lifetime of a few years for transport from the upper troposphere into the lower stratosphere, estimated from model calculations. Therefore, once HCl from the stratosphere reaches the upper troposphere, most of it will be removed by precipitation before it can be transferred back into the stratosphere. Similarly, HCl approaching the tropopause from below undergoes "precipitation filtering" before it can enter the stratosphere.

The removal of HCl in the troposphere affects the reduction of stratospheric ozone by ClO$_x$ in the following manner: if HCl is transported rapidly into the troposphere (if the vertical gradient in the HCl concentration for the lower stratosphere is large), then its ability to produce ClO$_x$ by reaction with HO in the stratosphere is less than would be the case if HCl is more slowly transported into the troposphere (small vertical gradient). The supposed rapid removal of HCl by precipitation causes large vertical gradients in the vicinity of the tropopause,

at least according to model simulations. Although the HCl
concentrations are difficult to measure near the tropopause, the estimated concentration gradient obtained from
those measurements in the lower stratosphere supports the
model calculations (Chapter 7). That is, the HCl concentration is low enough to have been undetectable in the
vicinity of 13 km (Lazrus *et al.*, 1975) and increases with
height above the tropopause (Figure 7.15). These observations are probably the best evidence to support the concept of rapid removal of HCl by precipitation. Chloride
ions are present in precipitation, but sea-salt nuclei
contribute in such large proportions that, even in central
continental areas, no anomaly due to HCl or other sources
could be discerned. [See Junge (1963) for Cl^- concentration distributions in precipitation.]

The problem of HCl concentration distributions in the
troposphere and removal by precipitation has been approached theoretically by Ryan and Mukherjee (1974) and
Stedman *et al.* (1975). They suggest that the HCl concentrations (and those of other very soluble tropospheric
gases such as HNO_3) are strongly controlled by the atmospheric water cycle of condensation and evaporation.
Their arguments are based on plausibility rather than
direct observations (which do not exist now). If they
are correct, then tropospheric removal of HCl is indeed
rapid.

Junge (1963) discusses some results obtained by
Georgii (1960), who measured atmospheric concentrations
of NH_3, NO_2, SO_2, and "Cl" near the ground before and
after rainstorms at Frankfurt/Main over the period June
1956-May 1957. The term "Cl" was used because the sampling method did not discriminate against possible contributions from chlorine-containing compounds other than HCl.
The results showed that Cl concentrations were decreased
an average of 61 percent and those for NH_3, NO_2, and SO_2
were decreased, respectively, by 48, 24, and 36 percent
after the rainstorms.

H. *Ionic Processes in the Stratosphere*

In discussing ionic processes in the troposphere (Section
II.C), it was only necessary to consider the removal of
the CFMs themselves. In the stratosphere, consideration
must also be given to processes that might involve reaction with the ClX from the CFMs and thereby affect the
chain decomposition of O_3.

Removal of CFMs or their photodissociation products by

direct ionization from cosmic rays can be ruled out by the same arguments that were applied to the troposphere. The local removal time, t_c, for this process for F-11 at 25 km is 10^5 yr.

In the stratosphere, the ambient positive ions are almost certainly oxonium ions of the type $H_3O^+(H_2O)_n$. Krankowsky and Arnold (1974) observed the dominant ions between 35 and 60 km to be $H_3O^+(H_2O)$ and $H_3O^+(H_2O)_2$. Below 35 km, even higher hydrates would undoubtedly be the more abundant ions. The times for removal of the CFMs by reactions with these ions will very likely be greater than 85 yr. But in this case the reaction products would almost certainly lead to ClO_x formation and thereby merely replace photolysis rather than constitute inactive removal.

There are no exothermic channels for reactions of $H_3O^+(H_2O)_n$ ions with Cl, ClO, or HCl.

It is unlikely that sufficient ammonia exists in the stratosphere to convert the oxonium ions to ammonium ions. But even if ammonium ions were present, no exothermic reactions can occur with CFM, Cl, ClO, or HCl.

There is a possibility that HCl can associate with positive ion clusters, e.g., to form $H_3O^+(H_2O)_n(HCl)_m$. However, such a process will not permanently remove HCl since neutralization with negative ions will most likely release the HCl again. The lifetime for neutralization is about 10^4 sec.

The ambient negative ions in the stratosphere are almost certain to be hydrates of NO_3^-. Reactions of these ions with CFMs, Cl, or HCl are endothermic. The reaction

$$NO_3^- + ClO \rightarrow Cl^- + NO_2 + O_2 \qquad (A.23)$$

is exothermic by 20 kcal/mole, but the reaction with the hydrated NO_3^- ions expected in the stratosphere will undoubtedly be endothermic. In addition, Cl^- has a large cross section for photodetachment, and the ions would be converted to Cl atoms in sunlight in a few seconds.

Thus, it appears that positive and negative ions in the stratosphere do not react with CFMs either in a manner or at a rate sufficient to constitute significant inactive removal. Moreover, the rates for reaction of the major ions with Cl, ClO, and HCl are also too slow to interfere with their reactions in the catalytic cycle, which destroys ozone. Even if some minor stratospheric ion were to react rapidly with ClO_x, the resulting density of product ions could never exceed the total ion density of

about 5×10^3 cm^{-3} so that only a negligible quantity of
Cl atoms could be tied up as ions at any time. Furthermore, these ions could play a significant role only if
they acted as a catalyst to produce a nonreactive Cl-containing compound. No such process has been proposed.
All the exothermic reactions that we have been able to
visualize or that have been proposed rapidly regenerate
ClO$_x$ molecules. For example, if Cl$^-$ were to be produced
rapidly in some manner, the reaction

$$Cl^- + HNO_3 \rightarrow NO_3^- + HCl \qquad (A.24)$$

is also fast so that HCl is regenerated and then parti-
tioned into active species. If hydration of the Cl$^-$
rendered this reaction endothermic, then the fate of the
Cl$^-$ hydrate would be neutralization

$$Cl^-(H_2O)_n + H_3O^+(H_2O)_m = HCl + (n + m + 1)H_2O \quad (A.25)$$

again with the release of the HCl.

   We can therefore safely conclude that *ionic processes
are unlikely to provide a significant "inactive" sink for
CFMs or to remove Cl, ClO, or HCl from the stratosphere.*

I.   *Heterogeneous Processes*

1. CFM Removal Stratospheric particles that may be of
importance in heterogeneous chemical reactions are those
primarily associated with the so-called Junge layer. This
is an aerosol layer between the tropopause and about 25 km
with peak concentrations about 5 or 6 km above the tropopause. The particles have radii that range from 0.05 to
1 μm and appear to consist largely of sulfuric acid and
water with variable and relatively minor proportions of
NH$_4^+$, silicates, and Na$^+$. CIAP Monographs 1 and 3 (1975)
provide summaries and review stratospheric aerosol properties and possible heterogeneous chemical processes associated with the aerosol but not those of halocarbons. A
recent article by Cadle and Grams (1975) reviews the properties of the aerosol that are important in radiation
transfer. In both of the above references, there is a
table of aerosol composition originally published by
Cadle (1972) in which chloride is reported as an important
constituent, having mass concentrations typically about
15 percent of the sulfate mass concentrations and having
mass mixing ratios in the range (0.2 to 1) $\times$ 10^{-9}.

The particles were collected by high-altitude aircraft on polystyrene filters. Cadle and Grams (1975) judge the Cl measurements to be reliable. However, Lazrus et al. (1975) state that the particulate Cl concentrations in the lower stratosphere collected in 1974 on balloonborne and aircraft filter samples were mostly below their limit of detection by colorimetric analysis (0.03 ppb). Neutron activation analyses of the filters gave an average Cl mixing ratio of 0.017×10^{-9} for the region between 15 and 18 km. In the same set of samples, the concentrations of gaseous Cl (probably HCl and ClO), collected on base-impregnated portions of the filters, were uniformly higher by factors >5 than the particulate Cl concentrations. It appears that the differences between the two sets of observations are more likely to be in analysis and interpretation rather than in natural variability.

If the recent observations of Lazrus et al. (1975) are representative of the average state of the stratosphere past and present, then it is reasonable to state that heterogeneous processes by which CFMs or other halocarbons are decomposed with attendant incorporation of products (Cl) in the particles are relatively unimportant (see Section II.D.2). This conclusion is supported by the studies conducted by Bigg et al. (1974), who have examined individual stratospheric particles, subjecting them to sensitive spot tests for various ionic components. While they have detected sulfate, persulfate, and ammonium ions, they have not found chloride, although it was readily detected in tropospheric particles.

The amount of CFM that may be absorbed by particles in the stratosphere and the associated removal time are estimated in the following calculation. Assume the solubility at one atmosphere pressure of CFM (F-11) is the same as it is in H_2O at 15°C; i.e., 0.08 percent by weight. The estimate of the amount of CFM absorbed is thus equal to 8×10^{-4} (0.08 percent) of the particle mass multiplied by the partial pressure of CFM in the stratosphere (about 10^{-11} atm). The mass concentration of aerosol in the vicinity of its peak value (at about 18 km) is typically 3×10^{-7} g m^{-3} [cf. Cadle et al. (1975)]. The mass concentration of F-11 corresponding to a typical volumetric mixing ratio of 10^{-10} at 18 km is 5.8×10^{-8} g m^{-3}. The concentration of F-11 absorbed is thus: 8×10^{-4} atm^{-1} × 1×10^{-11} atm × 3×10^{-7} g m^{-3} = 2.4×10^{-21} g m^{-3}. The fraction of F-11 absorbed is the concentration of F-11 absorbed divided by the mass concentration of F-11 or 4×10^{-14}.

The corresponding minimum removal time of F-11 is estimated by noting that the mean characteristic time to exchange all tropospheric air with the stratosphere is of the order of 15 yr (about five times the time to mix an inert substance from the troposphere into the stratosphere, the quality τ in Appendix B). If it is imagined that all the absorbed F-11 is removed from the atmosphere with the particles, then the F-11 removal time is the above exchange time divided by the fraction of F-11 absorbed, i.e., of the order of 10^{15} yr.

Solid particles covered by a monolayer would contain about 1 percent by weight of CFM for a mean particle radius of 0.2 μm. However, at low partial pressures (relative to the saturation vapor pressure) there is fractional coverage approximately proportional to the partial pressure of the absorbate. For the CFMs this would be $\lesssim 10^{-10}$, and the removal time for this process is estimated to be greater than 4×10^{11} yr. Physical adsorption from the stratosphere is therefore negligible. In addition, any physically adsorbed CFMs are likely to be released from the adsorbent in the troposphere, where approximately 90 percent of the particulate matter is removed by precipitation. As discussed previously, precipitation cannot remove significant amounts of CFMs.

There remains to be considered the possibility that adsorption is followed by chemical reaction with the aerosol, with subsequent release of the product gasses into the atmosphere. If γ is the probability that collision between a CFM molecule and a sulfuric acid aerosol leads to such a chemical reaction, then the reaction rate is γkcn, where k is the bimolecular rate constant and c and n are the numbers of CFM molecules and aerosol particles per unit volume, respectively. Insertion of the appropriate values in this expression leads to the time t_C for local removal of $\simeq (5 \times 10^5/\gamma)$ sec.

To estimate γ we may take advantage of the fact that CFMs are unreactive in bulk sulfuric acid solution and can therefore take the ultraconservative limit of 1 hour for the t_C of a CFM in such a solution. Now the collision frequency Z for a CFM in bulk liquid solution is of the order of 10^{13} sec^{-1}. Therefore γ cannot be greater than $(1/Zt_C) = 1/(3600 \times 10^{13}) = 3 \times 10^{-17}$. This is admittedly a crude estimate. However, it leads to an extremely long removal time of $(5 \times 10^5/3 \times 10^{-17})$ sec $\simeq 5 \times 10^{14}$ yr, so removal by this mechanism must be negligible.

All three of the processes considered in this section have removal times longer than 10^{11} yr. We can, therefore,

conclude that *inactive removal of CFMs from the stratosphere by heterogeneous processes is not at all significant.*

2. *ClO$_x$ Removal* If aerosol particles were able to remove
catalytic species such as ClO and render them unreactive
toward O or O$_3$, they would reduce the catalytic effectiveness of the ClO$_x$ cycle. The following simple calculation
sheds light on this possibility.

The gas-kinetic collision number, $Q\bar{v}$, where Q is the
collision cross section and $\bar{v}$ the relative velocity, for
collisions between ClO and spherical aerosol particles of
0.1-μm diameter, at 250 K, is 2×10^{-6} cm^3 sec^{-1}. For a
typical aerosol concentration of 1 per cm^3, this gives a
collision frequency of 2×10^{-6} sec^{-1}.

In laboratory experiments (Ogryzlo, 1961), it is
found that ClO survives 10^4 to 10^5 collisions with surfaces coated with aqueous acids such as phosphoric or
sulfuric acid. The corresponding ClO removal rate constant would thereby be decreased to less than 10^{-10} sec^{-1}.
This value is to be compared with the removal frequency
for ClO by reaction with atomic oxygen, $k_2(O)$, which was
shown to be about 10^{-2} sec^{-1} at 35 km altitude (see Section III.D, above). *It is therefore clear that heterogeneous removal of ClO is negligible when compared with
the homogeneous ozone catalysis processes.* Even at unit
probability of collisional removal, aerosol particles are
four orders of magnitude less effective than O atoms in
reaction with ClO. Moreover, a truly efficient removal
process for ClO by reaction with aqueous sulfuric acid is
difficult to postulate, since formation of Cl$_2$, HOCl, or
even HCl would not terminate the ClO$_x$ chain. Furthermore,
the aerosol concentration is known to decrease sharply
with increasing altitude above 25 km and is very much less
than 1 per cm^3 in the midstratosphere.

REFERENCES

Anderson, J. G. 1976. The absolute concentration of OH
($X^2\pi$) in the earth's stratosphere, *Geophys. Res. Lett.*
3:165-168.
Atkinson, R., D. A. Hansen, and J. N. Pitts, Jr. 1975.
Rate constants for the reaction of OH radicals with
CHF$_2$Cl, CF$_2$Cl$_2$, CFCl$_3$ and H$_2$ over the temperature range
297-434°K, *J. Chem. Phys.* 63:1703-1706.

Basco, N., and S. K. Dogra. 1971. Reactions of halogen oxides studied by flash photolysis. I. The flash photolysis of chlorine dioxide, *Proc. R. Soc. London* A323:29-68.

Basco, N., and R. D. Morse. 1973. The vacuum ultraviolet absorption spectrum of ClO, *J. Molec. Spectros.* 45:35-45.

Bigg, E. K., A. Ono, and J. A. Williams. 1974. Chemical tests for individual submicron aerosol particles, *Atmos. Environ.* 8:1-13.

Birks, J. 1976. University of Illinois. Unpublished results.

Broecker, W. S., and T. H. Peng. 1974. Gas exchange rates between air and sea, *Tellus* 26:21-35.

Cadle, R. D. 1972. Composition of the stratospheric sulfate layer, *EOS, Trans. Amer. Geophys. Union* 53: 812-820.

Cadle, R. D., P. J. Crutzen, and D. Ehhalt. 1975. Heterogeneous chemical reactions in the stratosphere, *J. Geophys. Res.* 80:3381-3385.

Cadle, R. D., and G. W. Grams. 1975. Stratospheric aerosol particles and their optical properties, *Rev. Geophys. Space Phys.* 13:475-501.

Calvert, J. G., and J. N. Pitts. 1966. *Photochemistry.* Wiley, New York, pp. 522-528.

Chang, J. S., C. Steen, and F. Kaufman. 1976. University of Pittsburgh. Unpublished results.

Chang, J. S., and F. Kaufman. 1976. University of Pittsburgh. Unpublished results.

CIAP Monograph 1. 1975. *The Natural Stratosphere of 1974.* A. J. Grobecker, ed. Final report, U.S. Dept. of Transportation, DOT-TST-75-51, Washington, D.C.

CIAP Monograph 3. 1975. *The Stratosphere Perturbed by Propulsion Effluents.* A. J. Grobecker, ed. Final report, U.S. Dept. of Transportation, DOT-TST-75-53, Washington, D.C.

Clyne, M. A. A., and J. A. Coxon. 1968. Kinetic studies of oxy-halogen radical systems, *Proc. R. Soc. London* A303:207-231.

Clyne, M. A. A., and W. S. Nip. 1976a. Reactions of chlorine oxide radicals. Part VI. The reaction O + ClO $\rightarrow$ Cl + O_2 from 220-426°K, *J. Chem. Soc. Faraday Trans.* In press.

Clyne, M. A. A., and W. S. Nip. 1976b. Study of elementary reactions by atomic resonance absorption with a nonreversed source. Part 1. The reaction of chlorine atoms with ozone giving chlorine monoxide and oxygen, *J. Chem. Soc. Faraday Trans. 2* 72:838-847.

Clyne, M. A. A., and R. T. Watson. 1974. Kinetic studies of diatomic free radicals using mass spectrometry. 2. Rapid bimolecular reactions involving the chlorine oxide ClO $X^2\pi$ radical, *J. Chem. Soc. Faraday I,* 70:2250-2259.

Clyne, M. A. A., B. J. McKenney, and R. T. Watson. 1975. Reactions of chlorine oxide radicals. Part 5. The reactions of ClO $(X^2\pi)\rightarrow$ products, *J. Chem. Soc. Faraday I,* 71:322-335.

Cole, R. K., and E. T. Pierce. 1965. Electrification in the earth's atmosphere for altitudes between 0 and 100 kilometers, *J. Geophys. Res.* 70:2735-2749.

Collis, M. J., F. P. Gintz, D. R. Goddard, E. A. Hebdon, and G. J. Minkoff. 1958. Nitryl chloride. I. Its preparation and properties of its solutions in some organic solvents, *J. Chem. Soc.* 438-445.

Coxon, J. A. 1976. Absorption cross sections for the $A^2\pi-^2\pi$ system of ClO. Report to The Manufacturing Chemists Association.

Crutzen, P. J. 1974. A review of upper atmospheric photochemistry, *Can. J. Chem.* 52(8):1569-1581.

Crutzen, P. J., and I. S. A. Isaksen. 1976. The impact of the chlorocarbon industry on the ozone layer, *J. Geophys. Res.* To be published.

Davis, D. D. 1975. Tropospheric hydroxyl radical measurement. *Chem. Eng. News* 53:22.

Davis, D. D. 1976. University of Maryland. Unpublished results.

Davis, D. D., and R. T. Watson. 1975. University of Maryland. Unpublished results.

Davis, D. D., E. S. Machado, R. L. Schiff, R. T. Watson. 1975a. Paper presented at 169th National Meeting of the American Chemical Society, Philadelphia, Pa.

Davis, D. D., J. F. Schmidt, C. M. Neeley, and R. J. Hanrahan. 1975b. Effect of wavelength in the gas-phase photolysis of carbon tetrachloride at 253.7, 184.9, 147.0, and 106.7 nm, *J. Phys. Chem.* 79:11-17.

Davis, D. D., R. T. Watson, T. McGee, W. Heaps, J. Chang, and D. Wuebbles. 1976. Tropospheric residence times for several halocarbons based on chemical degradation via OH radicals. Paper given at 171st American Chemical Society National Meeting, New York City, April 8.

DeMore, W. B., and C. L. Lin. 1975. Photochemistry of chlorine-ozone mixtures. Paper given at 169th National Meeting of the American Chemical Society, Philadelphia, Pa., April 9.

DeMore, W. B., and M. T. Leu. Jet Propulsion Laboratory. Unpublished results.

DeMore, W. B., and E. Tschuikow-Roux. 1974. Temperature dependence of the reactions of hydroxy and hydroperoxy radicals with ozone, *J. Phys. Chem.* 78:1447-1451.

Dixon-Lewis, G., J. B. Greenberg, and F. A. Goldsworthy. 1975. Reactions in the recombination region of hydrogen and lean hydrocarbon flames. Paper given at the 15th Symposium (International) on Combustion, the Combustion Institute, Pittsburgh, Pa., pp. 717-730.

E. I. du Pont de Nemours & Co. 1971. Organic Chemicals Dept., "Freon" Products Division, Wilmington, Del. "Solubility Relationships between Fluorocarbons and Water." Freon® Product Information B-43.

Fehsenfeld, F. C., P. J. Crutzen, A. L. Schmeltekopf, C. J. Howard, D. L. Albritton, E. E. Ferguson, J. A. Davidson, and H. I. Schiff. 1976. Ion chemistry of chlorine compounds in the troposphere and stratosphere, *J. Geophys. Res.* Submitted for publication.

Ferguson, E. E. 1971. D region ion chemistry, *Rev. Geophys. Space Phys.* 9:997-1008.

Fergusson, W. C., L. Slotin, and D. W. G. Style. 1936. The absorption spectrum of aqueous chlorine and hydrogen peroxide vapor, *Trans. Faraday Soc.* 32:956-962.

Gedanken, A., and M. D. Rowe. 1975. Magnetic circular dichroism spectra of methyl halides. Reduction of the $n \rightarrow \sigma *$ continuum, *Chem. Phys. Lett.* 34:39-43.

Georgii, H. W. 1960. Untersuchungen über atmosphärische Spurenstoffe und ihre Bedeutung für die Chemie der Niederschlage, *Geofis. Pura Appl.* 47:155-171.

Gurvich, L. V., G. V. Karachevtsev, V. N. Kondrat'yev, Yu. A. Lebedev, V. A. Medvedev, V. K. Potapov, and Y. S. Khodeev. 1974. *Bond Energies, Ionization Potentials and Electron Affinities* (Nauka, Moscow).

Hack, W., K. Hoyermann, and H. G. Wagner. 1975. Reaction of $NO + HO_2 \rightarrow NO + OH$ with $N_2O_2 \rightarrow HO_2 + N_2O$ as a hydroperoxy source, *Int. J. Chem. Kinet. 7 Symp.* 1:329-339.

Heicklen, J. 1969. Gas phase oxidation of perhalocarbons, *Advan. Photochem.* 7:57-148.

Herzberg, G. 1966. *Electronic Spectra of Polyatomic Molecules.* Van Nostrand, Princeton, N.J., p. 607.

Hester, N. E., E. R. Stephens, and O. C. Taylor. 1975. Fluorocarbon airpollutants II, *Atmos. Environ.* 9:603-606.

Hester, N. E. 1975. EPA, Las Vegas. Private communication.

Hochanadel, C. J., J. A. Ghormley, and P. J. Ogren. 1972. Absorption spectrum and reaction kinetics of the perhydroxyl radical in the gas phase, *J. Chem. Phys.* 56: 4426-4432.

Howard, C. J., and K. M. Evenson. 1976a. Rate constants for the reactions of hydroxyl with methane and fluorine, chlorine, and bromine substituted methanes at 296°K, *J. Chem. Phys.* 64:197-202.

Howard, C. J., and K. M. Evenson. 1976b. Rate constants for the reactions of OH with ethane and some halogen substituted ethanes at 296°K, *J. Chem. Phys.* 64:4303.

Johnston, H. S., E. D. Morris, and J. van den Bogaerde. 1969. Molecular modulation kinetic spectroscopy, $ClOO$ and ClO_2 radicals in the photolysis of chlorine in oxygen, *J. Am. Chem. Soc.* 91:7712-7727.

Junge, C. E. 1963. *Air Chemistry and Radioactivity*. Academic Press, New York.

Junge, C. E. 1976. The oceans as a sink for chlorofluoro-methanes, *Z. Naturforsch.* 31a:482-487.

Kaufman, E. D., and J. F. Reed. 1963. The vapour phase diffusion flame reaction of sodium with fluorinated chloromethanes, *J. Phys. Chem.* 67:896-902.

Kerr, J. A. 1966. Bond dissociation energies by kinetic methods, *Chem. Rev.* 66:465-500.

Knauth, H. D., H. Martin, and W. Stockmann. 1974. Determination of the enthalpy of formation of nitroxy chloride (chlorine nitrate, NO_3Cl) for the interpretation of the kinetics of the thermal decomposition of nitroxy chloride in the gas phase and in solution, *Z. Natur-forsch.* 29:200-210.

Kockarts, G. 1971. Penetration of solar radiation in the Schumann-Runge bands of molecular oxygen, *Mesospheric Models and Related Experiments*. G. Fiocco, ed. D. Reidel, Dordrecht, Holland, pp. 160-176.

Krankowsky, D., and F. Arnold. 1974. The nature of stratospheric positive ions. Paper given at XVII Meeting of COSPAR, Symposium on Solar-Terrestrial Physics, São Paulo, Brazil.

Kurylo, M. J., and W. Braun. 1976. Flash photolysis resonance fluorescence study of the reaction atomic chlorine + ozone → chlorine oxide (ClO) + molecular oxygen over the temperature range 213-298 K, *Chem. Phys. Lett.* 37:232-235.

Lazrus, A. L., B. W. Gandrud, and R. N. Woodard. 1975. Stratospheric halogen measurements, *Geophys. Res. Lett.* 2:439-441.

Leu, M.-T., and W. B. DeMore. 1976. Rate constants at 295 K for the reactions of atomic chlorine with H_2O_2, HO_2, O_3, CH_4 and HNO_3, *Chem. Phys. Lett.* In press.

Levy, H., II. 1973. Photochemistry of minor constituents in the troposphere, *Plant Space Sci.* 21(4):575-590.

Lin, C. L., and W. B. DeMore. 1975. Photochemistry of chlorine-ozone mixtures. Paper phys. 86, 169th American Chemical Society National Meeting, Philadelphia, Pa.

Lipscomb, F. J., R. G. W. Norrish, and B. A. Thrush. 1956. The study of energy transfer by kinetic spectroscopy. I. The production of vibrationally excited oxygen, *Proc. R. Soc. London* A233:455-464.

Liss, P. S., and P. G. Slater. 1974. Flux of gases across the air-sea interface, *Nature* 247:181-184.

Lovelock, J. E., R. J. Maggs, and R. J. Wade. 1973. Halogenated hydrocarbons in and over the Atlantic, *Nature* 241:194-196.

Lovelock, J. E. 1975. United Kingdom. Private communication.

Majer, J. R., and J. P. Simons. 1964. Photochemical processes in halogenated compounds, *Advan. Photochem.* 2:137-181.

Milstein, R., and F. S. Rowland. 1975. Quantum yield for the photolysis of CF_2Cl_2 in O_2, *J. Phys. Chem.* 79:669-670.

National Bureau of Standards. 1975. *Kinetic and Photochemical Data for Modeling Atmospheric Chemistry.* NBS Tech. Note 866.

Ogryzlo, E. A. 1961. Halogen atom reactions, Part I, *Can. J. Chem.* 39:2556-2562.

Parmelee, H. M. 1953. Water solubility of Freon refrigerants, Part I, *Refrig. Eng.* 61:1341-1345.

Perkins, H. C. 1974. *Air Pollution.* McGraw-Hill, New York, pp. 27 and 36.

Perry, R. A., R. Atkinson, and J. N. Pitts, Jr. 1976. Rate constants for the reaction of OH radicals with $CHFCl_2$ and CH_3Cl over the temperature range 298-423°K and with CH_2Cl_2 at 298°K, *J. Chem. Phys.* 64:1618-1620.

Phelps, A. V. 1969. Laboratory studies of electron attachment and detachment processes of aeronomic interest, *Can. J. Chem.* 47:1783-1793.

Rasmussen, R. A., and E. Robinson. 1975. Washington State University. Personal communication.

Rasmussen, R. A., K. J. Allwine, and W. H. Zoller. 1975. Analysis of halocarbons in Antarctica, *Antarctic J.* 10:231-236.

Rebbert, R. G., and P. J. Ausloos. 1975. Photodecomposition of $CFCl_3$ and CF_2Cl_2, *J. Photochem.* 4:419-434.

Rowland, F. S. 1975. University of California at Irvine. Unpublished data.

Rowland, F. S., and M. J. Molina. 1975. Chlorofluoromethanes in the environment, *Rev. Geophys. Space Phys.* 13:1-35.

232

Rowland, F. S., M. J. Molina, C. C. Chou, M. Vera Ruiz, and G. Grescentini. 1975. University of California at Irvine. Unpublished data.

Rowland, F. S., J. E. Spencer, and M. J. Molina. 1976. University of California at Irvine. Unpublished results.

Ryan, J. A., and N. R. Mukherjee. 1974. Estimate of tropospheric HCl cycle. Paper MDAC-WD-2290. McDonnell Douglas Astronautics.

Scheutzle, D., D. Cronn, A. Crittenden, and R. J. Charlson. 1975. Molecular composition of secondary aerosol and its possible origin, *Environ. Sci. Technol.* 9:838-845.

Schwartz, W. 1974. Chemical characterization of model aerosols, EPA-650-3-74-001, August.

Shepherd, J. 1961. *Aerosols: Science and Technology.* Interscience, New York.

Siegemund, G. 1973. Simple synthesis of carbonyl chloride fluoride and carbonyl bromide fluoride, *Angew. Chem. Internat. Edit.* 12:918-919.

Simonaitis, R., and J. Heicklen. 1975. Possible sink for stratospheric chlorine, *Planet. Space Sci.* 23(11): 1567-1569.

Smith, I. W. M., and R. Zellner. 1974. Rate measurements of reactions of hydroxyl radical by resonance absorption. 3. Reactions of hydroxyl radical with dihydrogen, dideuterium, and hydrogen and deuterium halides, *J. Chem. Soc. Faraday Trans. (II)* 70:1045-1056.

Stedman, D. H., W. L. Chameides, and R. J. Cicerone. 1975. The vertical distribution of soluble gases in the troposphere, *Geophys. Res. Lett.* 2:333-336.

Strehhlow, R. A. 1976. University of Illinois at Champagne-Urbana. Personal communication.

Taylor, O. C. 1975. University of California, Riverside. Private communication.

Wang, C. C., L. I. Davis, Jr., C. H. Wu, S. Japar, H. Niki, and B. Weinstock. 1975. Hydroxyl radical concentrations measured in ambient air, *Science* 189:797-800.

Watson, R. T. 1974. Chemical kinetics data survey VIII. Rate constants of ClO_x of atmospheric interest, NBSIR 74-516.

Watson, R. T., G. Machado, S. Fischer, and D. D. Davis. 1976. A temperature dependence kinetics study of the reaction of Cl ($^2P_{3/2}$) with O_3, CH_4, and H_2O_2, *J. Chem. Phys.* In press.

White, I. F. 1970. Ph.D. thesis, Queen Mary College, London.

Wittstruck, T. A., W. S. Brey, A. M. Buswell, and W. H. Rodebush. 1961. Solid hydrates of some halomethanes, *J. Chem. Eng. Data* 6:343-346.

Yost, D. M., and H. Russell. 1946. *Systematic Inorganic Chemistry*. Prentice-Hall, Inc., New York, p. 46.

Zahniser, M. S., F. Kaufman, and J. G. Anderson. 1974. Kinetics of the reaction of hydroxyl with hydrochloric acid, *Chem. Phys. Lett.* 27:507-510.

Zahniser, M. S., and F. Kaufman. 1976. University of Pittsburgh. Unpublished results.

Zahniser, M. S., F. Kaufman, and J. G. Anderson. 1976. Kinetics of the reaction atomic chlorine + ozone → chlorine oxide (ClO) + molecular oxygen, *Chem. Phys. Lett.* 37:226-231.

APPENDIX
B

DETAILS OF

TRANSPORT ANALYSIS

I. INTRODUCTION

The intent of this appendix is to summarize the theoretical considerations upon which we based our conclusions in
Chapter 5 regarding the sensitivity of the CFM-ozone problem to transport processes. First, mathematical formulation of transport in the mass conservation equation for a
chemical species is reviewed. Procedures for averaging
this equation over time and space are introduced, and the
form of the two-dimensional (2-D) and one-dimensional (1-D)
eddy-mixing models are obtained. For reasons to be discussed, only the 1-D global-mean models have yet been useful for predicting the reduction of ozone concentrations
due to CFM emissions. However, the previous theoretical
basis for deriving transport in these models has been
somewhat unsatisfactory. This transport has thus far
been obtained empirically to give agreement between observed and calculated profiles of certain chemical species.
   There has been no quantitative evaluation of the possible uncertainties in transport due either to uncertainties
in the chemical species data used to derive the transport
or in the methodology used in such a derivation. It is
difficult to evaluate the sensitivity of CFM-ozone reduction calculations to these various uncertainties from
information available in the literature. Consequently,
we have found it necessary to carry out an extensive theoretical and numerical study of the inference of transport
from a chemical profile, as summarized in Section III.
This study yields a number of possible transport

parameterizations consistent with the available data. These parameterizations together with an idealized model for the time-dependent distribution of a CFM provide a basis for determining the sensitivity of CFM-ozone reduction to uncertainties in transport. Theoretical and numerical results from such a study are described in Section IV. The prediction of ozone reduction due to CFM release is seen to be rather insensitive to uncertainties in transport, and the effects of transport are found to be describable in terms of a few simple concepts.

II. TRANSPORT IN THE CHEMICAL CONTINUITY EQUATIONS, AND AVERAGING PROCEDURES

If c_i is the number density of a chemical species, then it must satisfy a mass conservation equation, the "equation of continuity" of the form,

$$\frac{1}{n} \frac{\partial c_i}{\partial t} = S - L + T/n \qquad (B.1)$$

where S represents sources, L loss processes (both of which were discussed in Chapters 3 and 4), n is total air number density, and T transport, the term we are here interested in.* The transport term is of the form

$$T = - \text{div} (c_i \vec{V}) \qquad (B.2)$$

where div is the divergence operator in geometrical coordinates, and $\vec{V}$ is the three-dimensional (3-D) velocity vector. As written, Eqs. (B.1) and (B.2) are three dimensional and describe transport by motions over both large and small time and space scales. However, for the present problem, we are only interested in highly averaged answers for c_i, i.e., time scales of years and gross global patterns, if not global means. It is useful to recognize that variations of c_i with longitude are usually small compared with latitudinal variations, both because of the greater dependence of mean solar irradiance on latitude and the existence of stronger winds in the east-west direction.

---

*Equation (B.1) is the same as Eq. (7.1), except that for convenience in our later analysis, the source and loss terms are defined in terms of rates per unit air molecule.

Thus, if we apply an average over longitude and some span of time to Eqs. (B.1) and (B.2), denoting the average of a quantity ( ) by ($^-$) and deviations from this average by ( )', we get for the average of T,

$$\overline{T} = - \text{div } \overline{(c_i \vec{V})},$$

i.e.,

$$\overline{T} = - \text{div } \left[(\overline{c_i}\,\vec{\overline{V}}) + \overline{(c_i'\vec{V}')} \right] \tag{B.3}$$

With longitudinal averaging, T is independent of longitude and div and $\vec{V}$ now denotes 2-D quantities in the meridional plane. The term $\overline{c_i'\vec{V}'}$ in Eq. (B.3) involves the deviation or so-called "eddy" quantities, c_i' and $\vec{V}'$. In physical models, the correlation of c_i' and $\vec{V}'$ is derived from a detailed observation of velocity fields and concentration fields or a simulation of c_i' and $\vec{V}'$. The second kind of model that may be identified is a statistical model and would be derived from further relationships between this correlation term and other quantities including the mean terms, $\overline{c_i}$ and $\vec{\overline{V}}$.

The dynamic meteorologist distinguishes many different types of large-scale motion phenomena in the stratosphere. Improved understanding of the transport characteristics of those various observed motion types might be expected to lead to meaningful statistical transport theories. At present, the only theoretical guidance available is the 2-D extension of Prandtl's mixing-length hypothesis. In this approach (CIAP Monograph 3, Section 4.2) one replaces $\overline{c_i'\vec{V}'}$ by the spherical coordinate version of

$$\overline{c_i'\vec{V}'} = -n \left\{ \left[K_{yy} \frac{\partial(\overline{c_i}/n)}{\partial y} + K_{yz} \frac{\partial(\overline{c_i}/n)}{\partial z} \right] \hat{j} \right.$$

$$\left. + \left[K_{yz} \frac{\partial(\overline{c_i}/n)}{\partial y} + K_{zz} \frac{\partial(\overline{c_i}/n)}{\partial z} \right] \hat{k} \right\} \tag{B.4}$$

where $\hat{j}$ and $\hat{k}$ are unit vectors in the latitudinal (y) and (z) direction and K_{yy}, K_{yz}, K_{zz} are the so-called eddy-mixing coefficients. To use Eq. (B.3), 2-D eddy-mixing models need besides the K's a description of mean motion $\vec{\overline{V}}$ in the latitudinal and vertical directions. The eddy-mixing parameterization does not provide them. It is usually assumed that $\vec{\overline{V}}$ can be specified from observations.

However, the transport by the mean motion appears often to be highly correlated with the eddy transport such that the mean and eddy components of the transport tend to cancel (CIAP Monograph 3, Chapter 4, Section 5.1C). (This cancellation suggests that 2-D eddy transports alone do not generally depend on mean mixing ratio gradients.)

In the absence of meaningful treatments of this difficulty of relating 2-D transport to mean mixing ratio gradients and other problems such as a dearth of useful data on the latitudinal variation of chemical species, it would seem unlikely that 2-D models can give in the global mean much more reliable predictions of transport than do the 1-D models developed by a further averaging over latitude. The 2-D models are, furthermore, much more difficult to implement, but they might give a more accurate treatment of reactive species. No suitable 2-D models have to our knowledge been developed as yet for the CFM problem. Thus all reported calculations of the CFM problem up to now have used 1-D transport models, and this situation is likely to continue into the near future. The experience of the CIAP program was that 2-D models adequate for the NO_x problem did not give any results notably different from 1-D models that could be ascribed to the difference in dimensionality.

1-D models have a number of limitations in addition to their obvious inability to deal with latitudinal, longitudinal, and short-time-scale temporal variations. The vertical concentration profiles to be used in 1-D models should be global average profiles, but as a practical matter a profile from a single location may be all that is available in many cases. The eddy-mixing concept is simplistic in its formulation and by its very nature cannot accurately represent the sporadic nature and structure of mixing, especially the incursions of stratospheric air through the tropopause fold behind cyclonic storms. Further, different substances may have different mixing coefficients because of differences in the vertical distribution of sources and sinks (Mahlman, 1976). Even if two substances were to have equal mixing coefficients for the 2-D formulation [Eq. (B.4)], mixing coefficients appropriate to a 1-D model for global average concentration may differ if the distribution of the substances differs. In spite of such limitations, 1-D models have been used in many atmospheric transport problems involving constituents that do not by their presence significantly affect motions in the atmosphere. Until enough becomes known of atmospheric behavior to permit the application of more detailed models (i.e., 2-D and 3-D) yielding more detailed predictions (including at least latitudinal variations) with enough confidence to justify their use, reliance on 1-D models will continue. For all these reasons, consideration is restricted to 1-D models for the remainder of this appendix.

III. THE 1-D EDDY-TRANSPORT MODELS, THEIR DERIVATION AND UNCERTAINTIES

With the further averaging of Eq. (B.4) over latitude, the eddy-mixing hypothesis reduces to the calculation of global mean transport, as given by

$$\bar{T} = \frac{\partial}{\partial z} nK(z) \frac{\partial(\bar{c}_i/n)}{\partial z} \tag{B.5}$$

and $n = n(z)$ is now some mean gas density. By this stage all details of actual atmospheric motions have been averaged out. The vertical eddy-mixing coefficient $K(z)$ used in Eq. (B.5) is generally obtained without any explicit reference to these actual motions. It cannot correspond precisely to an average of the K_{zz} factor in Eq. (B.4) because all the terms in Eq. (B.4) can influence vertical transport either directly or indirectly through their effect on the latitudinal distribution of species. It must, however, be chosen to give approximate agreement between the predicted and observed distribution of some suitable trace species that depends on transport. Such models are thus almost entirely empirical. Only very rough estimates of transport time scales can be inferred from the scales of observed motions. The only *a priori* theoretical basis for estimating the accuracy with which such a model developed for one substance with certain sources and sinks will predict the distribution of another substance with different sources and sinks is the recognition that they necessarily involve similar transport time scales. Consequently, comparisons with independent data have to be used to establish the level of confidence to be placed in model predictions (e.g., Mahlman, 1976). For example, it is not possible to derive satisfactorily the global mean profile of water vapor using 1-D models based on observations of other substances.

It is evident from the above discussion that the correctness of the 1-D empirical model transport predictions depends in part on the validity of the data that have gone into deriving K and on the agreement between model predictions and reliable independent data. Since the models at best determine global mean vertical transport of a given species, both the data used to derive and the data used to verify the K's should in principle also be global. Unfortunately, in the past there have not been sets of observations suitable for deriving global means, so that global averages have in practice been replaced by averages over a few observational profiles obtained at one or more sites in the northern hemisphere. Thus one obvious question that must be addressed is how representative of global means such data are. This question is considered in more detail in Chapter 6 and Appendix C. It suffices to note here that methane (as given by Ehhalt et al., 1974) is the only chemical species observed over a sufficient range of altitudes to give K's at all the altitudes important for the CFM problem.

Data on nitrous oxide, as currently available, are lesser in vertical extent but have two other advantages. With the new data obtained by the National Oceanic and Atmospheric Administration (NOAA) and the older data obtained over Texas (as discussed in Appendix C) it is possible to estimate a global mean profile. Furthermore, the stratospheric lifetimes of N_2O are better known than those of CH_4. The sinks of N_2O in the troposphere are poorly known and may destroy more N_2O than the photodissociative stratospheric sink. This would not affect significantly the usefulness of N_2O for estimating transport into the stratosphere provided it remains nearly uniformly mixed in the troposphere. In conclusion, the data on N_2O are useful to validate transport parameterizations inferred from CH_4 data and even to provide another independent parameterization.

Even if the available data on chemical species distribution were impeccable, the procedures used to infer an eddy-mixing coefficient profile from these data still may not be entirely satisfactory. Such profiles have generally been assumed to be of a functional form with a few adjustable parameters. These parameters have then been tuned to provide a satisfactory fit to the data as evaluated subjectively by the investigator. It is difficult to estimate how much subjective bias and preconceived notions as to atmospheric structure may influence the results from such an approach. For example, several investigators have used the same data on methane to obtain somewhat differing K profiles. All the profiles so derived do, however, have significant similarities, such as decreases by one to two orders of magnitude from values at the top of the troposphere to those of the lower stratosphere. One would like to know whether the similarities between profiles result from constraints imposed by the data or, conversely, whether the similarities result from similar bias on the part of the various investigators and that there are drastically different profiles that equally well or better reproduce the data. Is it possible that the data satisfactorily define the mixing coefficient profile at some altitudes, but further data at other levels are needed to pin down the mixing coefficient elsewhere? Can large errors be introduced into the K profile by certain types of error or bias in the data?

To answer such questions, we have developed an objective procedure for deriving an eddy-mixing coefficient from a profile of a given chemical species. Assume global mean data on some species c_i is available at M different altitudes z_j and that the global mean sources and sinks of c_i can be adequately described. The procedure is then to choose the value of K such that when Eqs. (B.5) and (B.1) are solved for c_i the solution agrees with the data in a least-squares sense. Since we must select a continuous profile of K on the basis of a limited number of data points, this problem is in general underdetermined. Consequently, some additional constraints must be placed on the K's. One method for constraining the K's is to assume that they depend on only a small number of parameters. This approach is but a more systematic and objective application of the procedure previously used to infer mixing coefficients. However, it does readily admit investigation of a large number of assumed functional forms to see whether any form dissimilar to those previously considered can satisfactorily fit the available data. Another somewhat less arbitrary method is to impose some kind of smoothness criterion on the K profile.

The question of estimating a K profile from chemical data is somewhat analogous to that of the inference of an atmospheric temperature profile from remotely sensed infrared radiation. In that problem also the assumption is made that the temperature profile can be represented by a few degrees of freedom or be constrained in some other fashion, for example, by requiring a minimum departure from some prescribed profile. One way to obtain smooth profiles is by minimizing the departure from some constant value. This is known as the "minimum variance" procedure.

The temperature inversion problem (i.e., the use of infrared radiances measured from satellites to infer an atmospheric temperature profile) provides useful guidance for the present problem, but differs in being linear or nearly so. The differential equations given by (B.1) and (B.5) (or various integral equations that may be derived from them) relate the c_i and the $K(z)$ in a markedly non-linear fashion. Consequently, the least-squares fitting of the calculated to observed c_i must be done numerically for even the simplest of examples. A suitable procedure has been described by Marquardt (1963) and is available as a computer program in the IMSL mathematical software library. This program minimizes (in a least-squares sense) M functions of N variables. The M functions to be minimized are the differences between computed and observed c_i at M different points; the N variables are N parameters representing the K profile.

We restrict present consideration to use of a single chemical species profile in steady state. However, generalization to use of a time-dependent species or several species jointly for a best estimate of a K profile should not be difficult.

As mentioned earlier, both methane and nitrous oxide data now provide useful single steady-state species for deriving transport by employing the procedure developed here. The N_2O data of NOAA, however, only became available during the final revisions of this report, and the previously available N_2O data over Texas appeared by themselves to be less useful than the CH_4 data. Thus, most of the subsequent discussion is couched in terms of the methane data of Ehhalt *et al*. These data are available up to 50 km and are especially useful for evaluating the analysis of transport in past CFM studies. They have been used by several past investigators in deriving the K profiles they employed for evaluating the effect of CFMs on ozone. The data consist of 10 points between 15 and 31 km with 1- to 2-km spacing between points and with each point representing the average of several observations taken at different times but at essentially the same location. There are also two individual observations by a cryogenic rocket system sounding between 44 and 62 km and 40 and 50 km, respectively. The data points are placed at 50 and 44 km, corresponding to mass weighted averages of the sampled altitude.

Because all the methane data are measured at one location, there is no need to be concerned with the possible use of an altitude scale relative to the global mean tropopause (assumed here to lie at the 15-km level). However, with the use of the new NOAA data for N_2O (Appendix C) this problem does arise. The averaging for it was done as follows. First, profiles from Saskatoon and Antarctica were averaged together to give a high-latitude mean. The high-latitude data point at 14 km was assumed to represent the tropopause value and shifted to 15 km, whereas the data point at 18 km was shifted to 18.3 km. The Panama profile was assumed to be a representative equatorial profile and was adjusted as follows. The data point at 17 km was assumed to represent the tropopause and shifted to 15 km, whereas the data point at 18.5 km was shifted to 17.5 km. These shifts of data points near the tropopause follow a mathematical coordinate transformation such that the interval from the tropopause to 20 km is linearly stretched to map into the 15- to 20-km region. A similar transformation could be applied below the tropopause to, e.g., the 10-km level. Finally, a global mean profile for N_2O was derived by giving 1/4 weight apiece to the NOAA high-latitude and equatorial profiles as defined above and 1/2 weight to a subjectively smoothed version of the Ehhalt *et al.* average N_2O profile as obtained over Texas.

The transformation of measurements to follow the tropopause as described above has the advantage of measuring vertical distance relative to the surfaces along which relatively fast mixing occurs in the lower stratosphere. It is just above the tropopause in the region of slowest vertical transport where it seems most necessary and appropriate to use surfaces of relatively rapid mixing as coordinate surfaces. Above 20 km, there is little evidence that the concept that mixing parallels the tropopause is useful, and its application to define a vertical coordinate transformation appears an unnecessary complication.

Whether the tropopause averaging described is used or not in parameterizing transport is not expected to influence significantly the characteristic transport time scales for species such as the CFMs, which have no important sources or sinks near the tropopause. It would, however, make a major difference in the analysis of the residence time of NO_x introduced by SST's.

To evaluate the diffusion term, Eq. (B.5), we approximated the derivatives by standard centered finite differences using 200 layers 0.4 km apart. Boundary conditions consist of a prescribed value of methane at the ground and vanishing gradient of the mixing ratio at the top (at 80 km). The density in Eq. (B.5) is assumed to vary as $\exp(-z/7 \text{ km})$.

The methane profile to be fitted is determined by the balance between upward transport and loss processes, the latter primarily due to reaction of methane with HO and $O(^1D)$. The chemical loss rate of methane used for calculations, unless stated otherwise, is that given by Figure 1 of Wofsy and McElroy (1973). The values given there have the advantage of being readily available and having been used in the past by both Wofsy and Hunten to derive eddy-mixing coefficients. They, however, also have some drawbacks. In particular, the HO concentrations they used for determining lifetimes were much lower in the lower stratosphere than values calculated in current chemical models. Consequently, their methane lifetimes increase by more than an order of magnitude in descending from 25 km to the tropopause, whereas the current model lifetimes change but little over this region. The small methane lifetimes of Wofsy and McElroy in the lower stratosphere are found in our calculations to intensify significantly the depth of the minimum eddy-mixing coefficient region in the lower stratosphere. The depth of this minimum region is found to be an important factor in determining the characteristic time scales for the CFM problem. The mixing coefficient profiles derived without the dropoff of methane lifetimes below 25 km give nearly a factor of 2 faster transport time scales.

The current chemical models (as also Wofsy and McElroy) have copious concentrations of HO in the upper stratosphere in agreement with the recent observations of Anderson (Chapter 6 and Appendix C). Any downward revision of these HO concentrations would tend to slow down transport as inferred from CH_4, but less than proportionately, since reactions with $O(^1D)$ and Cl also contribute significantly to methane lifetimes. The rates of both these reactions are somewhat uncertain. In particular, recent measurements at NOAA (Schiff, 1976) indicate that $O(^1D)$ + CH_4 reacts at a rate factor of 4 slower than that recommended by CIAP. The value assumed by Wofsy and McElroy lies halfway between the NOAA and CIAP values.

The greatest uncertainty as to the Cl + CH_4 rate is the Cl concentration, which is poorly known due to modeling, observational, and chemical uncertainties. Methane lifetimes in Chang's model increase by 10 to 20 percent between 20 and 30 km with the introduction of the chlorine nitrate reservoir for chlorine. The primary present contributors to stratospheric chlorine are CCl_4 and CH_3Cl, whose tropospheric concentrations are not very well known (Chapter 6 and Appendix C). How much of these get to the stratosphere in turn depends on the transport model parameterization. Past models for stratospheric methane have generally not included any loss by the Cl reaction.

In summary, we ask the reader to keep in mind that all the results of this appendix unless otherwise stated use Wofsy and McElroy (1973) rather than more current methane lifetimes. Most of the resulting transport time scales are longer than the values regarded as most likely (cf. Table 5.1). We are primarily interested here in the relative comparisons.

To explore the sensitivity of derived K's to the methane data, several features of the data may be questioned. First, the lower 10 data points are somewhat noisy in that a smooth line cannot be drawn through them. A smooth line can, however, be drawn that falls within the error bars of the points. This we did to generate an alternate smoother set of data (referred to as the smoothed data). If the mean methane profile at the site of these observations is unrepresentative of the global mean, this event is likely a consequence of the downward subsidence of the Hadley cell and occurrence of minimum tropospheric exchange at that latitude. Consequently, global mean values may drop off less rapidly from the tropopause (at 15 km altitude where the data were taken) than does the observed profile. Another set of data points between 15 and 31 km (referred to as the stretched data) was inferred from the raw data by adding an extreme possible correction for this effect. Raw, smoothed, and stretched data points are shown in Table B.1.

One decision that must be made in applying a least-squares fitting procedure is the relative weight to be given the different data points. The lowest and top two points, in particular, have special significance and perhaps should be weighted more heavily. The bottom point should correspond to the mean tropospheric mixing ratio of methane so that errors in fitting it also represent departures from the observed complete mixing of methane within the troposphere; the top two points by themselves largely determine the K profile above 35 km. Since the top two points are obtained from single observations, they may depart further from actual mean values than do the lower points. After some experimentation, we adopted a definition of the error to be minimized of the form

$$E = \left\{ \sum_{j=1}^{M} w_j \left[\psi_0(z_j) - \psi_C(z_j) \right]^2 / M \right\}^{1/2} \qquad (B.6)$$

TABLE B.1   The Methane Data (ppm) Assumed for Calculating Eddy-Mixing Coefficients[a]

Altitude (km)	Raw Data	Smoothed Data	Stretched Data
15.0	1.32	1.32	1.32
16.0	1.21	1.21	1.30
18.0	1.17	1.13	1.25
20.5	0.92	1.04	1.20
22.5	1.02	1.00	1.15
24.0	1.00	0.96	1.10
25.5	0.99	0.92	1.05
27.0	0.97	0.88	0.97
29.0	0.80	0.80	0.80
31.0	0.70	0.74	0.70
44.0	0.37	0.37	0.37
50.0	0.25	0.25	0.25

[a]The raw data are taken directly from Ehhalt et al. (1974), the smooth data are generated by hand drawing a smooth line within the error bars of the data, and the stretched data by another hand-drawn line to represent a hypothetical extreme correction for possible departures of the data from global mean values.

where $\psi_c(z_j)$ is the calculated methane mixing ratio and $\psi_0(z_j)$ the observed methane mixing ratio, both determined at level z_j, and w_j is a weight function appropriate to that level and defined as $w_j = 1/\psi_0^2(z_j)$. This definition gives the fitting error as the sum of the root-mean-squared relative difference between observed and calculated values. Since the top two data points have magnitudes $\psi_0(z_j)$ of the order of 1/3 of the values of the lower points, or less, this definition gives them an order of magnitude more weight than would an error defined by constant w_j. The weight for the lowest data point at 15 km was then doubled to allow for its relative importance.

A large number of computer experiments have been performed to explore possible "best fits" to the raw, smoothed, and stretched methane profiles. The sensitivity of the results to an additional data point at 38 km and to changes in the ratio of the 44- to 50-km data points were also considered. The more informative results from these studies are now discussed. First, the studies with K

profiles given by small numbers of parameters are considered, then the studies involving a continuous K profile
are discussed.

In considering profiles specified by a few parameters,
we investigated various profiles depending on one to six
parameters. These results discussed here are selected
from the large number of experiments performed as representative of the best-fit profiles that are obtained. The
precise functional forms for which results are presented
here are given in Table B.2. Case A defines a constant K.
No very satisfactory fit could be obtained in this case.
For simplicity, we use the same constant value for the
raw, smooth, and stretched data examples. A slightly more
general case of exponential variation (with the same expo-
nent at all levels) gave little further improvement beyond
constant K and is not presented here. Case B is represented by a constant value up to 12 km, exponential decay to
18 km, and exponential growth above that level. The growth
and decay rates are the adjustable parameters. Similar
profiles but with the exponential decay beginning at other
levels were also considered. When the beginning level was
lower, slightly better fits could be obtained; and when it
was higher, the fits were notably poorer. Cases C and D
are patterned after the K profile suggested by Hunten
[described in the report of the Climatic Impact Committee
(1975), p. 116]. This profile assumes a constant value of
K to the tropopause, a rapid drop-off at the tropopause to
a constant low value in the lower stratosphere, and exponential growth above some level. The tropopause drop is
sufficiently smoothed to avoid difficulty with the finite
difference numerical procedures. For Case C we adjust
only the lower stratosphere K applied between 15 and 20 km
and the exponential growth above that level. For Case D,
we also allow the level at which exponential growth begins
to be adjustable and assume that the growth rate depends
on some (adjustable) power of the distance from its beginning. Such power dependence was not considered by Hunten.
Case E defines a smoothly varying K with one extremum.
Cases F and G provide more resolution by assuming, respectively, 6 and 40 segments defining the logarithmic slope
of K over adjacent intervals. Case G is to be used in
conjunction with the later to be described smoothness
criteria.

The methane and eddy-mixing coefficient profiles obtained by adjusting these parameters to obtain the best
root-mean-square (rms) fit to the data of Table B.1 using
Wofsy and McElroy (1973) methane lifetimes are shown in

TABLE B.2 Various Parametric Representations Assumed for the Eddy-Mixing Coefficients[a]

Case	Representation
A	$\ln (K) = -1$
B	$\ln (K) = \begin{cases} 1, & z \leq 1.2 \\ 1 + a_1(z - 1.2), & 1.2 < z \leq 1.8 \\ 1 + a_1(z - 1.2) + a_2(z - 1.8) & z > 1.8 \end{cases}$
C	$\ln (K) = \begin{cases} 1, & z \leq 1.5 \\ a_1, & 1.5 < z \leq 2.0 \\ a_1 + a_2\,(z - 2.0), & z > 2.0 \end{cases}$
D	$\ln (K) = \begin{cases} 1, & z \leq 1.5 \\ a_1, & 1.5 < z \leq a_3 \\ a_1 + a_2\,(z - a_3)a_4, & z > a_3 \end{cases}$
E	$\ln (K) = a_1 + a_2\,\mathrm{sech}^2\,[a_3(z - a_4)]$
F	$\ln (K) = 1.0 + \displaystyle\sum_{i=1}^{6} \mathcal{H}(z - z_i)a_i(z - z_i)$
	where $z_i = 1.2 + 0.4(i - 1)$, $\mathcal{H}(z - z_i) + \begin{cases} 0, & z < z_i \\ 1, & z > z_i \end{cases}$
G	$\ln (K) = 1.0 + \displaystyle\sum_{i=1}^{40} \mathcal{H}(z - z_i)a_i(z - z_i)$
	where $z_i = 0.9 + 0.12(i - 1)$

[a]Except for A, these parameters are adjusted to obtain best rms fits to the methane data given in Table B.1. Adjustable parameters are denoted a_i, and z is altitude in units of 10 km. Units of K are $10\ \mathrm{m}^2\ \mathrm{sec}^{-1}$. The letter labels of the cases are used to distinguish the various methane and K profiles shown in Figures B.1–B.3. Case G is used with additional smoothing criteria.

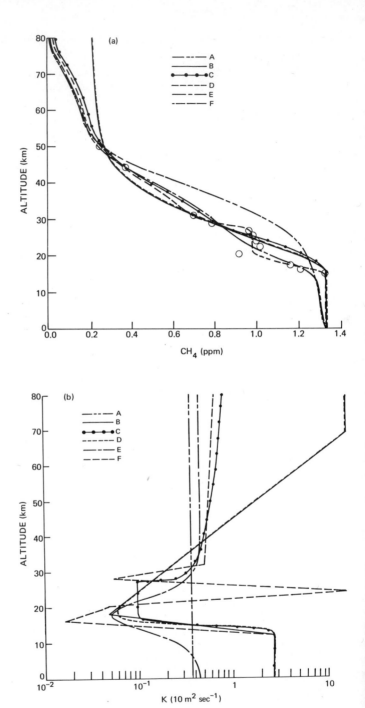

◄FIGURE B.1 (a) Methane profiles lettered according to and generated for the various K representations given in Table B.2. Except for profile A, the parameters of the representations were adjusted to give a best fit in the sense described in the text to the raw data of Table B.1 (also indicated on plot by dots with circles). (b) The various eddy-mixing coefficients that give the profiles of (a).

Figures B.1 to B.3. The rms error for each fit is given in Table B.3.* The fit is seen to depend mainly on the number of adjustable parameters for the models selected. The best fits are obtained with the six-parameter log linear profile (Case F). However, especially with the raw data, the inferred profile may be too jagged to be physically plausible. In Figure B.1, the K profile goes from values less than 0.2 m^2 sec^{-1} at 16 km to values greater than 10^2 m^2 sec^{-1} at 24 km. Consideration of Figure B.1 shows that the F profile allows good simulation of small-scale features of the raw methane data, in particular the rapid drop from 15 to 20 km and the apparently near constant region from 20 to 28 km. Of course, the reality of such features for the actual global mean vertical variation of methane is of considerable doubt. Figure B.2 shows that the six-piece K profile required to fit the smoothed data remains small at 25 km but does decrease to very small values, of the order of 0.1 m^2 sec^{-1}, at 16 km.

Indeed, the one common feature of all K profiles given by the raw or smoothed data are values less than 1 m^2 sec^{-1} between 16 and 20 km. This feature strongly suggests that any continuous K profile that would generate a methane profile similar to the observed data must have such a feature. Even in fitting the stretched data (Figure B.3), the calculations invariably give low values of K in the lower stratosphere, but in this case the region of minimum values is broader, lying between 18 and 30 km, and the extreme values are not as small. The most crucial feature in all the assumed versions of the data appears to be the 50 percent drop in CH_4 from 15 to 30 km. This relative dropoff is seen also to be present in the other two sources of data shown in Figure C.12. The lowest K values below 20 km in the raw and smoothed data examples depend on a 25

---

*Our interest here is primarily in the relative differences between different K profiles. Optimum K profiles will be discussed later.

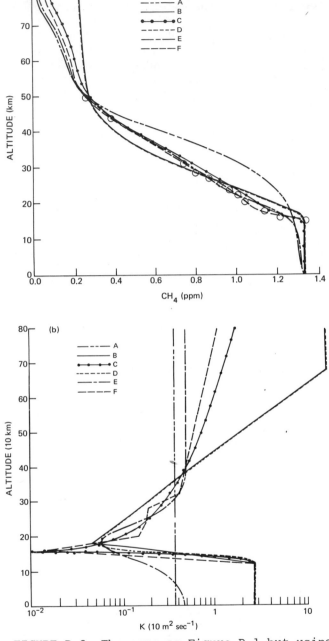

FIGURE B.2 The same as Figure B.1 but using the smoothed data of Table B.1.

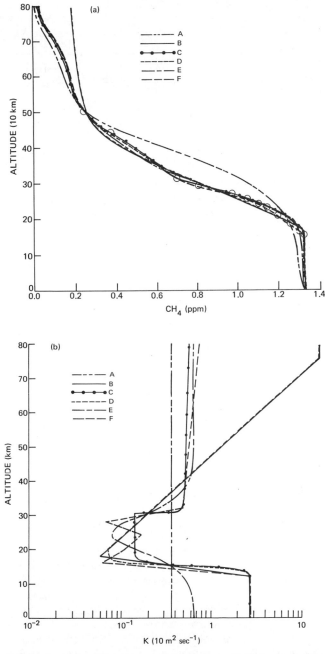

FIGURE B.3 The same as Figure B.1 but using the stretched data of Table B.1.

TABLE B.3   Root-Mean-Square Error as Defined by Eq. (B.6)
for the Optimum Fits to the Data of Table B.1 Obtained
Using the Functional Forms B-F Given in Table B.2[a]

Case	Data Raw	Smoothed	Stretched
B	0.10	0.08	0.05
C	0.10	0.07	0.05
D	0.10	0.04	0.02
E	0.07	0.03	0.03
F	0.03	0.01	0.003

[a]The significance of the smoothed and stretched data is
explained in Table B.1.

percent decrease of methane from 15 to 20 km.   The case A
of constant $K = 3.8$ m^2 sec^{-1} would only be consistent with
a less than 10 percent dropoff of CH_4 between 15 and 30 km.
     Another point to be noted is that the worst fit to the
top two data points at 44 and 50 km is given by profiles
B and D.   For these cases, the increase of K between 20
and 30 km requires large K values above 50 km because of
the restriction of K to log-linear variation.   All the
cases with independent flexibility above 30 km select a K
near 5 m^2 sec^{-1} at 50 km.   A study with the six-parameter
model of the dependence of K near 50 km on the cryogenic
rocket data indicated considerable sensitivity to the ratio
of these observed values.   In perturbing the observed values of 0.37 and 0.25 ppm so that their mean remained the
same, we found that K increased by a factor of 3 for each
0.02 ppm decrease of the larger and lower data point (and
compensating 0.02 ppm increase of the smaller and higher
point).   Very large values of K would evidently be required
at 50 km to give a near constant mixing ratio in the face
of the large methane destruction rates.   The sensitivity
to changes in the assumed value of tropospheric K were also
studied; changes in the tropospheric K (provided it remains large) do not seem to require significant changes in
K above the troposphere.   Likewise, changes in the methane
loss rates change the stratospheric K's by a proportional
amount although the tropospheric K's remained fixed.
     Finally, it should be noted that a crude fit to the data
as a whole can even be achieved with a constant eddy-mixing
coefficient as illustrated by Case A.   In general, the more

degrees of freedom allowed, the smaller the rms error achieved. However, achieving a close fit to noisy data by allowing a large number of degrees of freedom produces large oscillations in the K profile. Thus, probably the most significant decision that must be made in selecting a K profile is the degree of smoothness of the profile consistent with the quality of the data.

While the mean-square deviation is seen from Table B.3 to decrease with increasing number of adjustable parameters, as expected, the "oscillation" of K also increases (Figure B.1, cf. F), also as expected, probably largely because of the noise of the data. If one wishes to increase the number of adjustable parameters, we should (because of the noise) introduce into the error definition of smoothing criterion to avoid pathological oscillations of K.

A smoothness criterion can be objectively introduced into the K profile estimation, by requiring that some smoothness function be minimized as well as the fits to the data. That is, if we denote the error defined by Eq. (B.6) as E_0, we introduce a new error definition as

$$E = [E_0{}^2 + \lambda S(K)]^{1/2} \qquad (B.7)$$

where $S(K)$ is some function of K that decreases as the K profile becomes smoother, and λ is a parameter defining the degree of smoothness to be required. If $S(K)$ is chosen so that it goes to zero only in the limit of a constant K, then Eq. (B.7) cannot be exactly zero, even if K is allowed an infinite number of degrees of freedom. Thus, the criterion that E defined by Eq. (B.7) be minimum allows investigation of K's with so many degrees of freedom that arbitrariness in selecting a shape for K is no longer a problem. We initially attempted adjusting K at every grid point with various definitions of $S(K)$. However, such a procedure requires an exorbitant amount of computer time. Also, no smoothing criteria could be found that would eliminate discontinuous variations of K between grid points. An understanding of the reasons for such solutions would be needed to characterize fully the mathematical properties of the procedures being used. However, for present purposes, they are regarded as unphysical and hence unwelcome. A continuously varying K was achieved by assuming the functional form described by Case F of Table B.2 but using 40 linear segments for log (K), each 1.2 km in thickness and hence covering three grid points. That is,

$$\ln (K) = 1.0 + \sum_{i=1}^{4\Omega} \mathcal{H}(z - z_i) \cdot a_i (z - z_i) \qquad (B.8)$$

where $\mathcal{H}$ is the unit jump function and

$$z_i = 0.9 + 0.12(i - 1)$$

which together with z is measured in units of 10 km. For convenience, this expression is also listed as Case G in Table 5.2. The a_i parameters give the change of the gradient of ln (K) from one 1.2-km segment to the next between 9 and 57 km. After experimenting with several smoothness criteria, we discovered that satisfactory results could be achieved minimizing Eq. (B.7) with $S(K)$ defined by

$$S(K) = 0.02 \sum_{i=1}^{40} a_i^2 + 10a_{40}^2 \qquad \text{(B.9)}$$

This definition is essentially the average of the square of the a_i's with an additional factor of 10 more weight given to the top a_i to allow for its control of the gradient of log (K) from 57 km to the top of the model. The factor 0.02 is a normalization factor [i.e., $1/(40 + 10)$].

The best fit in the sense of minimum E defined by Eqs. (B.7) and (B.9) was obtained for the raw, smooth, and stretched methane data sets, using three values of the smoothing parameter λ, i.e., $\lambda = 10$, 0.1, and 0.001. These values of λ illustrate what we regard as heavy, moderate, and light smoothing. The fits so obtained are illustrated in Figures B.4-B.6, and the rms error in the fit to the data so obtained is given in Table B.4. The resulting eddy-mixing coefficients have profiles similar to those already discussed for the profiles derived assuming but a few degrees of freedom. Thus, the inferences made from Figures B.1-B.3 are generally confirmed here. In particular, the raw and smoothed data both generate a minimum K between 16 and 20 km, which becomes increasingly sharp and deep as less smoothing is applied. Furthermore, with only light smoothing applied to the raw data, a maximum K at 24 km greater than 10^2 m^2 sec^{-1} is generated, but the smoothed data give no suggestion of such a feature. Finally, the fit to the stretched data indicates a broad region

---

FIGURE B.4 Methane profiles generated by adjusting K's ▶ at each level to minimize the square difference between the calculated profile and the raw data of Table B.1 and the average change in the square logarithmic gradient of the profile, for various weights given to the latter term as described in text. A, constant K; B, heavy smoothing; C, medium smoothing; and D, light smoothing.

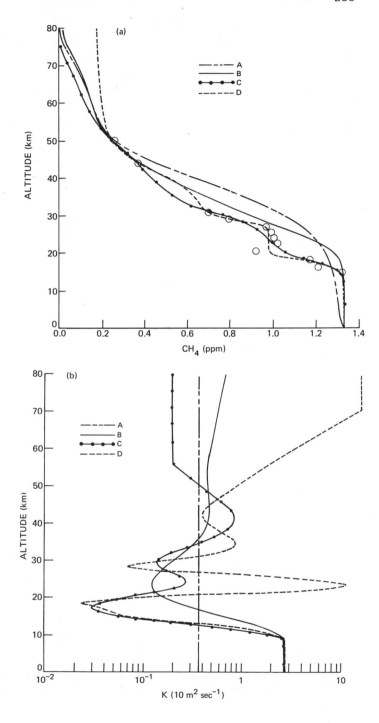

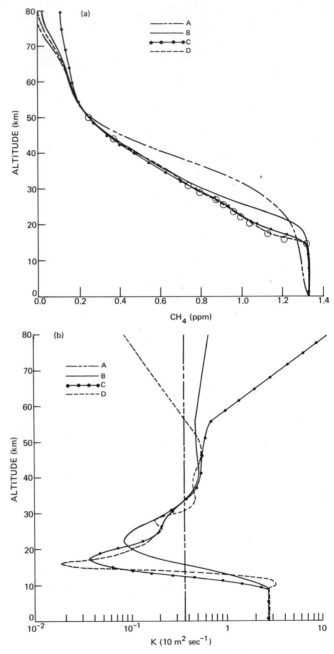

FIGURE B.5 Same as Figure B.4 but for
smoothed data of Table B.1.

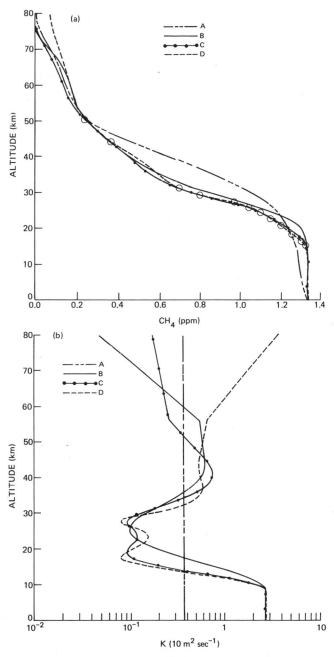

FIGURE B.6 Same as Figure B.4 but for the stretched data of Table B.1.

TABLE B.4   Root-Mean-Square Error Fit to Methane Data as Defined by Eq. (B.6) (and Case G of Table B.2) for Adjusting K at All Levels to Obtain the Profiles of Figures B.4-B.6[a]

| | Data | | |
Smoothing	Raw	Smoothed	Stretched
$\lambda = 10$ (heavy)	0.16	0.10	0.06
$\lambda = 0.1$ (moderate)	0.05	0.02	0.01
$\lambda = 0.001$ (light)	0.03	0.009	0.004

[a]The smoothness function to be minimized is defined by Eq. (B.7).

of minimum K between 18 and 30 km. In considering the sensitivity of the derived K profiles to possible inadequate data, we again adjusted the ratio of the 44- to 50-km data and obtained results similar to those earlier described. We also tested the sensitivity of the profile to absence of methane data between 32 and 46 km by assuming an additional data point at 38 km, which was varied from 0.3 to 0.5 ppm. This range of assumed values induced variation by factors of 3 in the derived K between 32 and 40 km; a 0.3 ppm value produced relatively small K below 36 km and relatively large values between there and 48 km.

Subjectively, we consider the K profiles generated with moderate smoothing as physically most acceptable. (The rms errors in achieving these fits to the raw, smoothed, and stretched data are, respectively, 5, 2, and 1 percent.) We have tabulated these values in Table B.5. Also given for comparison is a K profile derived using methane lifetimes from Chang's model (see Appendix D) and another derived to fit the global average N_2O profile as described earlier using N_2O lifetimes from Chang's model. Either of these eddy-mixing coefficient profiles and all the K profiles shown in Figures B.4-B.6 are regarded as within the realm of possibility. Thus, the range of uncertainty in predictions of 1-D transport models can be explored in terms of the differences obtained with the different profiles. The consequences of uncertainties in methane destruction rates are discussed further in the next section.

TABLE B.5   Optimum K Profiles Obtained by Adjusting K at
All Levels and with Moderate Smoothing Parameter (λ = 0.1)
in Units of m^2 sec^{-1}

z (km)	Data Fit to Methane Wofsy Lifetimes Raw	Smoothed	Stretched	Fit to Methane Chang's Lifetimes (No ClONO$_2$)	Fit to global N$_2$O Profile
≤ 8.8	27.2	27.2	27.2	27.2	27.2
10.0	18.6	18.6	24.9	20.8	19.3
12.0	5.02	5.06	17.7	8.1	5.81
14.0	1.08	1.14	10.7	2.47	1.38
16.0	0.37	0.42	5.89	0.98	0.50
18.0	0.33	0.38	3.24	0.88	0.50
20.0	0.68	0.64	1.90	1.79	0.93
22.0	1.77	1.20	1.27	4.61	1.58
24.0	2.67	1.76	1.02	6.65	1.87
26.0	2.24	2.07	0.99	5.23	1.97
28.0	1.51	2.24	1.14	3.34	2.53
30.0	1.43	2.60	1.50	3.18	3.88
32.0	1.95	3.18	2.09	4.52	6.07
34.0	3.17	3.88	2.93	7.70	8.02
36.0	4.99	4.54	3.92	12.7	8.84
38.0	6.99	5.05	4.91	18.3	10.1
40.0	8.30	5.35	5.70	21.9	14.0
42.0	8.35	5.50	6.14	21.7	22.7
44.0	7.38	5.56	6.26	18.5	39.7
46.0	5.98	5.63	6.16	14.3	71.9
48.0	4.72	5.75	4.98	10.9	
50.0	3.74	5.94	5.82	8.4	
52.0	3.00	6.24	5.69	6.7	

IV.   SENSITIVITY OF CFM-OZONE REDUCTION TO 1-D TRANSPORT
      MODELS

In this section we consider a hypothetical CFM with stratospheric destruction rates shown in Figure B.7, which lie
between those of F-11 and F-12, as given by Rowland and
Molina (1975). Upon stratospheric destruction, all of its
chlorine atoms are converted to "reactive chlorine," i.e.,
Cl, ClO, HCl, and ClNO$_3$ here designated ClX. The

260

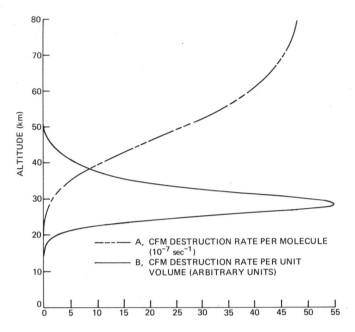

FIGURE B.7 The standard CFM loss rate assumed
for the transport sensitivity studies. A, inverse lifetime of CFM molecules; B, loss rate
per unit volume.

concentrations of the CFM and ClX are both defined as the
mixing ratio of their Cl atoms and normalized in a way
described below [i.e., Eq. (B.16)].

The CFM problem as here formulated is intended to iso-
late those aspects of transport that are insensitive to
particular model details. Let ψ_{CFM} = CFM mixing ratio and
ψ_{ClX} = the reactive chlorine mixing ratio normalized as
described below, so that the total Cl mixing ratio is
defined by

$$\psi = \psi_{CFM} + \psi_{ClX} \qquad (B.10)$$

The time-dependent evolution of ψ according to Eqs. (B.1)
and (B.5) is given by

$$\frac{\partial \psi}{\partial t} - \frac{1}{n} \frac{\partial}{\partial z} \left(nK \frac{\partial \psi}{\partial z} \right) = S - L \qquad (B.11)$$

where S represents the addition of chlorine atoms, assumed to be only in the form of CFMs, and L represents the Cl loss processes, assumed to involve only reactive chlorine (ClX) in the troposphere. The known sources and sinks for ψ occur in the troposphere. Consequently Eq. (B.11) in the stratosphere reduced to the equation for an inert tracer and has as the physically meaningful steady-state solution

$$\psi = \text{constant} \qquad (B.12)$$

On the other hand, the equations for ψ_{CFM} and ψ_{ClX} have, respectively, in the stratosphere a sink or source due to the photodissociation of CFM. In particular,

$$\frac{\partial \psi_{CFM}}{\partial t} - \frac{1}{n} \frac{\partial}{\partial z} \left(nK \frac{\partial \psi_{CFM}}{\partial z} \right) + J(z) \psi_{CFM} = S \qquad (B.13)$$

where $J(z)$ is the rate constant for dissociation of the CFM by solar radiation to ClX.

The CFM sources are assumed to occur entirely at the surface, i.e., CFM is inserted according to the boundary condition

$$n(0) K \frac{\partial \psi(0,t)}{\partial z} = n(0) K \frac{\partial \psi_{CFM}(0,t)}{\partial z} = -s(t) \qquad (B.14)$$

where $s(t)$ is the rate of CFM addition at the surface. The interior source S in Eq. (B.13), is assumed to vanish, that is,

$$S = 0 \qquad (B.15)$$

Putting all the CFM in at the lower boundary according to Eq. (B.14) is equivalent to specifying the source term $S(z,t) = s(t) \delta(z)$, where $\delta(z)$ is the Dirac delta function.

In discussing the results of our calculations, it is convenient to normalize ψ, ψ_{CFM}, and ψ_{ClX} by the total chlorine atoms in CFM added per atmospheric molecule, i.e., we define a normalization factor ψ^* by

$$\psi^* = \int_{-\infty}^{t_0} s(t)\,dt \Big/ \int_0^\infty n(z)\,dz \qquad (B.16)$$

where t_0 defines a time after which there is no more CFM addition. All further mixing ratios to be discussed will have been divided by the factor (B.16). [This normalization

precludes indefinite growth, and the steady-state case has
other complications, see, e.g., Eq. (B.27), but the conclusions we shall infer readily apply to these cases as
well.] A simple definition of mixing ratio at any altitude would be number of Cl atoms per total number of molecules at that z. Our ψ differs from such a definition
only by a constant of proportionality.

The mass averaged value of a quantity Y will be denoted
by a bar over it, i.e.,

$$\bar{Y} = \frac{\displaystyle\int_0^\infty n(z)\,Y(z)\,dz}{\displaystyle\int_0^\infty n(z)\,dz} \qquad (B.17)$$

This notation should not be confused with its earlier use
for longitudinal means. The quantity $\bar{\psi}(t)$ (as a consequence of our normalization) equals unity for $t > t_0$,
when there are no sinks, and when the integrated source
$\int_{-\infty}^{t_0} s(t)\,dt$ is finite. This limit is readily established
for $\bar{\psi}$ by integrating $n \times$ Eq. (B.11) over z and t under the
above-stated conditions.

We now introduce the two most important time scales of
the CFM-ozone reduction processes as inferred from our
instantaneous release analysis. These are: t_1 proportional to the time scale on which the atmosphere effectively
uniformly mixes up into the stratosphere, an inert constituent added at the surface (this time will be monitored
in numerical calculations by determining the time at which
maximum ClX mixing ratios are found); also t_2 = the time
scale for the loss term L in Eq. (B.11) to remove the CFM
from the atmosphere. This is also the time required to destroy the CFMs by solar photodissociation. It will be
monitored in numerical calculations by determining the time
required for ClX to decay to exp(-1) from the maximum concentrations. Both distinct time scales are determined in
large part by transport, and for problems of interest they
are well separated. That is, for realistic formulations,
$t_2 \sim 100$ yr and $t_1 \sim 10$ yr. Consequently, chlorine-containing molecules introduced instantaneously into the atmosphere undergo two separate processes in these two time
scales. First ψ becomes slowly mixed (i.e., approaches
unity) over t_1 without significant loss; afterward over
t_2, ψ slowly decays.

An important point to note from past model integrations
and those to be discussed is that at levels where ClX may
catalyze ozone destruction (i.e., above 25 km),

$$\psi_{ClX} \simeq \psi$$

so that a description of the evolution of ψ also applies
to the ψ_{ClX} in the region where the latter quantity affects the stratospheric ozone.

Several important conclusions for instantaneous releases
are immediately evident. First, if all chlorine atoms are
in the form of ClX for which ozone catalysis is important,
the significant role of transport in the CFM problem is
simply to determine the mixing ratio of total chlorine
atoms (CFM + ClX) in the stratosphere. Second, it follows
from near-complete mixing for times greater than t_1 [i.e.,
Eq. (B.12) prevails, and $t_1 \ll t_2$] that the maximum mixing ratio of total chlorine atoms (CFM + ClX) will be
insensitive to transport. In other words, if the transport mixes the total chlorine atoms (from all species)
throughout the troposphere and stratosphere before any significant losses occur, the maximum mixing ratio of chlorine
atoms (of all species) in the stratosphere will not depend
on the details of transport. Third, the total chlorine
atom concentration as it decays over t_2 is essentially uniformly mixed, hence the mixing ratio is independent of
altitude and so described completely by t_2.

These conclusions become much more powerful in the light
of a further result that we shall establish by numerical
calculation. That is, CFM releases over time scales of
10 yr or less are equivalent in the long run to instantaneous releases occurring at some particular release time.
Consequently, the effect of all the CFM releases that have
occurred up to now (because of their rapid past growth)
can be approximated by an equivalent instantaneous release
having occurred several years ago and with amount equal to
the total CFMs that have already been released to the atmosphere. Furthermore, the effect of additional CFM releases are essentially additive provided the diminution
of ozone by chlorine atoms is linear in the chlorine atom
concentrations (cf. Chapters 7 and 9).

We now proceed to establish more precisely the concepts
of the mixing and decay time scales for an instantaneous
release. Numerical calculations are then described, which
determine these time scales for the eddy-mixing coefficients inferred in the previous section and show that they
fully characterize the time evolution of ClX subsequent to
an instantaneous CFM release. Another series of calculations is described, which shows that over a wide range of
time scales various release scenarios can be approximated
by point releases. An analytic approximation to the time

evolution of an instantaneous release (depending only on t_1 and t_2) not only summarizes the time variation of the effect on ozone consequent to this particular scenario but also allows estimation of the effect of continuing exponential growth in terms of equivalent instantaneous releases.

Consider now the equations for mass averaged mixing ratios for ψ, ψ_{CFM}, and ψ_{ClX}. Averages are denoted by bars and defined as in Eq. (B.17). Assume the CFM surface emission has ceased. These equations then follow from Eqs. (B.10), (B.11), and (B.13).

$$\frac{\partial \bar{\psi}}{\partial t} + \bar{L} = 0 \tag{B.18}$$

$$\frac{\partial \bar{\psi}_{CFM}}{\partial t} + \overline{J(z)\ \psi_{CFM}} = 0 \tag{B.19}$$

$$\frac{\partial \bar{\psi}_{ClX}}{\partial t} + \bar{L} = \overline{J(z)\ \psi_{CFM}} \tag{B.20}$$

Simplification of these relationships follows from noting that at any given time most of the mass of the atmosphere and hence of chlorine atoms is in the troposphere, where the chlorine is mostly in the form of CFMs, whence

$$\left| \frac{\partial \bar{\psi}_{ClX}}{\partial t} \right| << \left| \frac{\partial \bar{\psi}}{\partial t} \right| \tag{B.21}$$

After $t = t_1$, this approximation with Eq. (B.18) reduces Eq. (B.20) to

$$\bar{L} \simeq \overline{J(z)\ \psi_{CFM}} \tag{B.22}$$

That is, there is a near balance between the mean photo-dissociation of the CFMs and the tropospheric loss of the ClX. This conclusion also applies in the presence of CFM sources. Since the mean source term for ClX is much larger than the mean time rate of change, there must also be a near steady state locally. That is, at the altitudes where most of the ClX resides and where its local loss is negligible, there should be a near balance between production of ClX and downward transport. Hence ψ_{ClX} is given above the troposphere approximately by

$$-\frac{1}{n}\frac{\partial}{\partial z} nK\frac{\partial \psi_{ClX}}{\partial z} \simeq J(z)\ \psi_{CFM} \tag{B.23}$$

This relationship also can be obtained by subtracting Eq. (B.13) from Eq. (B.11) neglecting $\partial \psi_{ClX}/\partial t$ and L (which is negligible above the tropopause). Integration of Eq. (B.23) with appropriate boundary conditions gives an approximate expression for ψ_{ClX} in terms of ψ. As a result of its quasi-steady state, the shape of the ψ profile should change but little so that this relationship establishes approximately a time-independent proportionality between ψ_{ClX} and ψ_{CFM} and hence between both of these and ψ. That is,

$$\left.\begin{array}{c} \psi_{CFM} = f(z)\,\psi \\[2mm] \psi_{ClX} = [1-f(z)]\psi \end{array}\right\} \tag{B.24}$$

where $f(z)$ is the proportionality factor to be determined from Eq. (B.23). Now assume that ψ has become fully mixed so that $\bar{\psi} = \psi$. With Eqs. (B.24) and (B.22) we can write Eq. (B.18) as

$$\frac{\partial \bar{\psi}}{\partial t} + \bar{\psi}/t_D = 0 \tag{B.25}$$

where

$$1/t_D = \overline{J(z)f(z)} \tag{B.26}$$

should be nearly equivalent to the empirically determined rate $1/t_2$.

In other words, the decay rate $(1/t_D)$ for chlorine concentrations in the stratosphere is defined by the mass weighted average of the product of CFM destruction rates and the (time-independent) fraction of chlorine in the CFM form.

Finally, note that the previous result [Eq. (B.25)] does not apply to steady-state conditions. With sources prescribed under steady-state conditions, ψ will be nearly uniformly mixed so that Eq. (B.25) is replaced by

$$\bar{\psi}/t_D = s \tag{B.27}$$

where s is now the average number of Cl atoms in CFM added to the atmosphere per unit time and per unit molecule of air. The previously assumed normalization is not applicable under these conditions, but other definitions are the same as before. Thus knowledge of the steady-state CFM emission rate and decay time are sufficient to establish

the steady-state concentration of total chlorine atoms,
hence, ClX, hence the effect of the CFMs on ozone for a
given chemical scheme.

   In the case of an instantaneous surface source, with
both n and K constant, and with no sinks, ψ at any altitude would be proportional to $\exp(-\tau/t)$, where $\tau = z^2/4K$.
Comparison of this time dependence with the exponential
decay implied by Eq. (B.25) suggests that ψ in the stratosphere should have time dependence of the form

$$\psi = \psi_{max} \exp[\phi(t_1) - \phi(t)] \tag{B.28}$$

where

$$\phi(t) = \frac{\tau}{t} + \frac{t}{t_D} \tag{B.29}$$

and t_1 is defined as the time at which ψ reaches the maximum value. The parameter τ in Eq. (B.29) is chosen so that
Eq. (B.28) reaches its maximum value when $t = t_1$. By setting the derivative of Eq. (B.29) equal to zero at $t = t_1$
we find $\tau = t_1^2/t_D$. The parameter τ rather than t_1 might
be interpreted as the mixing time scale. It is smaller
than t_1 by the ratio t_1/t_D and for realistic conditions is
of the order of 2 to 3 yr. This range of values is often
quoted as the time required to mix an inert substance from
the troposphere into the stratosphere.

   The validity of the above analysis is now explored with
numerical model integrations. For a standard model, we
use the optimum K profile for medium smoothing and fitting
the raw methane data with Wofsy's methane lifetimes as
given in Table B.4. Model atmosphere densities are taken
from Table 6 of Rowland and Molina (1975), and the standard CFM destruction assumed is that of Figure B.7. Two
forms for $s(t)$ in Eq. (B.14) are considered.

$$\left. \begin{array}{l} s_1(t) = \langle H \rangle \dfrac{c}{\sqrt{\pi}} \exp[-(ct)^2] \\[2mm] s_2(t) = \langle H \rangle \mathcal{H}(-t) c \, \exp(ct) \end{array} \right\} \tag{B.30}$$

The factor $\langle H \rangle$ is the mean scale height equal to the vertical integral of ρ and enters with the normalization Eq.
(B.16). The term $\mathcal{H}(t) = 1$ for $t < 0$ and is zero for $t > 0$.
The first of these expressions, Eq. (B.30), corresponds to
a smooth Gaussian growth up to a maximum value at $t = 0$
followed by decay of the same shape. The second corresponds to exponential growth up to $t = 0$, followed by an

abrupt cutoff. In both cases, c is adjustable to define the time scale of the source.

Differences in results depending on whether s_1 or s_2 is used as the source term and on what value of c is used should suggest the sensitivity of CFM-ozone reduction to details of source time dependence. The source terms have been normalized as described previously and so all give the same time-integrated release. An implicit numerical time-integration procedure was used for Eqs. (B.10), (B.11), and (B.13), with time and space steps sufficiently small that the solutions were reasonably insensitive to changes in either of these. A tropospheric loss term for ClX was prescribed below 10 km proportional to ClX and increasing linearly downward from 10 km to a surface inverse lifetime of $(10 \text{ day})^{-1}$. This loss term describes crudely the removal of HCl by rain-out.

An instantaneous source is defined by either s_1 or s_2 and $c >> (1 \text{ yr})^{-1}$. The results now discussed are obtained for such an emission and with the previously defined standard K and CFM lifetimes. The decay of chlorine atom concentrations after the time of maximum mixing ratio was monitored by measuring the time t_2 required for ψ_{ClX} at 30 km to decay to $\exp(-1)$ of its maximum value that is achieved at that level.

In Figures B.8 and B.9 we show the time history of ψ_{ClX} and ψ between 20 and 40 km. (The effect of ClX on ozone below 20 km is quite small compared with its effect above.) Several important points are obvious from these figures. First, the maximum values of ψ are nearly unity. In other words, the peak values of ψ could have been estimated by the assumption of complete mixing. Furthermore, $\psi_{ClX} = \psi$ above 30 km. Second, the time of peak concentrations is essentially the same for all altitudes shown and for both ψ and ψ_{ClX}. Third, the ratio of ψ_{ClX} to ψ does not seem to vary significantly with time. This result is seen dramatically in Figure B.10, which shows the ratio of these two quantities at the time of maximum ClX concentrations and at the time when these concentrations have decayed to $\exp(-1)$. The figure shows essentially no change in this ratio. The ratio of values of ClX at the latter and former times are also shown to illustrate that the relative decay is essentially independent of altitude. The time required for the ClX-to-total Cl ratio to approach a constant value at different altitudes is illustrated by Figure B.11. At the highest altitudes only a year or two is needed, but below 30 km it takes nearly 10 yr for ψ_{ClX}/ψ to reach its asymptotic values. However, this ratio is close to its asymptotic value at all altitudes in a few years.

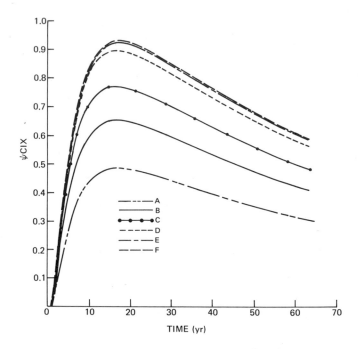

FIGURE B.8 Instantaneous CFM release scenario: time history of active chlorine mixing ratio (ψ_{ClX}) between 20 and 44 km. A value of 1.0 would be achieved with no sinks and total conversion of CFM to ClX. A, 20 km; B, 24 km; C, 28 km; D, 32 km; E, 36 km; F, 40 km.

The adequacy of expressions (B.28) and (B.29) for approximating the time history of ψ is illustrated by G in Figure B.11, which shows the ratio of this approximate expression to the ψ calculated at 32 km. Within one year after the addition of the CFM, the approximate expression (B.28) is seen to provide CFM mixing ratios reasonably close to those calculated numerically.

To see the effect of source time scale on the Cl concentrations, we integrated Eqs. (B.10), (B.11), and (B.13) for a large number of c's in Eq. (B.30). For the Gaussian source, we define source time scale as the time required for the Gaussian to drop to one half of its maximum value; for the exponential growth-sudden cessation, we define time scale as the doubling time. Solution time scales are again characterized by t_1, the time at which maximum

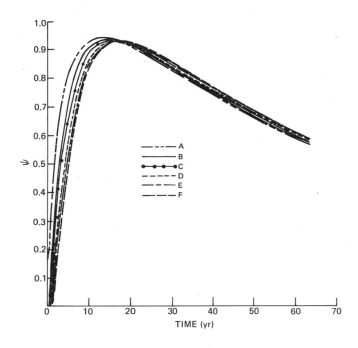

FIGURE B.9 Instantaneous CFM release
scenario: time history of total chlorine
mixing ratio (ψ) between 20 and 44 km. A
value of 1.0 would be achieved for an instantaneous release in the absence of sinks.
A, 20 km; B, 24 km; C, 28 km; D, 32 km; E,
36 km; F, 40 km.

concentrations are achieved, and t_2, the subsequent e-fold-
ing time, both determined by monitoring ClX at 30 km. We
have also determined the decay time calculated according
to Eq. (B.26) at $t = t_1$ and at $t = t_1 + t_2$. The results
are summarized in Tables B.6 and B.7. The major points
seen there are: (a) the decay times obtained theoretically from Eq. (B.26) are nearly independent of time, are
essentially independent of source time scale (i.e., the
width of the source in time), and agree reasonably with the
operational definition t_2; (b) the maximum mixing ratios
achieved remain within 10 percent of unity for time scales
less than 10 yr; (c) the time of maximum concentration
t_1 varies slowly with changing source time scale. For the
exponential growth-cutoff scenarios, this drift may be removed by adding some fraction (between 1/2 and 1) of the

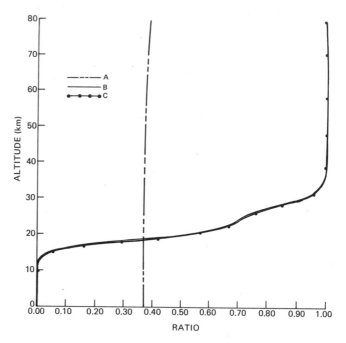

FIGURE B.10 Instantaneous CFM release scenario: A, ratio of ClX at e-fold time (t_2) to ClX at t_1, the time when peak ClX concentrations at 28 km were achieved; B, ratio of ClX to total Cl at t_1; C, ratio of ClX to total Cl at t_2.

doubling time to t_1 that is by measuring time of greatest mixing not from the cutoff time but more toward the middle of the source. For the Gaussian CFM emission scenarios, t_1 slowly increases with source width, indicating that time to maximum mixing should be measured from some time forward of the center of such a source.

Next, the dependence for an instantaneous source of CFM of t_1, t_2, and theoretical t_D on the 25 different values of K derived in the previous section are shown in Table B.8. The decay times are generally around 80-90 yr but with heavy smoothing applied to the derived K become as nearly small as 60 yr. Similarly, the time to maximum concentration is generally around 16 yr, but for the heavily smoothed K's it is about 10 yr. Maximum mixing ratios near unity are achieved in all cases.

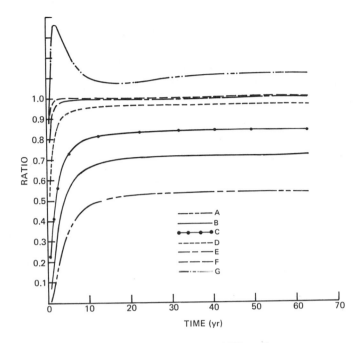

FIGURE B.11 Instantaneous CFM release
scenario: ratio of ClX to total Cl versus
time at various altitudes denoted by curves
A-F. Ratio at 32 km of approximate ψ to ψ,
using Eqs. (B.28) and (B.30) denoted by
curve G. A, 20 km; B, 24 km; C, 28 km; D,
32 km; E, 36 km; F, 40 km.

The extremely unrealistic example (Case A) of constant
$K = 3.8$ m^2 sec^{-1} is seen to give at most a factor of three
smaller time scales than those derived from the lightly
smoothed methane data. The characteristic time scales
appear to be largely controlled by the nature of the region of minimum K below 30 km. In particular, large uncertainties as to the maximum value of K as suggested by
F in Figure B.1 and D in Figure B.4 are found to be relatively unimportant as is the magnitude of K above 30 km.
The heavy smoothing condition was seen in Figures B.4-
B.6 to require what are probably excessive departures between the observed and theoretical methane profiles. We
conclude from the numerical results described here that
uncertainties in the measured methane profile and in the
fitting procedure employed can introduce errors no higher

TABLE B.6 Maximum Mixing Ratios in the Stratosphere near 30 km and Characteristic Times Calculated for CFM Emission with Gaussian Time Dependence of the Source and Various Time Scales as Defined in the Text[a]

Source Width (yr)	Max ψ	t_1	t_2	$t_D(t_1)$	$t_D(t_1 + t_2)$
1	0.98	17.3	98.8	95.8	100.5
3	0.97	17.9	98.8	95.8	100.5
5	0.96	19.0	98.8	95.8	100.5
7	0.94	20.7	98.8	95.8	100.5
9	0.92	22.6	99.2	95.8	100.5
11	0.90	24.5	99.6	95.8	100.5

[a] t_1 is the time measured from middle of source required to achieve maximum mixing ratios of ClX at 28 km; $t_1 + t_2$ is the time at which ClX has decayed to exp (-1) of its maximum value at that level; the t_D's are the analytically determined decay time evaluated at those times.

TABLE B.7 Same as Table B.6 but for Exponentially Growing Sources Cut off at $t = 0$

Width (yr)	Max ψ	t_1	t_2	$t_D(t_1)$	$t_D(t_1 + t_2)$
1	0.985	16.0	98.8	95.8	100.5
3	0.97	14.1	98.9	95.8	100.5
5	0.955	12.8	98.9	95.8	100.5
7	0.94	11.9	98.9	95.8	100.5
9	0.93	11.1	98.9	95.8	100.5
11	0.92	10.5	98.9	95.8	100.5

than 30 percent provided an optimum technique is employed. Both these uncertainties appear to make a comparable contribution to the total uncertainties.

In Table B.9, we show the sensitivity of the various characteristic times to relative changes in solar fluxes and CFM cross sections. The main point to be noted is that the times are rather insensitive to errors in solar fluxes and CFM cross sections. Rowland and Molina (1975) carried out a similar study and found a somewhat greater sensitivity to such changes.

In Table B.10, we show the sensitivity of various characteristic times to the approximately twofold increases or decreases of $K(z)$ resulting from changes in methane destruction rates by a factor of 2 in our derivation of $K(z)$. (The K's in the troposphere retain their previously assumed values in this calculation.) Adjusting methane destruction rates by the same fractional amount at all levels over a range of a factor of 4 leads to a factor of 3 change in the characteristic time scales of the CFM problem. There would be as much as a factor of 3 uncertainty in the methane lifetimes if the reaction with HO alone were responsible for the destruction of methane. The reactions of methane with $O(^1D)$ and Cl become more important than its reaction with HO (in the crucial region of 20-40 km) for smaller HO concentrations. Unfortunately, these rates are not much better known. Additional uncertainties result from the uncertainty in the concentration of $O(^1D)$ and the dependence of HO and $O(^1D)$ on somewhat uncertain solar fluxes. Uncertainty in the reaction rate of HO + CH_4 also introduces some uncertainty. Since it is highly temperature-dependent, errors in the appropriate temperature profile also introduce some uncertainty. In all the above processes, yet further uncertainty results from the global averaging inherent in 1-D models. On the basis of all of these mentioned uncertainties, we estimate the total range of the uncertainty in CH_4 lifetimes to be approximately a factor of 2 to 3.

During the latter stages of preparing this report, we revised our estimates of the CFM lifetimes to incorporate most recent data as follows. Eddy-mixing coefficients were derived independently from both CH_4 and N_2O data, using the latest available version of Chang's model without $ClONO_2$. The CH_4 lifetimes from the version of Chang'e model available for this calculation are if anything too short, for they include the fast (4×10^{-10}) $O(^1D)$ rate and a significant contribution from Cl.* On the other hand, for reasons discussed earlier, the methane data over Texas are likely to drop off more rapidly than the global mean and so give lifetimes that are too long. There is no known possible bias in the N_2O data, but random effects are expected to be larger because of the limited number of samples and the lack of data above 30 km. In particular, variation of the N_2O profile above 30 km

---

*For the calculations described in Chapter 7, this rate has been reduced to 1.3×10^{-10} cm^3 sec^{-1} (see Appendix D).

TABLE B.8 Characteristic Times in Years for the Time
Evolution of an Instantaneous Source of CFM for the
Different Eddy-Mixing Coefficients Derived in the Pre-
vious Section (based on Wofsy CH_4 lifetimes)[a]

	t_1	t_2	$t_D(t_1)$	$t_D(t_1 + t_2)$
K's derived by adjusting a limited number of parameters.[b] Case defined in Table B.2				
Case A[c]	5	33	36	38
Raw data Case B	15	94	91	94
Smoothed data Case B	15	89	90	95
Stretched data Case B	13	83	80	83
Raw data Case C[c]	15	92	88	92
Smoothed data Case C	15	91	95	95
Stretched data Case C	13	81	78	81
Raw data Case D[c]	12	91	78	80
Smoothed data Case D	15	d	87	90
Stretched data Case D	11	68	66	68
Raw data Case E	14	78	86	90
Smoothed data Case E	14	78	88	92
Stretched data Case E	11	69	73	76
Raw data Case F[c]	19	107	103	108
Smoothed data Case F	17	99	93	98
Stretched data Case F	13	84	76	78

K's adjusted at all levels with smoothing criterion [b]

Smoothing
factor

		t_1	t_2	$t_D(t_1)$	$t_D(t_1 + t_2)$
	Raw data	9	56	59	61
$\lambda = 10$	Smoothed data	11	70	72	75
	Stretched data	10	60	62	65
	Raw data[c]	17	93	94	99
$\lambda = 0.1$	Smoothed data[c]	16	90	93	97
	Stretched data	12	75	78	81
	Raw data	17	85	99	106
$\lambda = 0.001$	Smoothed data[c]	17	100	93	97
	Stretched data	12	74	76	79

TABLE B.9 Dependence of Characteristic Times on Variation of Solar Flux or CFM Cross Sections for an Instantaneous Release of CFM[a]

Perturbation in Solar Flux (or Cross Section)	Perturbation to t_1	Perturbation of $t_D(t_1 + t_2)$
-0.6	0.10	0.19
-0.2	0.02	0.04
0.2	-0.02	-0.03
0.6	-0.04	-0.08

[a]Values given are fractional increase (or decrease) over the standard case.

TABLE B.10 Fractional Change in CFM Time Scales from Standard Case G ($\lambda = 0.1$) Results for K's Derived Again for Same Case but with Methane Destruction Rates Halved and Doubled

	Fractional Change of Time Scale			
	t_1	t_2	$t_D(t_1)$	$t_D(t_1 + t_2)$
Methane destruction rate half as fast	2.0	1.7	1.8	1.8
Methane destruction rate twice as fast	0.5	0.6	0.6	0.6

[a]These are: t_1 = time required to achieve mixing ratio of ClX at 28 km; t_2 = time from t_1 for factor of e decay of ClX at 28 km; $t_D(t_1)$ and $t_D(t_1 + t_2)$ = decay time according to Eq. (B.26) with $f = \psi_{CFM}/\psi$ evaluated at t_1 and $t_1 + t_2$, respectively.
[b]Constant $K = 3.8$ m^2 sec^{-2}.
[c]Resolution of 0.02 used for this run.
[d]Doubtful results because of inadequate resolution.

within a likely range of uncertainty results in large variations in the inferred K profile. On the basis of some limited sensitivity calculations, we estimate a factor of 2 to 3 range of uncertainty in the transport time scales that can currently be inferred from N_2O measurements for this reason alone.

The CH_4 and N_2O derived transport would appear to give equally likely estimates of the actual transport. The eddy-mixing coefficients so derived for moderate smoothing are given in Table B.5. These coefficients were used with our hypothetical CFM to derive our t_1 and t_2 time scales. Additional calculations were made for F-11 and F-12. These indicate that the hypothetical CFM used in most of this appendix has lifetimes closer to F-11 than to F-12. Methane lifetimes provided by Cicerone (1976) for the Michigan chemical model were used to derive yet other estimates of the eddy-mixing profile and CFM transport. The t_1 and t_2 time scales derived from all these calculations are shown in Table B.11. The Chang CH_4 lifetimes give the fastest transport, the N_2O data the slowest transport, and the Cicerone CH_4 lifetimes intermediate values. Cicerone's CH_4 lifetimes are longer than Chang's, presumably in part because he uses the slow (1.3×10^{-10}) rate for $O(^1D)$ and does not include the reactions with chlorine.

On the basis of these and the earlier calculations, and the uncertainties in the procedure, we have determined the likely lifetimes and uncertainty ranges given in Table 5.1. Our upper limits correspond approximately to the time scales inferred from the Texas CH_4 profile with Wofsy's lifetimes. Our lower limits correspond nearly to the time scales that would be obtained if the NOAA equatorial N_2O data were taken as representative of global mean or if our global mean profile is assumed below 30 km but copious amounts of N_2O are assumed at higher levels, e.g., 0.4 ppm at 50 km a factor of 4 larger than the measured upper limit. In particular, our hypothetical CFM is given a $t_2 = $ 40-yr lower limit, whereas a calculation of K using the NOAA equatorial profile continued smoothly to higher altitudes and medium smoothing gave $t_2 = 32.3$ yr.

Several steps could be taken to reduce further the uncertainties due to transport in the chemical models predicting ozone depletion from CFMs. First, the chemical species data used to infer transport could be improved. The usefulness of methane data is somewhat limited by uncertainties in methane chemical lifetimes, due especially to uncertainties in HO concentrations. High-latitude and tropical data, however, would be helpful in ascertaining

TABLE B.11 Characteristic Times in Years for the Time Evolution of an Instantaneous Source of CFM Using Eddy-Mixing Coefficients from Most Recent Models and Data[a]

		t_1	t_2	t_D
CH₄ over Texas, CH₄ lifetimes from Chang				
F-11	$\lambda = 10$	5.09	31.1	33-34
	$\lambda = 0.1$	6.95	39.8	41-43
F-12	$\lambda = 10$	6.3	53.4	57-58
	$\lambda = 0.1$	8.5	61.5	65-67
Hypothetical	$\lambda = 10$	5.3	35.7	38-39
Appendix B				
CFM	$\lambda = 0.1$	7.3	42.6	45-47
CH₄ over Texas, CH₄ lifetimes from Cicerone				
F-11	$\lambda = 10$	6.4	38.3	40-41
	$\lambda = 0.1$	10.3	55.3	57-60
F-12	$\lambda = 10$	8.3	70.3	74-76
	$\lambda = 0.1$	12.6	87.0	91-94
Hypothetical	$\lambda = 10$	6.8	44.1	46-48
Appendix B				
CFM	$\lambda = 0.1$	10.7	60.5	63-66
N₂O global mean (based on Ehhalt and NOAA data)				
F-11	$\lambda = 10$	7.75	45.3	46-48
	$\lambda = 0.1$	12.5	68.0	70-74
F-12	$\lambda = 10$	9.9	82.7	87-89
	$\lambda = 0.1$	14.5	96.3	100-104
Hypothetical	$\lambda = 10$	8.2	51.9	54-56
Appendix B	$\lambda = 0.1$	12.9	73.9	77-81
CFM				
N₂O equatorial profile				
Hypothetical	$\lambda = 10$	4.4	29.5	31-32
Appendix B				
CFM	$\lambda = 0.1$	5.2	32.3	34-35

[a] The times are defined as in Table B.8. The K's are adjusted at all levels to give best fits within the heavy and medium smoothing constraints.

the representativeness of the existing data. Nitrous oxide should, ultimately, be more valuable than methane for determining transport because its stratospheric lifetime is more readily calculated. After global mean concentrations of CH_4 and N_2O can be satisfactorily estimated, their latitudinal and seasonal fluctuations should be ascertained and used in the development of latitudinally varying 2-D models. Even the ultimate refinements of the empirical models will have drawbacks and especially will not be able to predict large changes from present conditions properly. Physical and statistical models determining the motions responsible for transport from first principles represent the only potentially completely reliable approach.

The reader should again be reminded that the analysis of this appendix was not primarily intended to establish "best estimates" of the time history of CFMs but rather to determine the degree of uncertainty in state-of-the-art models due to uncertainties in the parameterization of transport. Furthermore, we have considered only the role of uncertainties in the transport of chlorine species. In reality, the 1-D models for ozone photochemistry involve transport of many other species, in particular that of NO_x and ozone as well. These transport processes have implicitly been assumed fixed in our analysis. Studies with the transport of all species varied are described in Chapter 7. The most notable differences reported there from the results of this appendix are some dependence on transport of the maximum ozone reduction after CFM cessation in contrast to the extreme insensitivity inferred here and less sensitivity of the final steady-state ozone reduction to transport than inferred from the analysis of this appendix. These differences are presumably due primarily to variations of the model NO_x concentrations with variations in transport.

In conclusion, the series of calculations that have been carried out indicate the degree to which various CFM scenarios can be approximated by instantaneous releases. Within such an approximation, the maximum effect of ClX on ozone is independent of transport. Only the time at which the maximum effect is achieved and the subsequent decay time depend on transport, but these can evidently be estimated to within a factor of 2 or 3 as indicated by calculation with the wide range of eddy-mixing coefficients derived in the previous section. The inclusion of other refinements such as increasing the dimensionality of the model would not be expected to give results outside the present range of possible results. The maximum effect of CFMs on ozone that can be achieved for steady-state emission is directly proportional to decay time and so has the same degree of uncertainty due to questions regarding transport as we found for instantaneous release. The concentrations of ClX during periods of rapidly increasing emission can be estimated with Eqs. (B.28) and (B.29) by assuming that all the emission occurred instantaneously at an earlier time, earlier by a period lying between the doubling time of the emission and half that amount.

REFERENCES

CIAP Monograph 3. 1975. *The Stratosphere Perturbed by Propulsion Effluents*. A. J. Grobecker, ed. Final report, Dept. of Transportation, DOT-TST-75-53, Washington, D.C.

Cicerone, R. J. 1976. University of Michigan. Personal communication.

Climatic Impact Committee. 1975. *Environmental Impact of Stratospheric Flight: Biological and Climatic Effects of Aircraft Emissions in the Stratosphere*. National Academy of Sciences, Washington, D.C.

Ehhalt, D. H., L. E. Heidt, R. H. Lueb, and N. Roper. 1974. Vertical profiles of CH_4, H_2, CO, N_2O, and CO_2 in the stratosphere, *Proc. Third Conf. on the Climatic Impact Assessment Program*, U.S. Dept. of Transportation, DOT-TSC-OST-74-15, pp. 153-160.

Mahlman, J. D. 1976. Some fundamental limitations of simplified transport models as implied by results from a three-dimensional, general-circulation/tracer model. To appear in *Proc. 4th CIAP Conference*.

Marquardt, D. W. 1963. An algorithm for least-squares estimation of nonlinear parameters, *J. Soc. Indust. Appl. Math.* 11:431-441.

Rowland, R. S., and R. J. Molina. 1975. Chlorofluoromethanes in the environment, *Rev. Geophys. Space Phys.* 13:1-35.

Schiff, H. I. 1976. York University. Personal communication.

Wofsy, S. C., and M. B. McElroy. 1973. Vertical mixing in the upper stratosphere and lower mesosphere, *J. Geophys. Res.* 78:2619-2624.

APPENDIX
C

DETAILS OF

ATMOSPHERIC MEASUREMENTS

I. INTRODUCTION

Measurements of atmospheric trace species are important
for several reasons. First, in some cases (e.g., CH_4
and N_2O) they provide the basis upon which the vertical
mixing coefficients used in one-dimensional (1-D) models
(or other models) are chosen. Second, measurements of
other species provide the means by which model calculations are validated. Finally, the results of some measurements are used directly as input to the calculations.
This appendix discusses the measurements, and the experimental techniques used in making those measurements, of
various atmospheric species relevant to our understanding
of the stratospheric impact of halogenated compounds.

II. OZONE AND ATOMIC OXYGEN

A. Ozone

The standard instrument for observation of atmospheric
ozone is the Dobson spectrophotometer. This instrument
is a specialized double-beam monochromator that measures
the ratio of the intensity in solar ultraviolet radiation
at two wavelengths. The wavelength pair is selected so
that one wavelength is absorbed strongly by atmospheric
O_3. The intensity ratio can be used to estimate the total
amount of O_3 in the optical path from the sun to the instrument. The Dobson instrument can also be operated in the

Umkehr mode, which measures scattered radiation and provides information on the vertical distribution of O_3 in
the stratosphere (Mateer and Dütsch, 1964; cf., Craig,
1965). If carefully maintained and calibrated, the
Dobson instrument is capable of measuring O_3 with an accuracy of a few percent in the direct-sun mode (Thomas
et al., 1974). About 60 Dobson instruments are distributed throughout the world. Gustin filter ozonometers,
which measure the ratio of a wavelength pair through
optical filters, are also widely used within the global
network, especially in the Soviet Union. However, they
are less reliable (Bojkov, 1969) than the standard Dobson
spectrophotometer (Dütsch, 1974). In addition to the
Umkehr method, which offers only limited vertical resolution, balloonborne electrochemical ozone sensors and
rocketborne optical sensors have been used to derive the
overall distribution in the stratosphere. Various types
of electrochemical *in situ* sensors have been used in the
past. Thomas *et al.* (1974) in an error analysis of measurements from an electrochemical detector described by
Komhyr and Harris (1971) estimate an accuracy of about 5
nbar or about 2-3 percent of the partial pressure of the
ozone maximum. They point out, however, that this error
analysis does not include possible errors in the cell reaction equation.

The major difficulty in determining the average total
O_3 content is, however, not caused by instrumental uncertainties but rather by the large natural fluctuations in
the O_3 content of the stratosphere, which occur on daily,
seasonal, and apparently also long-term, time scales.
Pittock (1974) has estimated that for an ideal network of
stations in order to establish a trend in the O_3 content
of 1 percent per decade to a 95 percent confidence level
it would take about 15 years of observation of global
ozone. Details on the observed trends can be found in the
IMOS (1975) and Climatic Impact Committee (1975) reports
(see also Chapter 9). Single O_3 soundings reveal a general
maximum in the O_3 concentration at 15-25 km, which, however, is often richly structured and broken up into layers,
especially at midlatitudes (Figure C.1, curve b). This
is caused by the flow pattern of air in the stratosphere,
which tends to transport O_3 quasi-horizontally in a strat-
ified manner from the main production region in the tropics.
If averaged, the fine structure disappears and a broad
maximum becomes the outstanding feature of the mean vertical O_3 profile at all latitudes (Figure C.2). These mean
profiles for different latitudes may be presented as a

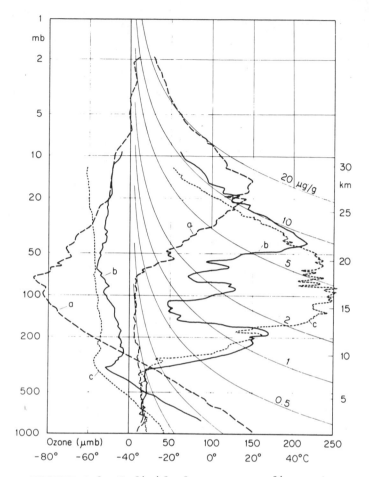

FIGURE C.1 Individual ozone soundings at different latitudes (the three right-hand curves, micromillibar scale). Simultaneous temperature profiles (left-hand curves, °C scale) are also given. Typically the equatorial distribution of ozone (curve a, Canal Zone, 9° N) is relatively smooth; the greatest variability with altitude is found at midlatitude (curve b); arctic soundings may show pronounced secondary maxima in addition to the low level single peak curve, c (from Dütsch, 1974).

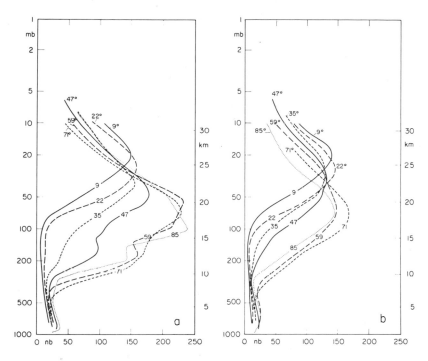

FIGURE C.2 Mean vertical ozone distributions at different
latitudes in spring (March/April, curves a at the left)
and fall (October/November, curves b at the right) (from
Dütsch, 1974). The concentration is given in nanobars.

"meridional cross section," a contour map of O_3 concentration versus latitude and altitude (in millibars pressure),
as done in Figure C.3 for the March/April data. The O_3
column above any latitude can be derived from that figure
by vertical integration. The total O_3 in a column has a
considerable seasonal variation, which depends on the
latitude (Figure C.4). Drawing a horizontal line through
Figure C.4 we can deduce the seasonal variation of total
ozone at any given latitude. It is obviously highest at
high latitudes. From a vertical line we can derive the
latitudinal dependence of total O_3 at any given month.
It turns out that the O_3 column is greatest at high latitudes at all times but especially during late winter and
early spring.

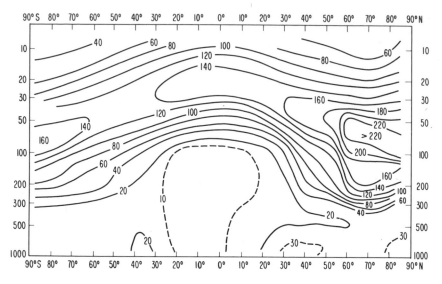

FIGURE C.3   Contour map of ozone concentration measured
in March/April for different altitudes and latitudes.
The altitude is in millibars.   The ozone isopleths are
given in units of nanobars (from Dütsch, 1974).

B.   Atomic Oxygen

Ground-state atomic oxygen $O(^3P)$ is both a precursor and
photolytic product of O_3; $O(^3P)$ and O_3 rapidly interconvert
in the sunlit atmosphere. $O(^3P)$ has been recently measured
in the stratosphere using a resonance fluorescence technique (Anderson, 1975).   The instrument was carried by a
balloon, and measurements were made during its descent on
a parachute.   The signal contributed from the resonance
fluorescence of $O(^3P)$ significantly exceeds that of background and allows one to assign an uncertainty of about
±30 percent to the measurements down to 29 km.   Below this
altitude the error increases to ±60 percent as O begins
to recombine appreciably with the O_2 within the instrument
flow system.   In Figure C.5 (taken from Anderson, 1975)
the data points represent average values obtained by integration over a 500 m above 30 km and over 1 km below 30 km.
The $O(^3P)$ mixing ratio depends on the O_3 concentration,
and it should be more variable than O_3 itself.

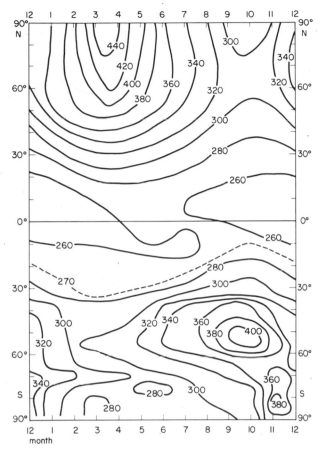

FIGURE C.4 Total ozone column as function of season and latitude (isopleths labeled in Dobson units $\equiv 10^{-3}$ cm ozone at normal temperature and pressure). Although the main features are presumably correct, there is still some uncertainty on details with respect to the southern hemisphere; the coverage is very poor between 50 and 60° S (from Dütsch, 1974).

III. NITROGEN COMPOUNDS

During the last 4 yr it has become clear that a catalytic cycle involving the nitrogen oxides NO and NO_2 is responsible for about 70 percent of the natural destruction of the stratospheric O_3. Thus, the measurement

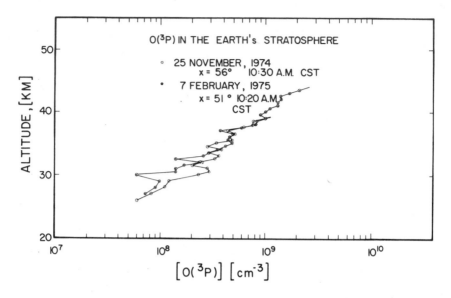

FIGURE C.5 Vertical concentration profile of ground-state atomic oxygen O(^{3}P) (from Anderson, 1975).

of nitrogen oxides, as well as that of the nitric acid product, HNO$_3$, and the nitrous oxide precursor, N$_2$O, are of primary importance. Most of the data are of recent origin, having been accumulated in connection with the Climatic Impact Assessment Program (CIAP).

A. NO

The vertical distribution of NO in the stratosphere is given in Figure C.6. *In situ* measurements of NO have been made using the chemiluminescence technique (Ridley *et al.*, 1974, 1975). Since on-board calibration of the instrument is employed and tests have been made to ensure the absence of serious sampling problems, the results should be reliable. The major objection that can be raised is that the method is not entirely specific. The spectrum of the NO—O$_3$ chemiluminescence is continuous; therefore, any other atmospheric species capable of producing chemiluminescence with O$_3$ in the spectral region viewed by the instrument may contribute to the signal. Although no such interfering species are likely to be present in the stratosphere, the method must, strictly

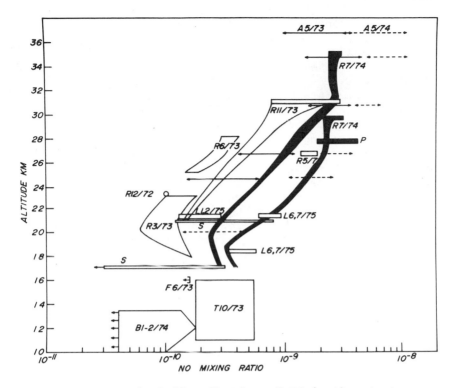

FIGURE C.6 Vertical distribution of NO in the stratosphere.

B1-2/74, Briehl *et al.* (1975); 25-37° N; chemiluminescence.
T10/73, Toth *et al.* (1973); 43-51° N; solar absorption.
S, Lowenstein *et al.* (1974); 25-49° N; chemiluminescence.
L6,7/75, Lowenstein *et al.* (1975); 25-49° N; chemilumines-
cence.
A5/73, Ackerman *et al.* (1973a); 44° N; solar absorption.
A5/74, Ackerman *et al.* (1974); 44° N; solar absorption.
P, Patel *et al.* (1974), Burkhardt *et al.* (1975); 33° N;
spin-flip laser absorption.
F6/73, Fontanella *et al.* (1974); 43-51° N; solar absorption.
R12/72, R3/73, Ridley *et al.* (1973, 1974); 33° N, chemilumi-
nescence.
R6/73, R11/73, Ridley *et al.* (1975); 33° N; chemilumines-
cence.
R7/74, Ridley *et al.* (1976a); 58° N; chemiluminescence.
R5/75, Ridley *et al.* (1976b); 33° N; chemiluminescence.

speaking, be considered to provide an upper limit for NO. The authors estimate the error to be ±50 percent. The spin-flip laser resonance method is a specific *in situ* technique, which should be capable of considerable accuracy (Patel *et al.*, 1974; Burkhardt *et al.*, 1975; Patel, 1976). Relatively few data have been presented to date with this technique, and there is no indication that on-board calibrations were made or that tests were made to ensure the absence of sampling errors. The estimated errors in the NO measurements by this method are about +20-30 percent (from the error bars in Figure C.6).

NO has also been measured using balloonborne infrared absorption technique (Ackerman *et al.*, 1973a). To obtain the required sensitivity, these measurements were made during twilight conditions for which long optical path lengths are available. Since the NO concentrations change rapidly during that time (cf. Figures C.7 and C.8), complications in the interpretation are to be expected.

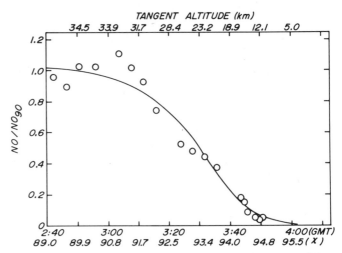

FIGURE C.7 The sunset decay of nitric oxide versus GMT and solar zenith angle, χ, at an altitude of 34.5 ± 0.5 km at 58° N, 95° W, on July 22, 1974. The circles represent the ratio of the NO mixing ratio to the average value of 2.7 ppbv measured under high sun conditions ($\chi < 90°$). Data from Ridley *et al.* (1976a). The smooth curve is the theoretical behavior calculated from a time-dependent 1-D model. Similar diurnal effects were shown by Patel *et al.* (1974).

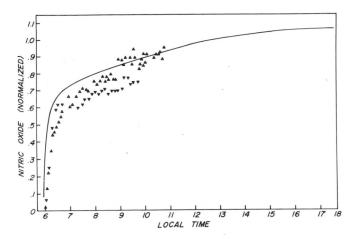

FIGURE C.8 The increase in nitric oxide following sunrise. Measurements were taken on May 15, 1975, at 33° N, 106° W at an altitude of 26 km. The NO mixing ratios are normalized to the noontime value. Data from Ridley *et al.* (1976b). The smooth curve is the theoretical behavior calculated from a time-dependent 1-D model.

Also, a comparison of *in situ* and remote measurements is of doubtful value. The vertical distribution of NO from these measurements is that given in Figure C.6. The volume mixing ratio curves for NO all show the same general behavior, viz., the mixing ratio increases with height in the 24- to 30-km region and then remains essentially constant up to 40 km. However, values reported with the same technique at the same location show considerable long-term variability. For example, the remote-sensing results of Ackerman *et al.* (1975), exactly 1 yr apart, differ by a factor of 5, while the *in situ* measurements of Ridley *et al.* (1975) with the same instrument at the same location have a spread of a factor of 7 at some altitudes. Lowenstein *et al.* (1975) conclude from their *in situ* measurements from U-2 aircraft that there is a seasonal effect, with NO being less in the winter.

There are also short-term fluctuations. Thus, Ridley *et al.* (1975) find changes up to a factor of 4 at a balloon float altitude of 30 km over a period of 3 hours for one flight (of six) and smaller changes for the other flights. Differences of a factor of 2 were observed at most altitudes between two flights made 6 days apart. Fluctuations of factors of 5 have been observed over a distance of 100 km during aircraft measurements. These fluctuations, observed with the same instrument, point to a large natural variability in the stratospheric NO concentrations.

A good part of the variability is caused by the fact that NO and NO_2 interconvert very rapidly on a time scale of about 100 sec in the stratosphere. The partitioning of "odd nitrogen" between NO and NO_2 is very sensitive to the prevailing O_3 concentration, the temperature, and the flux of solar radiation. Variation of any of these three parameters will cause fluctuations in the NO concentration (or NO_2 concentration). The dependence on solar flux is demonstrated in Figures C.7 and C.8. During sunset, solar radiation diminishes to zero, conversion of NO_2 to NO ceases, and the NO/NO_2 ratio becomes a sensitive function of time decreasing to zero. During sunrise, the ratio shows a rapid increase. Measurements of this kind are particularly valuable for testing models. Since they are made within a short time span, the measured time variation in the concentration becomes relatively independent of transport. In addition, chemical parameters like the concentration of O_3, HNO_3, and NO_x and the temperature should not change appreciably. Thus, since the temporal variation of NO is mainly determined by chemical reactions and the change in solar flux, a well-defined test situation is obtained.

B. NO_2

NO_2 so far has only been measured by remote-sensing techniques. The vertical profile shown in Figure C.9 shows an increase in the NO_2 mixing ratio in the lower stratosphere with a maximum around 28 km and a decrease above. Large fluctuations have been observed in the profiles, and Noxon (1975) found variations of more than a factor of 2 in the stratospheric NO_2 between successive nights. Fluctuations like this are probably not only due to chemistry but also to large-scale horizontal mixing. Since the formation of $ClONO_2$ is directly related to the NO_2 concentrations, the uncertainties in the NO_2 height profiles have taken on additional significance. It is likely that the average global height profile of N_2O as defined directly by observation is uncertain by a factor of 5. However, with the additional constraints gained by measurement of other species coupled with NO_2 by chemical reactions,

the NO_2 profile becomes significantly more certain,
i.e., within a factor of 2-3.

C. HNO_3

HNO_3 is formed in the stratosphere by reaction of NO_2
with the HO radical. It is converted back to NO_2 by
photodissociation and reaction with HO. Since that conversion is relatively slow, its chemical lifetime is about
5 weeks, and HNO_3 provides, so to speak, a holding reservoir for NO_x. *In situ* sampling of HNO_3 has been made
using filter-paper collection and subsequent analysis
(Lazrus and Gandrud, 1974). Although various tests have
been performed, the collection efficiency for these filter papers has not been unequivocally established. The
lowest values for HNO_3 are obtained from *in situ* filter-
paper measurements. This may be due to a collection efficiency less than unity. In addition, there are infrared
measurements both in absorption and in emission by Murcray
et al. (1973). The spread in the HNO_3 measurements shown
in Figure C.10 for this technique is considerably greater
than the 30 percent accuracy claimed by the various groups.
The data in Figure C.10 indicate an increase in the mixing
ratio in the lower stratosphere, which reaches a maximum
around 20 km and rapidly decreases above 20 km.

D. *Simultaneous Measurements of NO_x*

Simultaneous measurement of NO, NO_2, and possibly HNO_3
would eliminate some of the uncertainties due to the rapid interconversion between NO and NO_2. If, in addition,
O_3 and the temperature (T) could be measured simultaneously, the expected partitioning between NO and NO_2 could be
calculated and compared with the measured ratio. This
would provide a sensitive test of the chemical modeling.
Ackerman *et al.* (1975) reported simultaneous NO and NO_2
measurements but did not report measurements of T and O_3.
A joint balloon experiment coordinated by the Atmospheric
Environment Services of Canada provided *in situ* measurements of T, O_3, and NO and remote-sensing measurements
of NO_2 and HNO_3 by absorption and emission, respectively.
Not all the measurements were made at precisely the same
time of day and a time-dependent 1-D model was used to
compare the results. The calculated values agreed with
the measurements within 50 percent. With allowances for

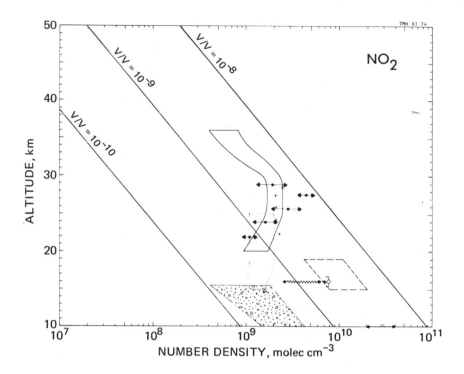

FIGURE C.9 Stratospheric NO$_2$ measurements.

⌐ Ackerman *et al.* (1974); balloon 44° N; sunset V 74;
solar absorption at 1597 and 1600 cm^{-1}; grille spectrometer; authors' uncertainty in measuring equivalent widths as amplified by inversion.

○ Ackerman and Muller (1972a, 1973); balloon 44° N; X 70;
2850-2925 cm^{-1} solar absorption; grating spectrometer;
authors' uncertainty at 12.5 km, upper limit at 16.1
km.

● Ackerman and Muller (1972a, 1973); based on spectra of
Goldman *et al.* (1970); balloon 33° N; sunset XII 67;
1612-1616 cm^{-1} solar absorption; grating spectrometer; authors' uncertainty.

□ Brewer *et al.* (1973); aircraft 44° N; VIII 73; 430-450
nm absorption in scattered sunlight; noon value with
authors' uncertainty.

◁ Brewer *et al.* (1973, 1974); aircraft 43-51° N; sunrise and sunsets VI, X, XI 73; 430-450 nm absorption
in scattered sunlight; envelope of several profiles.

⌁ Farmer *et al.* (1974); aircraft 43-51° N; sunsets VI,
IX, X 73; 2850-2925 cm^{-1} solar absorption; interferometer; authors' uncertainty.

the fact that the comparison involves a mix of *in situ* and remote-sensing techniques, this agreement is indicative of the uncertainties associated with the use of 1-D models, being well within the limits of accuracy discussed in Chapter 8 for the predicted reduction in stratospheric ozone.

E. N_2O

N_2O provides the major natural source of NO_x in the stratosphere. It is produced at the earth's surface by microbial activity and is oxidized in the lower stratosphere to form NO. A number of stratospheric measurements using different techniques have been published for N_2O. The results are shown in Figures C.11(a) and C.11(b). The oldest measurements are those of Schütz *et al.* (1970), who collected N_2O in a 5 Å molecular sieve by sucking known amounts of air through an absorption train. The amount of N_2O was determined by gas chromatography. For tropospheric samples with an N_2O mixing ratio of 0.2 ppm, a standard deviation of ±4.3 percent (or about ±0.01 ppm) could be obtained. Their stratospheric collection system provided smaller air samples, and the error should be somewhat bigger, possibly by a factor of 2. Their data [indicated by open squares in Figure C.11(a)] were collected on different flights in 1967 and 1968 over Germany at about 50° N. Murcray *et al.* (1972) measured an N_2O profile with a balloonborne infrared spectrometer by observing the atmospheric absorption of solar radiation at 2175-2250 cm^{-1}. The profile was obtained on January 18, 1972, over Alamogordo, New Mexico (about 32° N). The authors do not quote any

---

Fontanella *et al.* (1974); aircraft 43-51° N; sunset VII 73; 1604-1606 cm^{-1} solar absorption; interferometer; authors' uncertainty.

Harries *et al.* (1974a); aircraft 65-75° N; V 73; thermal emission at 38 cm^{-1}; interferometer; authors' uncertainty.

x Murcray *et al.* (1974); based on spectra of Goldman *et al.* (1970) and Murcray *et al.* (1969); balloon 33° N; sunset XII 67; 1604-1616 cm^{-1} solar absorption; grating spectrometer.

△ Murcray *et al.* (1974); based on spectra of Ackerman *et al.* (1973b); aircraft 43-51° N; sunset VII 73; 1604-1616 cm^{-1} solar absorption; grille spectrometer.

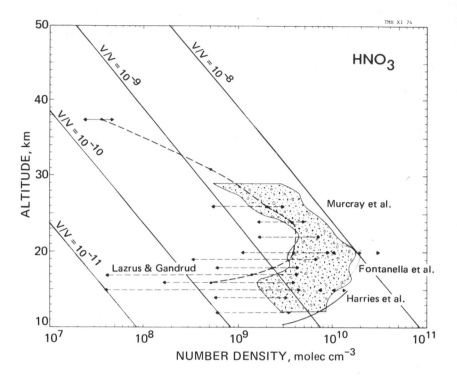

FIGURE C.10 Stratospheric HNO$_3$ measurements.

•••• Fontanella *et al.* (1974); aircraft 43-51° N; VII 73;
solar absorption at 1326 cm^{-1}; grille spectrometer;
authors' uncertainty.

—— Harries *et al.* (1974a, 1974b); aircraft 65-70° N; V
73; thermal emission between 9 and 31 cm^{-1}; interferometer.

–– Lazrus and Gandrud (1974); aircraft and balloon 75° N-
51° S; 71-73; paper filter capture. Dashed profile
above 16 km is average of 33° N and 34° S; IV-VI 73;
dashed arrows indicate extremes over all latitudes
and seasons.

Murcray *et al.* (1973, 1974); balloon 33° N; V-XI 70;
and 64° N; IX 71 and IX 72; 810-955 cm^{-1} thermal
emission; filter radiometer; hatched area is envelope of eight profiles. Concentrations at 64° N
are generally lower in altitude and larger than at
33° N.

errors, but, judging from the integrated absorption curves for CO and N_2O and the error for CO given below, the errors in the N_2O mixing ratio should be better than the 40 percent quoted for CO.

The measurements by Ehhalt *et al.* (1974) were made by separating CO_2 and N_2O from air samples collected cryogenically aboard a balloon and measuring the N_2O/CO_2 ratio in a mass spectrometer. The authors quote an error of ±0.01 ppm (Ehhalt *et al.*, 1975b). The profile collected on September 8, 1973, over eastern Texas (32° N) (full dots), which was strongly influenced by the peculiar wind pattern in the stratosphere on that day, shows a deep layer of constant N_2O concentration between 20 and 28 km altitude (Ehhalt *et al.*, 1975a). The open circles, which extend the balloon measurements to somewhat higher altitudes, were collected over Midland, Texas (32° N) on May 7, 1974. The full triangles represent vertical averages between 41 and 51 km altitude on May 23, 1973, and 44 to 62 km altitude on September 24, 1968. The samples were collected aboard an Aerobee rocket. At these altitudes the N_2O concentration is essentially zero with an error of ±0.007 ppm (Ehhalt *et al.*, 1975c). The most recent measurements are from Schmeltekopf *et al.* (1975) and Schmeltekopf (1976). They used a grab sampling system for sample collection and analyzed the N_2O content with a gas chromatograph equipped with an electron capture detector. They quote an error of ±0.02 ppm. The data [full squares in Figure C.11(a)] were collected on separate flights over Laramie, Wyoming (42° N) during July 1975. This technique was used to obtain height profiles over a wide latitude range [Figure C.11(b)]. These profiles show considerable dependence of the slopes of the curves on latitude. But they also show variability at the same location, which is clearly indicative of horizontal transport.

The tropospheric portion of the N_2O distribution, Figure C.11(a), is an average over six individual profiles that were collected mostly over the continental United States (Ehhalt *et al.*, 1975b). The error bars represent the standard deviation of the mean and indicate rather large natural fluctuations of N_2O in the troposphere. Large fluctuations have also been observed on a yearly basis by Schütz *et al.* (1970). On the other hand, the measurements of Schmeltekopf (1976) show remarkably similar concentrations just above and below the tropopause [Figure C.11(b)] for a wide range of latitudes. Since the values near the tropopause are used as input data

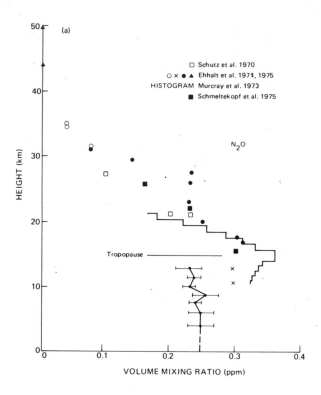

VOLUME MIXING RATIO (ppm)

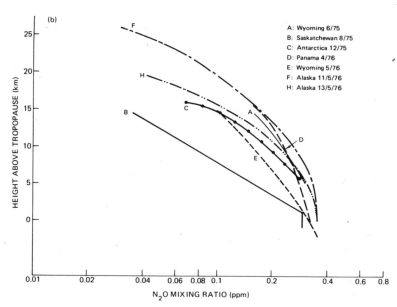

A: Wyoming 6/75
B: Saskatchewan 8/75
C: Antarctica 12/75
D: Panama 4/76
E: Wyoming 5/76
F: Alaska 11/5/76
H: Alaska 13/5/76

N_2O MIXING RATIO (ppm)

for the models in determining the stratospheric flux of NO_x, it is probably safe to say that this quantity is known to better than ±50 percent.

IV. HYDROGEN COMPOUNDS

The hydrogen compounds H_2O, CH_4, and H_2 are of interest, because they are the sources of HO and HO_2 radicals. Because of their high reactivity, the radicals are involved in many important reactions, including those that destroy O_3. Their total contribution to natural O_3 destruction is about 10 percent.

A. *Methane*

Of the hydrogen compounds mentioned above, CH_4 has gained particular importance. It is produced at the earth's surface and transported into the stratosphere. Since its chemical lifetime in the lower stratosphere is comparable with the transport time, it has been used as a tracer of vertical transport.

The CH_4 profiles have been measured using two techniques: balloonborne infrared spectrometry and air sampling with subsequent gas chromatographic analysis of the CH_4 content. There are two profiles measured by the former method: one by Ackerman and Muller (1972b) over southwestern France (about 45° N) and one by Cumming and Lowe (1973) over Quebec, Canada (45° N). Both profiles are shown in Figure C.12. Ackerman and Muller quote errors between 20 and 50 percent, which are indicated in Figure C.12. The data of Cumming and Lowe seem to have an error of about 30 percent. The profiles show a decrease of the CH_4 concentration in the stratosphere. The mean tropospheric profile appears to be a constant volume mixing ratio of 1.41 ppm (Ehhalt and Heidt, 1973b) as shown in Figure C.12. In view of

---

◄FIGURE C.11 Vertical distribution of the N_2O volume mixing ratio. (a) The stratospheric data were obtained by various authors in different years and locations (see text). They are normalized to the tropopause (assumed height 15 km) to facilitate comparison. The tropospheric profile is an average over six individual profiles (Ehhalt *et al.*, 1975b). (b) Vertical distributions of N_2O for different latitudes (Schmeltekopf, 1976).

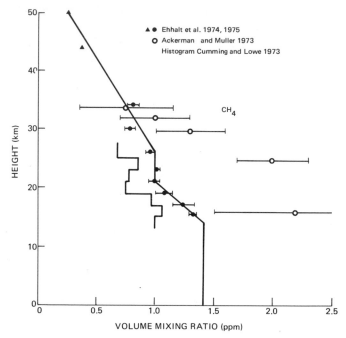

FIGURE C.12 Vertical distribution of the
CH₄ volume mixing ratio. The heavy line
and full dots represent the average of nine
profiles from nine different balloon flights
at 32° latitude. The error bars represent
the standard deviation of the data and are
mainly due to the natural variability in CH₄.
The other data are individual profiles; the
error bars represent the experimental uncertainties. The data are normalized to the
tropopause (assumed height 15 km) to facilitate comparison.

this, the CH₄ values of Ackerman and Muller appear to be
rather high in the lower stratosphere, perhaps because of
the interpretational difficulties associated with optical
techniques.

The full dots in Figure C.12 represent the profile obtained by averaging the profiles from nine balloon flights
made in different seasons. Most flights took place over
eastern Texas, 32° N (Ehhalt *et al.*, 1975a). The triangles
are results from rocket launches on September 24, 1968,
and May 23, 1973, and represent averaging between 44 and

62 km, and 41 and 51 km, respectively (Ehhalt *et al.*, 1975c). The error bars drawn for the balloon data are the standard deviations resulting from the natural variations between the individual profiles. Before 1973, a grab sampling system was used by the authors to collect 4 liters of ambient air, and a radio-frequency helium ionization detector was used in the gas chromatographic analysis of CH_4 and CO. The inherent difficulties in helium ionization detection required a long analysis time with repeated comparisons of the sample to a working standard. [Standard and sample runs were alternated four times (Heidt and Ehhalt, 1972).] The resulting error of the absolute CH_4 concentrations is stated as ±0.05 ppm. Contamination is reported to be small in the case of the balloonborne CH_4 measurements (Ehhalt and Heidt, 1973a) so that the error quoted for the analysis should represent the total experimental error.

The two rocket samples are particularly important, because the currently used profiles of the stratospheric vertical mixing coefficient rely heavily on measurements obtained from them. The sample collected on September 24, 1968, was analyzed for its CH_4 content 32 months after collection. However, that long storage apparently did not change the CH_4 concentration. Several individually kept aliquots of the sample all yielded close to 0.25 ± 0.02 ppm for the CH_4 concentration. The second rocket sample was measured in several aliquots within 2 weeks after collection and gave 0.35 ± 0.01 ppm, confirming the earlier results (see Figure C.12). No source of CH_4 contamination could be detected in the rocket sampler, and an estimate showed that at most 1.5 percent of the atmospheric CH_4 could be oxidized in the shock wave in the forward section of the rocket sampler during collection. Thus, the measured CH_4 values should closely represent the atmospheric value, and the quoted errors should represent the total experimental error. The natural variability at around 50 km is unknown. But, judging from the various data presented in Figure C.12, a standard deviation of at least ±20 percent has to be expected for a single measurement of the CH_4 concentration at any altitude because of natural variation.

The profile indicated by the solid points and line in Figure C.12 represents the latest average. Most calculations of the vertical mixing coefficient have been using a CH_4 profile published earlier, which included only seven individual profiles (Ehhalt *et al.*, 1974). The differences between the two are minor, and, for the

computation of the mixing coefficient, they are relatively
unimportant. The main drawback is the fact that the profile represents the average at one latitude, namely 32° N,
and might well reflect peculiarities that would not be observed at other latitudes. For example, the steeper decrease of the CH_4 mixing ratio immediately above the tropopause could partly be due to compression of the profile
by the average motion of air downward at 32° N. Conceivably a global average would be better represented by drawing a straight line through the data points.

B. Hydrogen

The available data on the stratospheric H_2 distribution
are summarized in Figure C.13. The full dots represent
the average of five independent profiles, mostly collected over eastern Texas (32° N) (Ehhalt et al., 1975a).
The error bars for the balloon data indicate the standard deviations due to the natural variability in the
single profiles. The triangles denote the H_2 measurements for the two rocket samples mentioned under CH_4. In
the latter case, the experimental error includes a correction for sample modification during collection (Ehhalt
et al., 1975c). The balloon flights prior to 1973 used
a grab sampling system for sample collection; in the
subsequent flights much larger samples (5 liters STP)
were collected by a cryogenic system. The samples were
analyzed by gas chromatography using a radio-frequency
helium ionization detector (Heidt and Ehhalt, 1972). The
authors quote an analysis error of ±3 percent. For the
early grab sampling flights in 1967 and 1970 and the
highest sampling altitudes (Ehhalt and Heidt, 1973a), the
contamination was considerable. For those flights 75 percent of the observed H_2 was due to contamination. The
contamination could be corrected for, but this introduced
a substantial error. Therefore, the two earliest profiles
were omitted from the average in Figure C.13. In the subsequent flights, contamination was greatly reduced; for
the cryogenically collected samples it is negligible. It
seems safe to assume an experimental error of ±5 percent
for the stratospheric H_2 data after 1970. Most of the
individual stratospheric profiles exhibit a slight maximum in the H_2 concentration varying in altitude between
25 and 30 km, which is still visible at 30 km in the average profile of Figure C.13. On the basis of an average
tropospheric H_2 concentration of 0.55 ppm, as measured in

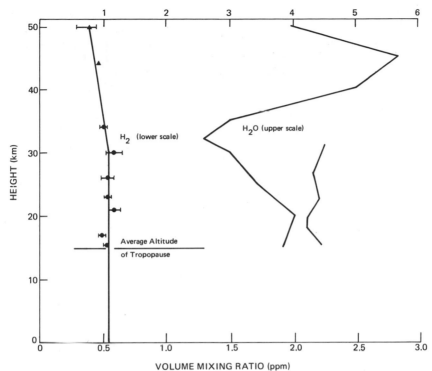

FIGURE C.13 Vertical profiles of H_2 (lower scale) and
H_2O (upper scale). The H_2 data are from Ehhalt *et al.*
(1975a) and represent an average over five balloon
flights all at 32° latitude. The H_2O profile on the
right is an average over nearly 100 individual profiles
obtained over Washington, D.C., by Mastenbrook (1974).
The line at the left represents an individual H_2O profile
obtained by Evans (1974) with a rocket-launched radiometer.

aircraft samples, a constant H_2 concentration up to 30 km
altitude has been drawn in Figure C.13.

C. H_2O

Figure C.13 also contains some of the available data on
the vertical H_2O distribution in the stratosphere. The
line on the left represents a profile obtained by Evans
(1974) with a rocket launched radiometer over Churchill
(58° N). The author quotes an error of ±1 ppm for the
higher and ±0.5 ppm for the lower altitudes. This single

profile indicates the possibility of rather large vertical variations in the stratospheric H_2O concentration. In the lower stratosphere, up to 30 km, a series of water vapor profiles has been measured on a monthly basis for 10 yr over Washington, D.C. (39° N) by Mastenbrook (1974). The mean profile (from Table 1, Mastenbrook, 1974) averaged over nearly 100 balloon flights is also shown in Figure C.13. The measurements were made by a carefully calibrated frost-point hygrometer during descent of the balloon to minimize contamination from the instrument and flight train. Since water is such a ubiquitous substance, contamination that is difficult to estimate is the most likely source of error in that measurement. It would, most likely, modify the data at the highest altitudes; below 25 km, 0.5 ppm at most should be contributed by contamination.

D. Hydroxl Radical

A few measurements also exist for the hydroxl radical, HO. The first measurements were made by Anderson (1971). He used a rocketborne spectrophotometer to measure light at 3064 Å resulting from solar radiation resonantly scattered by HO. His measurements obtained on April 22, 1971, 18:16 MST over White Sands, New Mexico (32° N) extend between 45 and 70 km altitude (Figure C.14). The errors, mostly due to calibration and uncertainties in the solar flux, are estimated to be approximately ± a factor of 2 to 3. Anderson (1976a) has made additional measurements between 30 and 43 km using the same technique previously used to detect $O(^3P)$. The results shown in Figure C.14 are from two flights launched from the National Scientific Balloon Facility, Palestine, Texas (32° N). The data point at 40 km was obtained on July 18, 1975, whereas the others were obtained on January 12, 1976. A few tropospheric measurements also exist. Wang et al. (1975) reported HO measurements during July and August 1974 near Dearborn, Michigan, made by laser-induced fluorescence. They observed a diurnal variation near the ground of between 5×10^6 and 5×10^7 molecules/cm^3, as indicated in Figure C.14. However, their detection limit is 5×10^6 cm^{-3}, and the calibration error is estimated to be about a factor of 3. Another source of uncertainty results from the fact that HO produced by the experiment itself is also observed. The amount of HO generated in the experiment depends on O_3 and HNO_2, whose concentration is

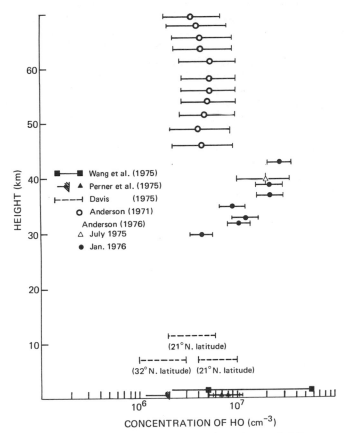

FIGURE C.14 Vertical distribution of HO concentration (in cm^{-3}). Experimental uncertainties are indicated by bars. The tropospheric measurements were both made close to the surface.

not known. Thus, the authors state that the true HO concentration could be lower than the ones observed. Further HO measurements close to the earth surface were made by Perner *et al.* (1975) using optical absorption along a 7.8-km path. The measurements were made over Julich, West Germany, during August and September 1975. The concentrations remained usually below 2×10^6, molecules/cm^3, the detection limit. Only in two instances could HO be observed. The error is estimated to be ±30 percent.

Davis (1976) has recently measured HO in the upper troposphere using laser-induced fluorescence. The

measurements were obtained during October at two sites
(32° N and 21° N) and at two altitudes (7 and 11.5 km).
After scaling their results for seasonal and diurnal effects, Davis (1976) obtained a value of 9×10^5 molecules/
cm^{-3} for the globally averaged HO concentration.

Like atomic oxygen, the concentration of HO depends on
a number of parameters that make it highly variable. Its
proper interpretation requires auxiliary measurements of
species such as H_2O, O_3, CH_4, H_2, CO, NO, and NO_2, as
well as of solar uv. Although such measurements can be
expected in the troposphere within the next year, it
will be much more difficult to extend them into the
stratosphere.

V. CARBON MONOXIDE

Carbon monoxide is of minor importance, but since it is
produced in the stratosphere during CH_4 oxidation, it
permits a useful check on the validity of the CH_4 chemistry employed in the 1-D calculations. The available
carbon monoxide data are summarized in Figure C.15. There
are, at present, three profiles that are quite different
in character. Seiler and Warneck (1972), on aircraft
flights over Europe at 50° N, observed constant values in
the upper troposphere, a rapid decrease across the tropopause (at 9 km altitude) and a much lower constant value
in the lower stratosphere. This type of profile was observed on three occasions, all associated with a cyclonic
weather situation. The authors use an airborne chemical
detection method in which the CO-containing air is passed
over hot mercuric oxide. This results in an additional
release of mercury vapor, whose concentration is monitored
optically. They claim an error of ±3 percent.

Goldman *et al.* (1973) with an infrared spectrometer
carried aboard a balloon observed a more gradual decrease
starting in the upper troposphere and continuing in the
lower stratosphere. These measurements were taken over
Alamogordo, New Mexico (32° N) in January 1972. The authors quote an error of ±40 percent for the absolute concentration and 5 to 10 percent for the concentration values
relative to each other.

The other data are from Ehhalt *et al.* (1974, 1975a).
The data were measured by collecting 5 liters STP of
air by a cryogenic sampler carried aboard a balloon. The
samples were analyzed with a gas chromatograph; after
separation the CO was converted to CH_4 over a hot nickel

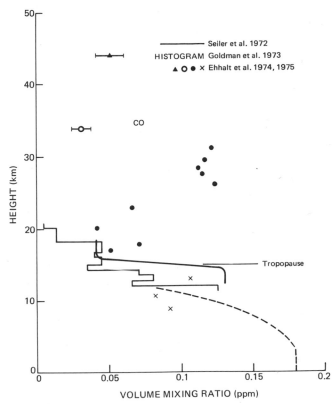

FIGURE C.15 Vertical distribution of the CO
volume mixing ratio in the stratosphere.
The stratospheric data were obtained by various authors in different years and locations (see text). They are normalized to
the tropopause (assumed height 15 km) to
facilitate comparison. The tropospheric
portion is an interpolation from a crude
meridional cross section (Seiler, 1974).

catalyst and detected by a flame ionization detector.
The measurement errors are 0.005 to 0.01 ppm. The
crosses and the full dots for data obtained on a flight
on September 9, 1973, over eastern Texas (32° N) also
indicate a decrease starting in the troposphere and
continuing into the stratosphere. This profile has a
very peculiar feature, however, in that it exhibits an
increase of 0.05 to 0.1 ppm between 22 and 26 km. This
is apparently because of the complicated stratospheric

wind pattern referred to above in the discussion of N_2O (Ehhalt *et al.*, 1975a). Measurements from a balloon on May 7, 1974, and a rocket on May 23, 1973, give low values for the upper and middle stratosphere. Apparently there are large natural latitudinal variations of CO, and it is presently difficult to draw an average CO profile from the individual experimental profiles. The vertical distribution of CO in the troposphere also depends on the latitude, mainly because the surface source distribution depends on latitude. The dashed curve in Figure C.15 represents the presumed tropospheric CO profile at 30° N. It is derived from a meridional cross section published by Seiler (1974), which, however, is somewhat speculative in nature.

VI. HALOGEN COMPOUNDS

Measurements of the halocarbons $CFCl_3$, CF_2Cl_2, CCl_4, and CH_3Cl, as well as those of other chlorine-containing compounds like HCl, are most pertinent to the present problem. Most of the stratospheric data summarized in Chapter 6 and the concentration profiles presented in Figures 6.1-6.6 became available during 1975.

Several groups have measured one or more halocarbons. Among them are included Wilkniss *et al.* (1973, 1975a, 1975b), Su and Goldberg (1973), Lillian *et al.* (1975), Hester *et al.* (1974), Zafonte *et al.* (1975), Hanst *et al.* (1975), and Williams *et al.* (1976b). Aircraft have been used for sample collection by Lovelock (1974), Hester *et al.* (1975), Krey and Lagomarsino (1975) and Rasmussen (1975), which restricts their data coverage to the lower stratosphere below about 20 km. Heidt *et al.* (1975) using a balloonborne cryogenic sampling system and Schmeltekopf *et al.* (1975) using a balloonborne grab sampling system have provided data from the middle stratosphere. While most authors used gas chromatography with an electron capture detector for the analysis of the halocarbons, some measurements have been made using an infrared detection technique. For calibration of the measured concentrations, most authors relied on standard gas mixtures. However, Lovelock has used a coulometric calibration. The reproducibility of the actual measurements with a given instrument under particular conditions (the precision) is much better than the absolute accuracy of the concentrations determined. The major difficulty is proper calibration. Krey and Lagomarsino (1975) quote a precision of better than 10 percent, while Rasmussen (1975) gives 3 percent but estimates his accuracy to be ±10 percent. Heidt et al. (1975) quote an absolute error of ±10 percent, and so do Schmeltekopf et al. (1975). All errors quoted represent the standard deviation (1σ) of the particular group's determinations. A cross calibration between all the laboratories is planned and partly completed. At the 95 percent confidence limits, the uncertainties are ±30 percent for $CFCl_3$ and CF_2Cl_2, ±50 percent for CH_3Cl, and a fourfold range for CCl_4.

A. F-11

The most extensive data are available for $CFCl_3$ (F-11). Two measurements of the latitudinal dependence of its concentration (meridional cross section) in the lower stratosphere have been made by Krey and Lagomarsino (1975), one in spring 1974 and the other in the fall; they are shown in Figure 6.1. Relatively extensive measurements also exist up to 34 km (Figure 6.2); for clarity only the average of Rasmussen's results are shown. For the same reason, Hester et al.'s values and those of Krey and Lagomarsino have been omitted; they agree well with the data points plotted. Rasmussen's flights were made during May 1975 over Alaska. Lovelock's data were obtained in June 1974 over central England. The data by Schmeltekopf et al. (1975) and Schmeltekopf (1976) were obtained on flights over Laramie, Wyoming (42° N); Yorkton, Saskatchewan (51° N); Panama (9° N); and Fairbanks, Alaska (64° N). The data from Heidt et al. also come from separate flights. The samples indicated by the open and crossed circles were collected on September 9, 1973; the full triangles represent samples from May 7, 1974; and the open squares samples from June 2, 1975. All were collected over Texas at about 32° N. The hatched area between 40 and 50 km refers to a rocket sample that was collected in that altitude interval in May 1973 and in which $CFCl_3$ could not be detected. [The sensitivity was 2×10^{-13} (Heidt et al., 1975).] These samples were analyzed during 1975, and the authors checked that the samples could be stored in stainless steel containers without deterioration for long periods of time. The profile shows a rather constant mixing ratio in the troposphere and a rapid decrease in the stratosphere.

B.   F-12

Figure 6.3 shows the data for CF_2Cl_2 (F-12). The symbols
refer to the same sampling dates as in Figure 6.2, and
the same comments apply. The electron capture detector
is somewhat less sensitive to CF_2Cl_2 than to $CFCl_3$, and
thus the upper limit for the rocket sample is somewhat
higher. At present, the limit for the detection of the
mixing ratio for CF_2Cl_2 has been improved to 5×10^{-12}.
No concentration gradient is observed in the troposphere,
and the stratospheric gradient is much smaller than that
observed for $CFCl_3$ because of the slower photodissociation of CF_2Cl_2.

C. CCl_4 and CH_3Cl

To date, the measurements for CCl_4 and CH_3Cl are restricted to the troposphere and lower stratosphere. Again,
both gases seem to be well mixed in the troposphere, and
both decrease in the stratosphere. For CCl_4, the measurements over central England by Lovelock indicate a much
steeper decrease than those over Alaska by Rasmussen
(Figure 6.4). The measurements of CH_3Cl by Rasmussen
were made over the State of Washington (Figure 6.5).
Sampling analyses and calibration errors for these compounds are much greater than for the CFMs.

D.   Hydrogen Chloride

Another chlorine-containing compound, which is quite important for the understanding of stratospheric chlorine
chemistry, is HCl (Figure 6.6). A number of stratospheric
profiles now exist for HCl. Several have been measured
by Lazrus et al. (1975, 1976) using base (alkali) impreg-
nated filters carried aboard balloons or aircraft (WB57f
or U-2) to collect HCl. The data points below 20 km
were all obtained by the aircraft. The HCl collected is
quantitatively analyzed in the laboratory using a colorimetric method. The profiles shown in Figure 6.6 represent
the measurements from flights over Alamogordo, New Mexico
(32° N) in 1974, 1975, and 1976. Another profile was measured by Farmer et al. (1976) using an infrared spectrometer carried aboard an aircraft. These data provide a

vertical profile between 14 and 21 km altitude and the total column density of HCl above 21 km (see Figure 6.6). Both profiles (Farmer and Lazrus) show an increase in the HCl concentrations from rather low values in the upper troposphere to high values at 20 km (0.5 ppb). Lazrus *et al.* (1975) showed that the contribution of chlorine from solid particles is negligible and estimate a total measurement error of ±30 percent (Lazrus *et al.*, 1976). Farmer *et al.* (1976) assign an error of 15 percent to their HCl measurement at 21 km. Two additional measurements of HCl at 20 and 30 km have been made over southwest France by Ackerman *et al.* (1976). Their results are indicated by the solid triangles on Figure 6.6. Williams *et al.* (1976a) have obtained a height profile from 13 to 30 km from balloonborne infrared measurements at Alamogordo, New Mexico (32° N). Their results are indicated by the open triangles in Figure 6.6.

Farmer *et al.* also derived tropospheric profiles from ground-based observations. The observed zenith column abundances varied between 6.5×10^{15} and 1.1×10^{16} molecules cm^{-2}. From those total column values they derived the two tropospheric profiles in Figure 6.6, assuming an exponential decrease of HCl with increasing altitude in the troposphere and a surface volume mixing ratio for HCl of 1 ppb. The latter value is based on a measurement by Junge (1957) of gaseous chlorine compounds over Hawaii. Farmer *et al.* assign an absolute error of 25 percent to their total column abundance for HCl. Another possible source of error in the calculated profiles is introduced through the assumed surface concentration of HCl. Junge (1957) used a 0.01 normal KOH solution to collect gaseous chlorine compounds and subsequently measured all Cl^- in the solution. Thus other gases such as Cl_2 or CH_3Cl could have contributed to his gaseous chlorine determination, a fact that was pointed out by Junge himself. However, Cl_2 is believed to have a low concentration in the atmosphere, so it probably is not a serious contaminant. Even CH_3Cl, whose atmospheric abundance may reach 1 ppb, is not a likely contaminant, although its solubility is relatively high (400 cm^3 gas/100 ml H_2O at NTP) because its hydrolysis rate in the KOH solution is too slow (Moelwyn-Hughes, 1949). Nevertheless, the hatched area, which allows for a surface concentration of HCl as low as 0.5 ppb, seems to provide appropriate limits for the tropospheric HCl profile.

E. ClO

There are three reports on the measurement of this important free radical. Two of these, using remote-sensing
techniques from the ground have made only tentative identification. Carlson (1976) obtained photoelectric spectra
of the sun in December 1975 using the main beam of the
Solar Telescope of Kitt Peak Observatory, Arizona, and a
vacuum spectrometer operated in the ninth order. Observations were made at 303.5 nm corresponding to the 3-0
band of the $A^2\Pi \rightarrow X^2\Pi$ transition. Since the band is
diffuse because of predissociation (Coxon and Ramsay,
1976) no rotational structure could be observed, which
makes the identification difficult. Interference from
other atmospheric absorbers (notably O_3) in this spectral
region adds to the difficulty in interpreting the very
weak absorption, which is barely detectable above the
noise level. Therefore, positive identification of ClO
must be considered tentative. The maximum absorption observed, if attributed to ClO, would correspond to an
upper limit of about 2 ppbv at 30 km.

The Kitt Peak Radio Telescope was used in December
1975 by Ekstrom (1976) to obtain the atmospheric microwave spectrum near 93 GHz, which was tentatively identi-
fied with the $^2\Pi_{3/2}$, $J = \frac{5}{2} \rightarrow \frac{3}{2}$ transition of ClO.
Observations were made both in absorption against the sun
or in emission against the cold sky. The data contained
strongly curved base lines and indications of interference
from unidentified species. For these reasons, no claim
can be made for a positive identification of ClO. Even
if the spectra are interpreted in terms of ClO, concentrations can only be derived on the basis of an assumed
temperature profile. Therefore, the suggestion that the
ClO concentration may be 50 times the value calculated
from the models must be viewed as highly speculative.

In June 1976, Anderson (1976b) succeeded in making an
in situ measurement at Palestine, Texas. A flow reactor
was dropped by parachute from 45 km, which provided an
airflow through the apparatus of about 200 m/sec. Measurements were made down to 25 km. The reactor had been
used previously to measure O atoms and HO radicals (see
Appendix C, Sections II and IV), the species being detected by resonance fluorescence. ClO does not fluoresce
and is therefore converted to atomic chlorine, which is
then measured by resonance fluorescence at 118.8 nm. This
conversion of ClO to Cl is accomplished by addition of
NO (cf. Appendix A, Section II.D, Reaction 3) to the

sampled air upstream of the detection region. The detection limit is better than 10^5 atoms/cm^3, and the altitude resolution better than 0.5 km. One possible ambiguity
in the method is that the solar blind detector will respond to longer wavelengths, which may have other unknown
origins. Future experiments will incorporate an appropriate filter to make the detection more definitive. Nevertheless, extensive laboratory testing of the apparatus
has convinced Anderson that he has indeed measured some
species in the atmosphere that is capable of producing
Cl atoms upon reaction with NO; ClO appears to be the
only candidate. The measured concentrations were from
20 to 100 times the values calculated by the models over
the 25 to 45 km altitude range. The measured concentrations are also inconsistent with the HCl measurements
and with the limits placed on the ClONO$_2$ concentration.
Most disturbing, however, is that the amount of ClO
measured exceeds the total Cl content of all the chlorine-
containing constituents believed to be present in the
stratosphere. Further confirmation of the technique is
indicated. Furthermore, it is worth emphasizing once
again that, because of the known variability of the
stratosphere, too much significance should not be placed
on a single measurement. Additional measurements by
this and other techniques of this important constituent
are obviously required and are likely to be forthcoming
in the near future.

F. Chlorine Nitrate

Because of the possible role of chlorine nitrate, direct
spectroscopic evidence for its presence in the atmosphere
has been sought. This substance has a relatively intense
infrared absorption with a distinctive *PQR* structure and
with the *Q* branch centered at 1292.4 cm^{-1}. There is
a second, broad absorption with two extremely sharp lines
at 780.3 and 809 cm^{-1}.* These characteristic features
can be sought in some of the excellent reference solar

---

*These frequencies are taken from a reference spectrum of
ClONO$_2$ kindly provided by M. Molina. The spectrum was
recorded with spectral slit width near 0.02 cm^{-1}.

spectra recorded in recent years. For example, in the spectra recorded by Migeotte *et al.* (1971) at 10,000 feet in the Swiss Alps, there is absorption at 1292.48 cm^{-1} attributable to H_2O, which would obscure weaker absorption by $ClONO_2$. In contrast, near 780 cm^{-1} the spectrum is relatively clear. There are weak absorptions due to CO_2 at 780.53 and 779.80 cm^{-1} but no sharp feature at 780.3 cm^{-1}. Near 809 cm^{-1} there is distinct absorption at 809.4 cm^{-1} attributable to CO_2, and an extremely weak absorption is noted at 808.7 cm^{-1} but not assigned. Thus, absorption due to $ClONO_2$ is not apparent unless the marginal feature at 808.7 cm^{-1} is so identified. Assuming that the resolution in these spectra of Migeotte is about the same as that used in the spectra furnished by Molina, a crude upper limit can be placed on the $ClONO_2$ atmospheric mixing ratio--less than a few parts per billion. This upper limit is not definitive, since model calculations indicate that $ClONO_2$ concentrations would be expected at this time to be only a few tenths of a part per billion in the altitude range of interest, 20 to 30 km (cf. Figure 7.16).

More useful, perhaps, are the balloon spectra recorded in September 1975, by Murcray and colleagues (Murcray, 1976). These were recorded at sunset, providing a long absorbing path with tangential altitudes for the line of sight in the range 15 to 21 km. Murcray finds that the spectral region near 809 cm^{-1} overlapped with absorption due to CO_2. The region near 780 cm^{-1} seems more useful, and marginal evidence of absorption near 781 cm^{-1} might be present in the spectra (Murcray, 1976). He concludes that his spectra place an upper limit of 1 ppb on the $ClONO_2$ mixing ratio.

Hanst (1976) has also attempted to analyze Murcray's data. He tries to estimate an expected $HNO_3/ClONO_2$ absorption intensity ratio and on this basis concludes that $ClONO_2$ is not important in the stratospheric chemistry. His analysis is probably less significant than that of Murcray, both because of difficulty in estimating the apparent extinction coefficient of $ClONO_2$ from reference spectra at higher resolution and different pressure and because of a strong altitude dependence of the $HNO_3/ClONO_2$ number ratio.

In summary, the available data do not prove that $ClONO_2$ is present in the stratosphere at the presently anticipated level, but they do not rule it out.

G. *Sulfur Hexafluoride*

Because it is released at the surface and is very chemically stable, SF$_6$ can serve as a tracer for atmospheric motions in the lower stratosphere. Two cross sections of the distribution in latitude of SF$_6$ have been measured by Krey and Lagomarsino (1975) (see Figure C.16). The errors are the same as those discussed for CFCl$_3$. The cross sections can help to estimate the meridional distribution of other long-lived species.

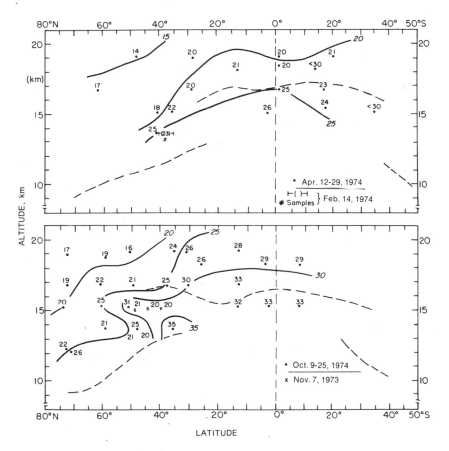

FIGURE C.16 Latitude-altitude cross section of SF$_6$ in hundredths of parts per trillion. Dashed lines denote the observed tropopause positions during sampling periods. Solid lines are subjectively drawn between observation points.

VII. DISCUSSION

The data presented in this appendix show that only for O_3 do we have a true globally averaged stratospheric profile. For some constituents, such as N_2O and NO, some profiles have been obtained at a number of latitudes. There exist at best average vertical profiles at a given latitude for the other trace gases, and it requires some judgment to estimate how closely those profiles represent a global average. Nevertheless, the available experimental trace gas profiles provide certain guidelines from which the modeled profiles cannot deviate too much. Just how such deviation occurs depends on the trace gas considered. Probably for the longer-lived trace gases such as CH_4, H_2, and N_2O, the global mean profiles will not deviate significantly from the ones measured so far. Since their destruction processes are slow, and the sources are diffuse, they stay around long enough to be well mixed throughout the troposphere. Stratospheric N_2O concentrations do show considerable variability indicative of horizontal transport, and it is likely that CH_4 and H_2 will have similar behavior. Additional measurements are obviously indicated to improve the reliability of the vertical mixing coefficients derived from these profiles.

It is far more difficult to establish global averages for the shorter-lived constituents. As discussed in Section III.A on NO, their concentrations strongly depend on a number of parameters and, in fact, vary considerably in the stratosphere, both in time and space. Thus, the experimental data define a broad bandwidth for the global average and still allow a certain latitude in the 1-D calculations. To establish their atmospheric roles more precisely, simultaneous measurements of several chemically related species are very important, more so than those of single compounds.

So little is known about the very short-lived reactive species like atomic oxygen, HO and HO_2 radicals, or ClO that even single, isolated measurements must be regarded as important. Again, the concentrations of these compounds depend strongly on a number of parameters so that simultaneous measurements are required to be able to interpret properly the measured concentrations of the radicals. This experimental knowledge is still insufficient for proper comparisons with model calculations. Since suitable, new techniques are at hand, further O and HO profiles can be expected soon. To obtain

simultaneous measurements of all parameters required for
the interpretation of HO will be more difficult and correspondingly further off in the future. For NO_x, simultaneous measurement of all parameters involved is equally
desirable, and experiments are planned by several groups.
Both sets of measurements, for HO and NO_x, would help to
clarify the details of their important roles in stratospheric chemistry.

Of immediate concern for the chlorine problem are measurements of HCl, ClO, and $ClONO_2$ in the stratosphere.
Some HCl height profiles are available, but additional
measurements are required to determine both the magnitude
and shape of the profiles. Upper limits only are available for $ClONO_2$ concentrations, but the sensitivity of
the measurements must be improved before the significance
of this compound can be fully evaluated. ClO measurements are under way, and the results should be available
in the near future. Measurements of total Cl in the
stratosphere would also be extremely useful.

There is an obvious need to increase the number and
accuracy of CFM measurements in the troposphere in order
to establish the existence or absence of tropospheric
sinks for these compounds by comparison of the total
atmospheric content with the total amount released (cf.
Appendix E). Improved accuracy in measuring tropospheric
CH_3Cl and CCl_4 is also desirable.

REFERENCES

Ackerman, M., and C. Muller. 1972a. Stratospheric ni-
    trogen dioxide from infrared absorption spectra, *Nature*
    240:300-301.
Ackerman, M., and C. Muller. 1972b. Stratospheric meth-
    ane from infrared spectra, *Proc. Symp. Sources, Sinks
    and Concentrations of Carbon Monoxide and Methane in the
    Earth's Environment,* August 15-17, St. Petersburg, Fla.
Ackerman, M., and C. Muller. 1973. Stratospheric meth-
    ane and nitrogen dioxide from infrared spectra, *Pure
    Appl. Geophys.* 106-108:1325-1335.
Ackerman, M., J. C. Fontanella, D. Frimout, A. Girard,
    N. Louisnard, C. Muller, and D. Nevajans. 1973a.
    Stratospheric nitric oxide from infrared spectra,
    Nature 245:205-206.
Ackerman, M., D. Frimout, C. Muller, D. Nevejans, J. C.
    Fontanella, A. Girard, L. Gramont, and N. Louisnard.

1973b.   Recent stratospheric spectra of NO and NO_2, *Aeron. Acta* A120.

Ackerman, M., J. Fontanella, D. Frimout, A. Girard, N. Louisnard, and C. Muller. 1974. Simultaneous measurements of NO and NO_2 in the stratosphere, *Aeron. Acta* A133.

Ackerman, M., J. C. Fontanella, D. Frimout, A. Girard, N. Louisnard, and C. Muller. 1975. Simultaneous measurements of NO and NO_2 in the stratosphere, *Planet. Space Sci*. 23:651:660.

Ackerman, M., D. Frimout, A. Girard, M. Gottignier, and C. Muller. 1976. Stratospheric HCl from infrared spectra, *Aeron. Acta* A158.

Anderson, J. G. 1971. Rocket measurement of OH in the mesosphere, *J. Geophys. Res*. 76:1820-1824.

Anderson, J. G. 1975. The absolute concentrations of $O(^3P)$ in the earth's stratosphere, *Geophys. Res. Lett*. 2:231-234.

Anderson, J. G. 1976a. The absolute concentration of OH $(X^2\pi)$ in the earth's stratosphere. *Geophys. Res. Lett*. 3:165.

Anderson, J. G. 1976b. University of Michigan. Private communication.

Bojkov, R. D. 1969. Differences in Dobson spectrophotometer and filter ozonometer measurements of total ozone, *J. Appl. Meterol*. 8:362-368.

Brewer, A. W., C. T. McElroy, and J. B. Kerr. 1973. Nitrogen dioxide concentrations in the atmosphere, *Nature* 246:129-133.

Brewer, A. W., C. T. McElroy, and J. B. Kerr. 1974. Spectrophotometric nitrogen dioxide measurements, *Proc. Third Conf. on the Climatic Impact Assessment Program*, DOT-TSC-OST-74-15, U.S. Dept. of Transportation, pp. 257-263.

Briehl, D. C., E. Hilsenrath, B. A. Ridley, and H. I. Schiff. 1975. *In situ* measurements of NO, H_2O and O_3 from aircraft, *NASA Technical Memorandum* TM-X-3174, January.

Burkhardt, E. G., C. A. Lambert, and C. K. N. Patel. 1975. Stratospheric nitric oxide: measurements during daytime and sunset, *Science* 188:1111-1113.

Carlson, R. W. 1976. Investigation of atmospheric chlorine oxide through solar absorption spectroscopy. Preprint.

Climatic Impact Committee. 1975. *Environmental Impact of Stratospheric Flight: Biological and Climatic Effects of Aircraft Emissions in the Stratosphere*. National Academy of Sciences, Washington, D.C.

Coxon, J. A., and D. A. Ramsay. 1976. Reinvestigation of the absorption spectrum of ClO, *Can. J. Phys.* 54:1034-1042.

Craig, R. H. 1965. *The Upper Atmosphere*. Academic Press, New York, pp. 184-189.

Cumming, C., and R. P. Lowe. 1973. Balloon-borne spectroscopic measurement of stratospheric methane, *J. Geophys. Res.* 78:5259-5264.

Davis, D. D. 1975. Tropospheric hydroxl radical measurement, *Chem. Eng. News* 53:22.

Davis, D. D. 1976. University of Maryland. Private communcation.

Dütsch, H. U. 1974. The ozone distribution in the atmosphere, *Can. J. Chem.* 52:1491-1504.

Ehhalt, D. H., and L. E. Heidt. 1973a. The concentration of molecular H_2 and CH_4 in the stratosphere, *Pure Appl. Geophys.* 106-108:1352-1360.

Ehhalt, D. H., and L. E. Heidt. 1973b. Vertical profiles of CH_4 in the troposphere and stratosphere, *J. Geophys. Res.* 78:5265-5271.

Ehhalt, D. H., L. E. Heidt, R. H. Lueb, and N. Roper. 1974. Vertical profiles of CH_4, H_2, CO, N_2O and CO_2 in the stratosphere, *Proc. Third Conf. on the Climatic Impact Assessment Program,* DOT-TSC-OST-74-15, U.S. Dept. of Transportation, pp. 153-160.

Ehhalt, D. H., L. E. Heidt, R. H. Lueb, and W. Pollock. 1975a. The vertical distribution of trace gases in the stratosphere, *Pure Appl. Geophys.* 113:389-402.

Ehhalt, D. H., N. Roper, and H. E. Moore. 1975b. Vertical profiles of nitrous oxide in the troposphere, *J. Geophys. Res.* 80:1653-1655.

Ehhalt, D. H., L. E. Heidt, R. H. Lueb, and E. A. Martell. 1975c. Concentrations of CH_4, CO, CO_2, H_2, H_2O and N_2O in the upper stratosphere, *J. Atmos. Sci.* 32:163-169.

Ekstrom, P. A. 1976. Stratospheric ClO abundance: a tentative microwave emission measurement. Preprint.

Evans, W. J. F. 1974. Rocket measurements of water vapor in the stratosphere. *Proc. Int. Conf. on Structure, Composition and General Circulation of the Upper and Lower Atmospheres and Possible Anthropogenic Perturbations,* Melbourne, Australia, January, pp. 249-256.

Farmer, C. B., O. F. Raper, R. A. Toth, and R. A. Schindler. 1974. Recent results of aircraft infrared observations of the stratosphere, *Proc. Third Conf. on the Climatic Impact Assessment Program,* DOT-TSC-OST-74-15, U.S. Dept. of Transportation, pp. 234-245.

318

Farmer, C. B., O. F. Raper, and R. H. Norton. 1976. Spectroscopic detection and vertical distribution of HCl in the troposphere and stratosphere, *Geophys. Res. Lett.* 3:13-16.

Fontanella, J. C., A. Girard, L. Gramont, and N. Louisnard. 1974. Vertical distribution of NO, NO_2 and HNO_3 as derived from stratospheric absorption infrared spectra, *Proc. Third Conference on the Climatic Impact Assessment Program*, DOT-TSC-OST-74-15, U.S. Dept. of Transportation, pp. 217-233.

Goldman, A., D. G. Murcray, F. H. Murcray, and W. J. Williams. 1970. Identification of the v_3 NO_2 band in the solar spectrum observed from a balloon-borne spectrometer, *Nature* 225:443-444.

Goldman, A., D. G. Murcray, W. J. Williams, J. N. Brooks, and C. M. Bradford. 1973. Vertical distribution of CO in the atmosphere, *J. Geophys. Res.* 78:5273-5283.

Hanst, P. E. 1976. Evidence against chlorine nitrate being a major sink for stratospheric chlorine. Part II. Examination of stratospheric infrared spectra to establish limits for the $HNO_3/ClNO_3$ ratio. Preprint.

Hanst, P. L., L. L. Spiller, D. M. Watts, J. W. Spence, and M. F. Miller. 1975. Infrared measurement of fluorocarbons, carbon tetrachloride, carbonyl sulfide, and other atmospheric trace gases. *J. Air Pollut. Cont. Assoc.* 25:1220-1226.

Harries, J. E., J. R. Birch, J. W. Fleming, N. W. B. Stone, D. G. Moss, N. R. W. Swann, and G. F. Neill. 1974a. Studies of stratospheric H_2O, O_3, HNO_3, N_2O, and NO_2 from aircraft, *Proc. Third Conf. on the Climatic Impact Assessment Program*, DOT-TSC-OST-74-15, U.S. Dept. of Transportation, pp. 197-212.

Harries, J. E., N. W. B. Stone, J. R. Birch, N. R. Swann, G. F. Neill, J. W. Fleming, and D. G. Moss. 1974b. Submillimeter wave observations of stratospheric composition, *Proc. Int. Conf. on Structure, Composition, and General Circulation of the Upper and Lower Atmospheres and Possible Anthropogenic Perturbations*, Vol. 1, pp. 275-291.

Heidt, L. E., and D. H. Ehhalt. 1972. Gas chromatographic measurement of hydrogen, methane and neon in air, *J. Chromatog.* 69:103-113.

Heidt, L. E., R. Leub, W. Pollock, and D. H. Ehhalt. 1975. Stratospheric profiles of CCl_3F and CCl_2F_2, *Geophys. Res. Lett.* 2:445-447.

Hester, N. W., E. R. Stephens, and O. C. Taylor. 1974. Fluorocarbons in the Los Angeles Basin, *J. Air Pollut. Cont. Assoc.* 24:591-595.

Hester, N. E., E. R. Stephens, and O. C. Taylor. 1975. Fluorocarbon air pollutants. Measurements in lower stratosphere, *Environ. Sci. Technol.* 9:875-876.

IMOS. 1975. Report of the Federal Task Force on Inadvertent Modification of the Stratosphere. *Fluorocarbons and the Environment*. Council on Environmental Quality, Federal Council for Science and Technology.

Junge, C. E. 1957. Chemical analysis of aerosol particles and of trace gases on the island of Hawaii, *Tellus* 9:528-537.

Komhyr, W. D., and T. B. Harris. 1971. Development of an ECC Ozonesonde, NOAA Tech. Rep. ERL-APCL 18, Boulder, Colo.

Krey, P. W., and R. J. Lagomarsino. 1975. Stratospheric concentrations of SF_6 and CCl_3F. Health and Safety Laboratory, *Environ. Quart.* HASL-294.

Lazrus, A. L., and B'. W. Gandrud. 1974. Distribution of stratospheric nitric acid vapor, *J. Atmos. Sci.* 31:1102-1108.

Lazrus, A. L., B. W. Gandrud, R. N. Woodard, and W. A. Sedlacek. 1975. Stratospheric halogen measurements, *Geophys. Res. Lett.* 2:439-441.

Lazrus, A. L., B. W. Gandrud, R. N. Woodard, and W. A. Sedlacek. 1976. National Center for Atmospheric Research. Private communication.

Lillian, D., H. B. Singh, A. Appleby, L. Lobban (8 other authors). 1975. Atmospheric fates of halogenated compounds, *Environ. Sci. Technol.* 9:1042.

Lovelock, J. E. 1974. Atmospheric halocarbons and stratospheric ozone, *Nature* 252:292-294.

Lowenstein, J., J. R. Paddock, I. G. Poppoff, and H. F. Savage. 1974. NO and O_3 measurements in the lower stratosphere from U-2 aircraft, *Nature* 249:817-819.

Lowenstein, M., H. F. Savage, and R. C. Whitten. 1975. Seasonal variations of NO and O_3 at altitudes of 18.3 and 21.3 km, *J. Atmos. Sci.* 32:2185-2190.

Mastenbrook, H. J. 1974. Water vapor measurements in the lower stratosphere, *Can. J. Chem.* 52:1527-1531.

Mateer, C. L., and H. U. Dütsch. 1964. Uniform evaluation of "Umkehr" observations from the world network. I. Proposed standard Umkehr evaluation technique, Tech. Rep. NCAR, Boulder, Colo.

Migeotte, M., L. Nevin, and J. Swensson. 1971. The solar spectrum from 2.8 to 23.7 microns, Part I, *Photometric Atlas*. (University of Liège. (See *Handbook of Lasers*, R. J. Pressley, ed. The Chemical Rubber Co., Ohio, pp. 91, 112

Moelwyn-Hughes, E. A. 1949. The kinetics of certain reactions between methyl halides and anions in water,
Proc. R. Sóc. 196:542-553.

Murcray, D. G. 1976. University of Denver. Private
communication.

Murcray, D. G., F. H. Murcray, W. J. Williams, T. G.
Kyle, and A. Goldman. 1969. Variation of the infrared
solar spectrum between 700 cm^{-1} and 224 cm^{-1} with altitude, *Appl. Opt.* 8:2519-2536.

Murcray, D. G., A. Goldman, F. H. Murcray, W. J. Williams,
J. N. Brooks, and D. B. Barker. 1972. Vertical distribution of minor atmospheric constituents as derived
from air-borne measurements of atmospheric emission and
absorption infrared spectra, *Proc. Second Conf. on the
Climatic Impact Assessment Program,* DOT-TSC-OST-73-4,
U.S. Dept. of Transportation, pp. 86-98.

Murcray, D. G., A. Goldman, A. Csoeke-Poekh, F. H.
Murcray, W. J. Williams, and R. N. Stocker. 1973.
Nitric acid distribution in the stratosphere, *J. Geophys.
Res.* 78:7033-7038.

Murcray, D. G., A. Goldman, W. J. Williams, F. H. Murcray,
J. N. Brooks, J. Van Allen, R. N. Stocker, J. J. Kosters,
D. B. Barker, and D. E. Snider. 1974. Recent results
of stratospheric trace gas measurements from balloonborne spectrometers, *Proc. Third Conf. on the Climatic
Impact Assessment Program,* DOT-TSC-OST-74-15, U.S. Dept.
of Transportation, pp. 184-192.

Noxon, J. F. 1975. Stratospheric and Tropospheric Composition in Research, Aeronomy Laboratory, Environmental
Research Laboratories, NOAA, Boulder, Colo., Nov.

Patel, C. K. N. 1976. Proceedings of the Conference on
Laser Spectroscopy of the Atmosphere, June 1975,
Norway. To be published in *Opt. Quantum Electron.*

Patel, C. K. N., E. G. Burkhardt, and C. A. Lambert. 1974.
Spectroscopic measurements of stratospheric nitric oxide
and water vapor, *Science* 184:1173-1176.

Perner, D., D. H. Ehhalt, W. W. Paetz, U. Platt, E. P.
Roth, and Q. Volz. 1975. OH radicals in the lower
troposphere. Unpublished manuscript.

Pittrock, A. B. 1974. Ozone climatology, trends and the
monitoring problem, *Proc. Int. Conf. on Structure, Composition and Circulation of the Upper and Lower Atmospheres and Possible Anthropogenic Perturbations,
Melbourne, Australia,* January, pp. 455-466.

Rasmussen, R. 1975. Data presented to Panel on Atmospheric Chemistry, Snowmass, Colo., July.

Ridley, B. A., H. I. Schiff, A. W. Shaw, L. Bates, L. C. Howlett, H. Levaux, L. R. Megill, and T. E. Ashenfelter. 1973. *In situ* measurements of nitric oxide in the stratosphere, *Nature* 245:310-311.

Ridley, G. A., H. I. Schiff, A. W. Shaw, L. Bates, L. C. Howlett, H. Levaux, L. R. Megill, and T. E. Ashenfelter. 1974. Measurement of NO in the stratosphere between 17.4 and 22.9 km, *Planet. Space Sci.* 22:19-24.

Ridley, B. A., H. I. Schiff, A. W. Shaw, and L. R. Megill. 1975. *In situ* measurements of NO using a balloon-borne chemiluminescent instrument, *J. Geophys. Res.* 80:1925-1929.

Ridley, B. A., J. T. Bruin, H. I. Schiff, and J. C. McConnell. 1976a. Altitude profile and sunset decay measurements of stratospheric nitric oxide, *Can J. Meteorol.* In press.

Ridley, B. A., M. McFarland, J. T. Bruin, H. I. Schiff, and J. C. McConnell. 1976b. Diurnal measurements of stratospheric NO. Paper given at the AGU annual meeting, San Francisco, December.

Schmeltekopf, A. L. 1976. National Oceanic and Atmospheric Administration, Boulder, Colo. Private communication.

Schmeltekopf, A. L., P. D. Goldan, W. R. Henderson, W. J. Harrop, T. L. Thompson, F. C. Fehsenfeld, H. I. Schiff, P. J. Crutzen, I. S. A. Isaksen, and E. E. Ferguson. 1975. Measurements of stratospheric $CFCl_3$, CF_2Cl_2 and N_2O, *Geophys. Res. Lett.* 2:393-396.

Schütz, K., C. Junge, R. Beck, and B. Albrecht. 1970. Studies of atmospheric N_2O, *J. Geophys. Res.* 75:2230-2246.

Seiler, W. 1974. The cycle of atmospheric CO, *Tellus* 26:116-135.

Seiler, W., and P. Warneck. 1972. Decrease of the carbon monoxide mixing ratio at the tropopause, *J. Geophys. Res.* 77:3204-3214.

Su, C. W., and E. D. Goldberg. 1973. Chlorofluorocarbons in the atmosphere, *Nature* 245:27.

Thomas, R. W., K. Guard, A. C. Holland, and J. F. Spurling. 1974. Ozone measurement systems improvement studies, NASA Tech. Note NASA TN D7758.

Toth, R. A., C. B. Farmer, R. A. Schindler, O. F. Raper, and P. W. Schaper. 1973. Detection of nitric oxide in the lower stratosphere, *Nature* 244:7-8.

Wang, C. C., L. I. Davis, Jr., C. H. Wu, S. Japar, H. Niki, and B. Weinstock. 1975. Hydroxyl radical concentrations measured in ambient air, *Science* 189:797-800.

Wilkniss, P. E., R. A. Lamontagne, R. E. Larson, J. W. Swinnerton, C. R. Dickson, and T. Thompson. 1973. Atmospheric trace gases in the southern hemisphere, *Nature* 245:45-47.

Wilkniss, P. E., J. W. Swinnerton, R. A. Lamontagne, and D. J. Bressan. 1975a. Trichlorofluoromethane in the troposphere, distribution and increase 1971 to 1974, *Science* 187:832-834.

Wilkniss, P. E., J. W. Swinnerton, D. J. Bressan, R. A. Lamontagne, and R. E. Larson. 1975b. Carbon monoxide, carbon tetrachloride, Freon-11, methane, and radon 222 concentrations at low altitude over the Arctic Ocean in January 1974, *J. Atmos. Sci.* 32:158:162.

Williams, W. J., J. J. Kosters, A. Goldman, and D. G. Murcray. 1976a. Measurement of the stratospheric mixing ratio of HCl using infrared absorption techniques, *Geophys. Res. Lett.* To be published.

Williams, W. J., J. J. Kosters, A. Goldman, and D. G. Murcray. 1976b. Measurements of stratospheric fluorocarbon distributions using infrared techniques. To be published.

Zafonte, L., N. E. Hester, E. R. Stephens, and O. C. Taylor. 1975. Background and vertical atmospheric measurements of fluorocarbon-11 and fluorocarbon-12 over southern California, *Atmos. Environ.* 9:1007-1009.

THE LLL ONE-DIMENSIONAL STRATOSPHERIC MODEL

I. INTRODUCTION

Most of the predictions that are used in this report for the effects of CFMs on stratospheric ozone have been derived from the LLL (Lawrence Livermore Laboratory) one-dimensional model. This model is designed to include many of the characteristics of other 1-D calculations; it is capable of reproducing the principal results of other workers. For example, by using the same chemical reaction rates and eddy-mixing coefficients, this model can generate results virtually identical to those first reported by Wofsy *et al.* (1975) on the potential effect of CFMs on ozone (Figure D.1). Comparisons of the concentration profiles calculated for various trace species are also in close agreement. For the sake of completeness, the basic structure of the LLL model is described in this appendix.

For a vertical one-dimensional description of the atmosphere, the conservation equation governing the temporal variation in the concentration of the ith constituent, $c_i = c_i(z,t)$ is

$$\frac{\partial c_i}{\partial t} = P(c) - L(c)\ c_i - \frac{\partial}{\partial z}\left[K\rho\ \frac{\partial}{\partial z}\left(\frac{c_i}{\rho}\right)\right] \qquad (D.1)$$

where $P(c)$ is the production of c_i due to photochemical interactions of the other c_j species; $L(c)c_i$ is the loss of c_i due to chemical interaction of c_i with the other c_j species; K is the vertical transport coefficient (often

323

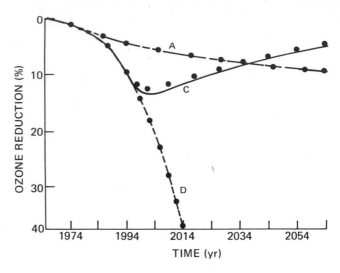

FIGURE D.1 Reductions of global ozone calculated at LLL (●) compared with results from Wofsy *et al.* (1975) (solid line) for various cases of CFM use. Three test cases were simulated: Case A corresponds to a constant release rate after 1975; Case D corresponds to a release rate doubling every 7 years after 1975; Case C is same as Case D with release stopped after 1995.

referred to as the "eddy"-mixing or diffusion coefficient); $\rho = \rho(z)$ is the ambient air density; and t and z are, respectively, time and altitude. The concentrations and air density are given usually in molecules per unit volume, i.e., the number density. The physical domain of the model extends from the ground to 55 km, comprising most of the troposphere and stratosphere. Primary emphasis has been placed on the calculation of stratospheric distributions of the trace chemical species, in particular, ozone.

At present, the atmospheric concentrations of 28 trace species are included in the model. These species are listed in Table D.1. For each of 20 of these species, we solve an equation of the form of Eq. (D.1) to obtain its vertical profile. Three other species, $O(^1D)$, H, and N, are assumed to be in instantaneous equilibrium. The vertical distributions of five constituents, N_2, O_2, CO, H_2O, and H_2, are presently assumed to be time-independent and are held constant. The distributions of these species are

TABLE D.1 Atmospheric Species Included in the 1-D
Calculations

Species Calculated	Calculated Assuming Instantaneous Equilibrium
$O(^3P)$	$O(^1D)$
O_3	H
NO	N
NO_2	
N_2O	
CH_4	
HNO_3	
HO	
HO_2	
H_2O_2	*Assumed Constant*
Cl	N_2
ClO	O_2
$OClO$	CO
$ClONO_2$	H_2O
$ClNO_2$	H_2
HCl	
CCl_4	
$CFCl_3$	
CF_2Cl_2	
CH_3Cl	

based on published measurements. Ambient temperature and
air density distributions are taken from the U.S. Standard
Atmosphere [see, e.g., CIAP Monograph 1 (1975)].

II. THE SYSTEM OF CHEMICAL AND PHOTOCHEMICAL REACTIONS

As shown in Tables D.2 and D.3, there are presently 91 chemical and photochemical reactions in the calculations. Most of the reaction rate constants are the values recommended in the reviews of Hampson and Garvin (1975) and Watson (1974). The rate constants are continuously updated, and the values adopted here include recent measurements of ClX reaction rates by Anderson *et al.* (1975) and Davis *et al.* (1975). Discussion of the most important rate constants is included in Chapters 4 and 8 and Appendix C. The computer program is structured so that reaction rates can be changed or reactions added with a minimum of effort.

The solar flux and absorption cross-section data used to derive the photodissociation rates* in the model are also based on measured data, as described by Gelinas *et al.* (1973). Recent one-dimensional calculations have utilized a constant sun condition; that is, the solar zenith angle χ has been assumed to be constant throughout the calculations. In this type of calculation it is assumed that χ represents the average solar conditions in the atmosphere. For the present, a value of cos χ equal to 0.707 has been adopted. Studies have also been performed (Wuebbles and Chang, 1975) to test the effect of seasonal and diurnal variations in χ. Also, as described in Chapter 7, a diurnal sun was employed in determining the role of ClONO$_2$ formation.

In an attempt to account for the diurnal variation of photodissociation, the solar flux is halved in the calculations. Numerical experiments have indicated that using half of the solar flux has little effect on the perturbations of the ozone concentration, except for the effect of ClONO$_2$, but it does have a significant effect on the

---

*The photodissociation rates are calculated at every altitude using the expression

$$J = I(\lambda)e^{-A(\chi)}\sigma(\lambda)\phi(\lambda)\,d\lambda$$

where λ is the wavelength of impinging light; I is the intensity of solar radiation; A is the attenuation factor due to atmospheric absorption, which depends on the solar zenith angle χ; σ is the photoabsorption cross section of the molecule under consideration; and ϕ is the quantum yield for dissociation.

TABLE D.2 Chemical and Photochemical Reactions of O_x, NO_x, and HO_x

Reaction	Rate[a]
$O_2 + h\nu \rightarrow O + O$	1.54×10^{-9}
$O_3 + h\nu \rightarrow O + O_2$	5.59×10^{-4}
$O_3 + h\nu \rightarrow O(^1D) + O_2$	9.57×10^{-3}
$O + O_2 + M \rightarrow O_3 + M$	$1.07 \times 10^{-34} \exp(510/T)$
$O + O_3 \rightarrow 2O_2$	$1.9 \times 10^{-11} \exp(-2300/T)$
$NO_2 + h\nu \rightarrow NO + O$	1.01×10^{-2}
$O_3 + NO \rightarrow NO_2 + O_2$	$9.0 \times 10^{-13} \exp(-1200/T)$
$O + NO_2 \rightarrow NO + O_2$	9.1×10^{-12}
$N_2O + h\nu \rightarrow N_2 + O(^1D)$	8.22×10^{-7}
$N_2O + O(^1D) \rightarrow N_2 + O_2$	0.7×10^{-10}
$N_2O + O(^1D) \rightarrow 2NO$	0.7×10^{-10}
$NO + h\nu \rightarrow N + O$	2.70×10^{-6}
$N + O_2 \rightarrow NO + O$	$1.1 \times 10^{-14} \, T \exp(3150/T)$
$N + NO \rightarrow N_2 + O$	2.7×10^{-11}
$O(^1D) + H_2O \rightarrow 2HO$	2.1×10^{-10}
$O(^1D) + CH_4 \rightarrow HO + CH_3$	1.3×10^{-10}
$HNO_3 + h\nu \rightarrow HO + NO_2$	1.27×10^{-4}
$O_3 + HO \rightarrow HO_2 + O_2$	$1.6 \times 10^{-12} \exp(-1000/T)$
$O + HO \rightarrow O_2 + H$	4.2×10^{-11}
$O_3 + HO_2 \rightarrow HO + 2O_2$	$1.0 \times 10^{-13} \exp(-1250/T)$
$O + HO_2 \rightarrow HO + O_2$	$8.0 \times 10^{-11} \exp(-500/T)$
$H + O_2 + M \rightarrow HO_2 + M$	$2.08 \times 10^{-32} \exp(290/T)$
$O_3 + H \rightarrow HO + O_2$	$1.23 \times 10^{-10} \exp(-562/T)$
$HO_2 + HO_2 \rightarrow H_2O_2 + O_2$	$1.7 \times 10^{-11} \exp(-500/T)$
$HO_2 + HO \rightarrow H_2O + O_2$	2.0×10^{-11}
$HO + NO_2 + M \rightarrow HNO_3 + M$	$\dfrac{2.76 \times 10^{-13} \exp(880/T)}{1.17 \times 10^{18} \exp(222/T) + M}$

TABLE D.2   Continued

Reaction	Rate[a]
$HO + HNO_3 \rightarrow H_2O + NO_3$	8.9×10^{-14}
$H_2O_2 + h\nu \rightarrow HO$	1.65×10^{-4}
$H_2O_2 + HO \rightarrow H_2O + HO_2$	$1.7 \times 10^{-11} \exp(-910/T)$
$N_2 + O(^1D) + M \rightarrow N_2O + M$	2.8×10^{-36}
$N + NO_2 \rightarrow N_2O + O$	1.4×10^{-12}
$NO + O + M \rightarrow NO_2 + M$	$3.96 \times 10^{-33} \exp(940/T)$
$NO + HO_2 \rightarrow NO_2 + HO$	2.0×10^{-13}
$H_2 + O(^1D) \rightarrow HO + H$	1.4×10^{-10}
$HO + HO \rightarrow H_2O + O$	$1.0 \times 10^{-11} \exp(-550/T)$
$N + O_3 \rightarrow NO + O_2$	5.7×10^{-13}
$NO_2 + O_3 \rightarrow NO_3 + O_2$	$1.2 \times 10^{-13} \exp(-2450/T)$
$HO_2 + h\nu \rightarrow HO + O$	3.97×10^{-4}
$HO + CH_4 \rightarrow H_2O + CH_3$	$2.36 \times 10^{-12} \exp(-1710/T)$
$HO + HO + M \rightarrow H_2O_2 + M$	$2.5 \times 10^{-33} \exp(2500/T)$
$H_2O_2 + O \rightarrow HO + HO_2$	$2.75 \times 10^{-12} \exp(-2125/T)$
$O(^1D) + M \rightarrow O + M$	$1.63 \times 10^{-11} \exp(215/T)$
$NO_3 + h\nu \rightarrow NO_2 + O$	2.36×10^{-3}
$CO + HO \rightarrow H + CO_2$	1.4×10^{-13}

[a]Photodissociation coefficients are calculated at 55 km (top boundary of the model) at noontime.

distribution of some of the trace constituents, e.g., $O(^3P)$ and $O(^1D)$. The model calculates the photodissociation rates, as well as the vertical ozone column needed for their derivation, at every time step. Therefore, any perturbations of the ozone column have an immediate effect on the photodissociation rates employed subsequently; this takes into account any "self-healing" photochemical processes involving ozone.

TABLE D.3 Chemical and Photochemical Reactions of ClX

Reaction	Rate
$Cl + O_3 \rightarrow ClO + O_2$	$2.97 \times 10^{-11} \exp(-243/T)$
$Cl + OClO \rightarrow 2ClO$	5.9×10^{-11}
$Cl + O_2 + M \rightarrow ClO_2 + M$	1.7×10^{-33}
$Cl + CH_4 \rightarrow HCl + CH_3$	$5.4 \times 10^{-12} \exp(-1133/T)$
$Cl + ClO_2 \rightarrow Cl_2 + O_2$	5.0×10^{-11}
$Cl + ClO_2 \rightarrow 2ClO$	1.4×10^{-12}
$Cl + NO + M \rightarrow ClNO + M$	$1.7 \times 10^{-32} \exp(553/T)$
$Cl + ClNO \rightarrow Cl_2 + NO$	3.0×10^{-11}
$Cl + NO_2 + M \rightarrow ClNO_2 + M$	$6.9 \times 10^{-34} \exp(2115/T)$
$Cl + ClNO_2 \rightarrow Cl_2 + NO_2$	3.0×10^{-12}
$ClO + O \rightarrow Cl + O_2$	$3.38 \times 10^{-11} \exp(75/T)$
$NO + ClO \rightarrow NO_2 + Cl$	$1.13 \times 10^{-11} \exp(200/T)$
$ClO + O_3 \rightarrow ClO_2 + O_2$	$1.0 \times 10^{-12} \exp(-2763/T)$
$ClO + O_3 \rightarrow OClO + O_2$	$1.0 \times 10^{-12} \exp(-2763/T)$
$ClO + NO_2 + M \rightarrow ClONO_2 + M$	see text (Chapters 4 and 8)
$ClO + ClO \rightarrow Cl + OClO$	$2.0 \times 10^{-12} \exp(-2300/T)$
$ClO + ClO \rightarrow Cl_2 + O_2$	$2.0 \times 10^{-13} \exp(-1260/T)$
$ClO + ClO \rightarrow Cl + ClO_2$	$2.0 \times 10^{-13} \exp(-1260/T)$
$HCl + O(^1D) \rightarrow Cl + HO$	2.0×10^{-10}
$HO + HCl \rightarrow H_2O + Cl$	$2.0 \times 10^{-12} \exp(-310/T)$
$O + HCl \rightarrow HO + Cl$	$1.75 \times 10^{-12} \exp(-2273/T)$
$ClO_2 + M \rightarrow Cl + O_2 + M$	$1.5 \times 10^{-8} \exp(-4000/T)$
$O + OClO \rightarrow ClO + O_2$	5.0×10^{-13}
$NO + OClO \rightarrow NO_2 + ClO$	3.4×10^{-13}
$NO + OClO \rightarrow NO_2 + ClO$	3.4×10^{-13}
$N + OClO \rightarrow NO + ClO$	6.0×10^{-13}
$H + OClO \rightarrow HO + ClO$	5.7×10^{-11}
$Cl + HO \rightarrow HCl + O$	$2.0 \times 10^{-12} \exp(-1878/T)$

TABLE D.3 Continued

Reaction	Rate
$Cl + HO_2 \rightarrow HCl + O_2$	3.0×10^{-11}
$Cl + HNO_3 \rightarrow HCl + NO_3$	$4.0 \times 10^{-12} \exp(-1500/T)$
$ClO_2 + HO_2 \rightarrow HCl + 2O_2$	3.0×10^{-12}
$ClONO_2 + h\nu \rightarrow ClO + NO_2$	1.12×10^{-3}
$HCl + h\nu \rightarrow H + Cl$	7.49×10^{-7}
$ClO_2 + h\nu \rightarrow ClO + O(^1D)$	5.52×10^{-3}
$ClO + h\nu \rightarrow Cl + O$	7.46×10^{-3}
$ClO + h\nu \rightarrow Cl + O(^1D)$	1.29×10^{-3}
$ClNO_2 + h\nu \rightarrow Cl + NO_2$	2.09×10^{-3}
$OClO + h\nu \rightarrow ClO + O(^1D)$	5.79×10^{-3}
$CF_2Cl_2 + h\nu \rightarrow 2Cl$	1.98×10^{-6}
$CFCl_3 + h\nu \rightarrow 2.5\ Cl$	1.45×10^{-5}
$CCl_4 + h\nu \rightarrow 2Cl$	3.31×10^{-5}
$CFCl_3 + O(^1D) \rightarrow 2Cl$	5.8×10^{-10}
$CF_2Cl_2 + O(^1D) \rightarrow 2Cl$	5.3×10^{-10}
$Cl + H_2 \rightarrow HCl + H$	$5.7 \times 10^{-11} \exp(-2400/T)$
$Cl + H_2O_2 \rightarrow HCl + HO_2$	$1.0 \times 10^{-11} \exp(-810/T)$
$ClONO_2 + O \rightarrow ClO + NO_3$	2.0×10^{-13}
$CH_3Cl + h\nu \rightarrow Cl + CH_3$	4.31×10^{-7}
$CH_3Cl + HO \rightarrow Cl + H_2O + CH_2$	$1.58 \times 10^{-12} \exp(-1049/T)$

III. TRANSPORT COEFFICIENTS

As was pointed out before, the vertical transport in the one-dimensional model is parameterized through the so-called "eddy"-mixing coefficient, K. The LLL computer program for one-dimensional calculations has been designed such that the profiles of K utilized by other groups can be easily incorporated. Such profiles (see Figure 7.1) have been utilized to test the sensitivity of the results to the transport parameters (Chang, 1974).

IV. NUMERICAL METHODS AND BOUNDARY CONDITIONS

The numerical technique used to solve Eq. (D.1) in the
calculations is that described by Chang *et al.* (1974).
The main advantage of using this method, which is a
variable order, multistep, implicit method, is its ability to solve differential equations containing mathematical stiffness such as those resulting from the chemical
kinetic terms in Eq. (D.1). The boundary conditions are
assumed to be fixed or time-varying source-dependent
concentrations for each species at the surface and zero
flux conditions at 55 km, the upper boundary. Numerical
studies have indicated that these boundary conditions
do not introduce significant error into the calculations.

REFERENCES

Anderson, J. C., F. Kaufman, and M. S. Zahmiser. 1975.
Kinetics of some stratospheric reactions of chlorine
species. Paper presented at the American Chemical
Society Meeting, Philadelphia, Pa.

Chang, J. S. 1974. Simulations, perturbations and inter-
pretations, *Proc. Third Conf. on the Climatic Impact
Assessment Program,* DOT-TSC-OST-74-15, U.S. Dept. of
Transportation, pp. 330-341.

Chang, J. S., A. C. Hindmarsh, and N. K. Madsen. 1974.
Simulation of chemical kinetics transport in the strato-
sphere, *Stiff Differential Systems,* R. Willoughby, ed.
Plenum Press, New York.

CIAP Monograph 1. 1975. *The Natural Stratosphere of 1974.*
A. J. Grobecker, ed. U.S. Dept. of Transportation, DOT-
TST-75-51, Washington, D.C.

Davis, D. D., E. S. Machado, R. L. Schiff, and R. T.
Watson. 1975. The temperature dependence of some Cl
atom reactions of stratospheric interest. Paper presented at the American Chemical Society Meeting, Philadelphia, Pa.

Gelinas, R. J., R. P. Dickinson, and K. E. Grant. 1973.
Solar flux and photodissociation calculations for LLL
atmospheric physics programs, Lawrence Livermore Laboratory Rep. UCRL-74944.

Hampson, R. F., and D. Garvin. 1975. Chemical kinetic
and photochemical data for modeling atmospheric chemistry, NBS Tech. Note 866.

Watson, R. J. 1974. Chemical kinetics data survey VIII,
rate constants of ClO_x of atmospheric interest, NBSIR
74-515.

332

Wofsy, S. C., M. B. McElroy, and N. D. Sze. 1975. Freon
    consumption: implication for atmospheric ozone, *Science*
    187:535-537.
Wuebbles, D. J., and J. S. Chang. 1975. Sensitivity of
    time-varying parameters in stratospheric modeling, *J.*
    Geophys. *Res*. 80:2637-2642.

APPENDIX   DETECTION OF A

E   TROPOSPHERIC SINK FOR CFMs

   BY MATERIALS BALANCE

The removal of some fraction f of a CFM before it can be
transported from the troposphere to the stratosphere will
cause a proportional decrease in the magnitude of the
ozone reduction by the CFM, the reduction becoming approximately $(1 - f)$ of what it would otherwise have been.
The photolytic ozone-destroying removal of CFMs in the
stratosphere is slow, the removal time τ_S being about 50
to 100 yr, respectively, for F-11 and F-12 (Table
4.1). Therefore, inactive removal does not have to be
rapid to have an important effect on the ozone reduction.
For example, a tropospheric sink with a removal time τ_T
of 50 years (see Chapter 4) would remove as much F-11 as
stratospheric photolysis ($f = 1/2$) and in the case of
F-12, twice as much ($f = 2/3$). Thus, such a sink would
decrease the ozone reductions by 1/2 and 2/3 of the respective values for F-11 and F-12 in the absence of the
sink.

Because of this sensitivity to weak tropospheric sinks,
we have examined as quantitatively as possible the wide
variety of suggested possibilities (Chapter 4 and Appendix A). Three processes were found to have estimated
inactive removal times for F-11 and F-12 that are short
enough to warrant further, more detailed study. Lower
limits of $\sim 10^2$ (70 and 200), 10^3, and 5×10^3 years have
been placed, respectively, on the removal times for solution in the surface waters of the oceans (followed by some
unknown degradation process) and by ion-molecule reactions
and photodissociation in the troposphere. If each of
these processes actually removed F-11 and F-12 in the

333

time corresponding to the *lower limit* set for it, the (maximum) combined effect would be a decrease in the predicted ozone reductions by about two fifths of what they would be in the absence of such inactive removal. However, we expect the effect to be no more than 20 percent (a decrease by one fifth), based on the limited data available for the oceanic sink in the case of F-11.

In principle, conclusive evidence for the actual amount of such removal or for the presence or absence of an otherwise unidentified sink can be sought by comparing the amount of CFM released with the amount in the atmosphere, i.e., by making some kind of materials balance to see if any CFM is unaccounted for. However, the release of CFMs has been increasing exponentially and most has been in the atmosphere for only a few years. Thus, the atmospheric concentrations expected for even a relatively short removal time τ_T of, say, 30 yr do not differ as yet by a readily measurable amount from those for an indefinite τ_T (zero tropospheric removal).

Nonetheless, two types of materials balance have been examined. Chapter 6 mentions comparison of the estimated release rates with the rates at which the CFMs have been observed to accumulate over a period of time at particular tropospheric locations. These comparisons are subject to localized influences difficult to evaluate, and the time span of the measurements available is limited to a few years, which is short for the purpose. Therefore, the global type of materials balance (see, e.g., Rowland and Molina, 1975a) seems to us to be more promising, and we discuss it in more detail at this point.

The total amount of a CFM released up to a given time must either be in the atmosphere *at that time* or have been removed from it in some way. Also, it is highly unlikely that F-11 or F-12 has a natural source of any consequence, for thermodynamic reasons, and none are known. Therefore, we can write

$$\begin{array}{c} \text{CFM} \\ \text{released} \end{array} = \begin{array}{c} \text{CFM in} \\ \text{troposphere} \end{array} + \begin{array}{c} \text{CFM in} \\ \text{stratosphere} \end{array}$$

$$+ \begin{array}{c} \text{CFM} \\ \text{photolyzed} \end{array} + \text{Sink(s)} \qquad \text{(E.1)}$$

Division of this equality by the total amount released gives

$$1 = B_t + B_s + F_p + \Delta \qquad \text{(E.2)}$$

or

$$\Delta = 1 - B_t - B_s - F_p \qquad \text{(E.3)}$$

where B_t and B_s are the fractions of total CFM release
still present as the tropospheric and stratospheric burdens, F_p is the fraction photolyzed in the stratosphere,
and Δ is the fraction otherwise unaccounted for (sinks).
The presently available atmospheric observations of F-11
and F-12 are neither widely enough dispersed nor sufficiently accurate to provide a true global average for
the tropospheric burden B_t. However, we will use the
estimates of B_t in Chapter 6 to illustrate the procedures
involved.

The MCA data of Chapter 3 indicate that the total
world production of F-11 ($CFCl_3$) through August 1975 was
3.32 million metric tons, with an estimate that 85 percent of the production has been released to the atmosphere.
This corresponds to 1.24×10^{34} molecules. Of them, 5.3
percent has been photolyzed in the stratosphere, as obtained from calculations of the type described in Chapter
7. The stratospheric burden is not in steady state with
the troposphere, and both the experimental data and 1-D
calculations show it to be about 5 percent of the total.
In Chapter 6, the global tropospheric average was *estimated* from Rasmussen's data to be 110 ppt (±40 percent)
for F-11 as of September 1, 1975. The troposphere contains 90 percent of the atmosphere's 1.1×10^{44} molecules,
so 110 ppt corresponds to 1.09×10^{34} molecules (110 ×
$10^{-12} \times 0.9 \times 1.1 \times 10^{44}$) or 88 percent of the 1.24×10^{34}
molecules of F-11 released. Substitution of these values
in Eq. (E.3) gives

$$\Delta = 1 - 0.88 - 0.05 - 0.05 = 0.02 \qquad \text{(E.4)}$$

This result suggests that Δ is close to zero, corresponding to little or no tropospheric removal. However, the
estimated ±40 percent uncertainty in the tropospheric global burden and ±5 percent in release rates (Chapter 3)
produce a combined uncertainty* in Δ of ±0.37, i.e., Δ =
0.02 ±0.37.

The significance of this uncertainty range depends on
the relationship between Δ and the magnitude of a tropospheric sink. Approximate analytical expressions have
been developed to determine the relation (Rowland and

---

*Combined as the square root of the weighted sum of the
squares of the (2σ) deviations.

Molina, 1975b; Johnston, 1975), and, more recently, numerical methods have been used (Rowland and Molina, 1976; Jesson et al., 1976). Typical results we have obtained are given in Table E.1, which shows the effects of different tropospheric removal rates on the *tropospheric* burden at different times. The calculations were based on the 1974 MCA estimates of annual world production through 1973 (IMOS, 1975) followed by a constant production rate at the 1973 levels. The release rate was taken to be 90 percent of production. The results are generally insensitive to the details of the growth curves for release; for example, the values for F-11 and F-12 agree within about 0.005.

Since there are no known natural sources for F-11 of any significance, the negative side of the uncertainty range in Δ (±0.37) is excluded for physical reasons. The positive side (+0.39) is found via the data in Table E.1 to correspond to a tropospheric removal rate of 13 percent per year, or a steady-state removal time of 8 yr compared with about 50 yr for stratospheric photolysis. Thus, although the analysis points to a tropospheric removal rate of close to zero ($\tau_T \sim 250$ yr), the upper limit of the uncertainty range corresponds to a large removal rate of 13 percent per year. The difficulty is that when Δ is a small difference between two large numbers, even modest percentage errors in the latter give magnified uncertainties in the corresponding removal rate.

TABLE E.1   Reductions in the Tropospheric Burden B_t Caused by Different Tropospheric Removal Rates[a]

τ_T (yr)	k (yr^{-1})	1975	1977	1979	1983
100	0.01	0.05	0.04[b]	0.06	0.07
50	0.02	0.08	0.09	0.11	0.13
30	0.033	0.17			
20	0.05	0.24			
10	0.10	0.34	0.36	0.39	0.45

[a]The values given are the fraction of the tropospheric burden that would be missing, i.e., $(B_t^{\,o} - B_t)/B_t^{\,o}$, where $B_t^{\,o} = 1 - B_s - F_p$.
[b]The low value is a computational artifact produced by lumping early release and assuming constant release after 1973.

As pointed out above, the global tropospheric burden was *estimated* from the sparse data on hand. The latitudinal averaging used in Chapter 6 is based on the ^{85}Kr radiotracer results, a procedure that assumes that the release schedule of ^{85}Kr has been similar to that of the CFMs. Also, the CFM measurements are concentrated near the surface at midlatitudes, which could bias the results. For such reasons, it is possible to estimate the global burden in a different manner, obtain a different value, and arrive at a different conclusion. Jesson *et al.* (1976), noting that most measurements of F-11 concentrations have been made at 40-52° N latitude, have used a latitude distribution profile obtained by Lovelock in 1972 to estimate a global average. Lovelock's data showed a particularly large difference (almost twofold) between concentrations at 40-50° N latitude and in the southern hemisphere, so global averages estimated from them are lower (80-93 ppt) and correspond to a 10- to 15-yr tropospheric sink. However, it is possible that Lovelock's early data may be less reliable than his more recent measurements (1975). Furthermore, they lead to a distribution that is 16-27 percent lower for the southern hemisphere than that indicated by the ^{85}Kr distribution.

In the case of F-12 (CF_2Cl_2), which is more difficult to measure, the *estimated* global tropospheric burden of 208 ppt given in Chapter 6 corresponds to 2.06×10^{34} molecules. However, only 1.9×10^{34} molecules have been released (Chapter 3). Thus, with allowance for the 10 percent or so that either is in the stratosphere or has been destroyed there, the apparent tropospheric burden of F-12 is nearly 20 percent higher ($\Delta = -0.18$) than is physically possible. This result is included to emphasize the present rudimentary state of the materials balance approach to the question of tropospheric sinks. Also, it should be noted that sinks for F-12 may differ in importance from those for F-11. If reliance is to be placed on this method, it must be applied independently to the more difficult case of F-12 as well as to F-11.

This Panel believes that present data are inadequate to draw any firm conclusions about the presence or absence of a significant tropospheric sink on the basis of materials balance. While global release data may now be reliable to ±5 percent (Chapter 3), our knowledge of the amount of F-11 and F-12 in the atmosphere is inadequate for the purpose in its time and global coverage as well as in its accuracy. Although it is ineffective at present, this materials balance approach could be of some use in the future. If the amounts released and the global burden can both be determined to ±5 percent, then, as may be shown by analysis of Table E.1, it should be possible to distinguish between a tropospheric removal time of >50 yr and one of <20 yr. Further, if rates of atmospheric release decline or remain constant, the spread in calculated tropospheric concentrations with varying lifetimes will increase from those expected at present, making it easier to obtain definitive results at a future date.

REFERENCES

IMOS. Report of Federal Task Force on Inadvertent Modification of the Stratosphere. 1975. *Fluorocarbons and the Environment*. Council on Environmental Quality, Federal Council for Science and Technology.

Jesson, J. P., P. Meakin, and L. C. Glasgow. 1976. The fluorocarbon-ozone theory II: tropospheric lifetimes. An experimental estimate of the tropospheric lifetimes of CCl_3F. Submitted to *Atmos. Environ.*

Johnston, H. S. 1975. University of California, Berkeley. Private communication.

Lovelock, J. E. 1975. Natural halocarbons in the air and in the sea, *Nature* 256:193-194.

Rowland, F. S., and M. J. Molina. 1975a. Chlorofluoromethanes in the environment, *Rev. Geophys. Space Phys.* 13:1-35.

Rowland, F. S., and M. J. Molina. 1975b. University of California, Irvine. Private communication.

Rowland, F. S., and M. J. Molina. 1976. Estimated future atmospheric concentrations of CCl_3F (fluorocarbon-11) for various hypothetical tropospheric removal rates, *J. Phys. Chem.* In press.

APPENDIX F

RECOMMENDED ATMOSPHERIC

AND

LABORATORY STUDIES

Since this Panel began its work in April 1975, a large amount of new information and data have become available on stratospheric chemistry. Clearly, interest in the area and the amount of research in progress has increased rapidly. On the one hand, the willingness of investigators to provide the Panel with preliminary results and unpublished manuscripts has been most helpful. On the other, the fragmentary and preliminary nature of the data and often contradictory interpretations proposed have made the information difficult to assess. Looking to the future, it is plain that the quality and reliability of new data, rather than quantity, will be critical in determining future progress in our understanding.

Scattered throughout the report are many suggestions for improving our ability to predict the perturbation of stratospheric chemistry by CFMs and other pollutants. Most of the proposals fall within the general rubric of the studies recommended in the interim report of the panel (July 1975). However, in the course of our work since then, some studies have been completed, new aspects have emerged, there is a clearer sense of urgency in others, and we now have a better perspective for what is known and what is needed. In this Appendix, we have collected these thoughts into two groups--those studies that might make substantial improvements within a year or two in our ability to deal with the CFM problem and the broader-gauge, longer-range activities that will better enable us to manage stratospheric pollution in the long run.

In general terms, *the greatest need is for verification of stratospheric ozone chemistry through measurements of trace atmospheric constituents, both stable and reactive.* The dearth of such measurements is a serious limitation to the establishment of the reliability of calculations and predictions of stratospheric effects. A carefully planned program of such measurements, with special attention to correlation of the measured species, space and time distributions, and attainable accuracy, carries the highest priority. These measurements should include NO_x, HO_x, and ClX species, as well as the CFMs, CH_4, N_2O, HF, HCl, COF_2, and others. The measurements will serve a variety of specific purposes, as described below.

I. SPECIFIC IMMEDIATE OBJECTIVES OF CFM RESEARCH*

1. *Establish the role of $ClONO_2$ in the stratosphere*--This might be accomplished by comparing infrared measurements of its concentration (Chapter 6) with calculated results (see Figure 7.16). However, particularly if disparities are revealed by such comparisons, there will need to be further study of the stratospheric processes that might convert the $ClONO_2$ back into catalytically active species (Appendix A).

2. *Verify the reliability of predicted ozone reductions*--By this we mean a more direct means than has been available so far for estimating and narrowing the *total* uncertainty of the ozone reductions predicted for the CFMs (Chapters 1 and 8), including the possibility of as yet unidentified processes. Observations of ClO and Cl profiles in the stratosphere and their comparison with calculated values (Figure 7.16) seem to be the most likely means of accomplishing this objective. However, the analysis of any disparities probably will be difficult and requires comparisons with simultaneous measurements of other species such as HCl and O_3. Preferably, the observations of ClO and Cl should be by two independent methods, such as resonance fluorescence and microwave emission in the case of ClO (Chapter 6). Direct observation of a decrease in the total ozone column and its attribution to the CFMs will take longer (Chapters 6 and

---

*The items are listed roughly in order of priority and probable length of time required (shorter, more critical first).

9); changes in the ozone profile may be a more sensitive indication of CFM perturbations (Figure 8.6).

3. *Develop an accurate materials balance*--Inasmuch as a materials balance can detect, in principle, otherwise unidentified sinks for the CFMs (Appendix E), a detailed analysis should be made of how best to obtain a materials balance accurate enough to be meaningful. The study should define a minimum set of observations (including its time span) sufficient to establish a true global burden and develop standards of sensitivity, calibration, and accuracy necessary for the results to be useful. If such a study supports its feasibility, there should then be a coordinated observational program and data analysis for both F-11 and F-12.

4. *Reduce errors in rate constants*--Considerable improvement in the rate constants for the $Cl + CH_4$ and $HO + HO_2$ reactions should be sought and may be expected within the next year or two. Further work on several of the other reactions in Table 8.3 would serve to reduce their lesser contributions to the uncertainties. A strong effort should be made to reduce the uncertainty factor of the $HO + HO_2$ reaction to 1.5 and those of the other reactions to 1.3, or less, including the effects of temperature dependences.

5. *Detailed evaluation of identified sinks*--As a complement to the materials balance approach, inactive removal processes of the CFMs, once identified, should be characterized in quantitative terms. In particular, oceanic removal of F-11 should be investigated more thoroughly by additional measurements of its concentration gradient in surface waters, by observations of its transport across the air-water interface, by redetermination of its solubility in seawater, and by a search for mechanisms that might contribute to its degradation in the surface waters. The same types of studies should be made for F-12. Also, efforts should be made to place narrower limits on the removal of F-11 and F-12 by photolysis and ion-molecule reactions in the troposphere (Appendix A).

6. *Improve other aspects of atmospheric chemistry*--As pointed out in Chapter 8, photochemical processes and concentrations of natural species (NO_x and HO_x) contribute appreciable uncertainties to the prediction of ozone reduction by the CFMs. These aspects should be studied further, to place better limits on their importance and to seek the most productive ways of reducing their contributions.

Attainment of the six objectives listed above should eliminate whatever possibility there is of subsequently finding an unidentified factor that has a major effect on predictions of ozone reduction by the CFMs. Also, it would reduce the overall (identified) uncertainty of the predictions from a tenfold to a fourfold or a fivefold range.

II. A LONGER-RANGE, MORE GENERAL PROGRAM

The research program outlined in Section I would provide information that is needed specifically and urgently for the CFM problem; it includes studies for which there is a reasonable expectation of accomplishment within one to two years. Besides this there are other ways in which the assessment of the CFMs could be improved. Furthermore, there is the need to strengthen our ability to deal with problems of stratospheric pollution in general. These two main concerns overlap to some extent and are considered jointly in this section. Although we have tried to give a comprehensive list of the capabilities needed to deal effectively with problems of stratospheric pollution, this synopsis should not be viewed as all inclusive. The field is too complex and developing too rapidly for that.

A. *Sources and Sinks*

The reliable prediction of ozone reduction by halogen-containing compounds and other pollutants requires accurate data about the composition and amounts of the compounds released into the atmosphere (sources) and knowledge of any inactive removal processes (sinks) that compete appreciably with the ozone destroying processes in the stratosphere.

1. *Monitoring of releases*--All major releases of volatile halocarbons (and other potential stratospheric pollutants) on a global scale should be monitored via data on production and use and, when possible, by observations of the actual atmospheric concentrations. This is needed not only to determine the possible magnitude of their effects on the stratosphere but also to enable materials balance studies of tropospheric removal to be made.

For the CFMs, the least certain part of past production is the small contribution (5 percent) from Eastern Bloc countries. However, their production may be rising rapidly; if so, it will become increasingly important to have accurate data from them (Chapter 3). Further analysis will also be needed of uses that lead to delayed release of CFMs, such as sealed refrigeration units and foaming agents as well as inventory stocks, to ensure that errors in the delayed release rates do not affect the accuracy of a materials balance.

2. *Determination of sinks*--For every substance identified as a potential or actual hazard to the stratosphere, there should be a careful search for tropospheric sinks that might materially reduce its environmental impact. As a minimum, this requires accurate measurements of its photodissociation properties and of the rate constants for its reactions with the various species in the troposphere, including temperature dependences for both. Furthermore, the concentrations of the tropospheric species need to be known more accurately. If necessary, verification of sinks or their absence might be sought in a materials balance (Appendix E).

3. *Products of atmospheric reactions*--In assessing the effects of a particular pollutant, it is highly desirable to know the complete sequence of events from injection, through the various chemical and/or photochemical processes, to removal of the final products from the atmosphere. For example, relatively little is known about the fate of the halocarbon radicals remaining after photodissociation of the CFMs, e.g., CF_2Cl from CF_2Cl_2. Subsequent reactions convert them into HF and compounds like COFCl. It is important to detect decomposition products such as COF_2, $COCl_2$, and COFCl (in addition to the HF and HCl) and to measure their atmospheric abundances to determine whether, for example, COFCl releases additional Cl and how much of the product is finally removed from the stratosphere in what forms. Also, one might ask what effects will be produced by the HF and HCl in rain and snow.

B. *Transport Models*

4. *Improvement in reliability of 1-D calculations*--The accomplishment of the objectives in Section I would leave the approximations of the 1-D model as the major source of uncertainty in predicting ozone reduction by the CFMs. That uncertainty, which would also be important in deal-

ing with other pollutants, could be reduced by further "fine tuning" of the calculations via comparisons with more extensive and more accurate global distribution data for various trace gases with globally distributed steady-state sources and chemical sinks localized in various regions of the atmosphere (see Appendix B). Examples include CH_4 and N_2O with surface sources and stratospheric (20 to 50 km) sinks and HF and O_3 with stratospheric sources (20 to 50 km) and tropospheric sinks. Another aspect of 1-D calculations that requires further analysis is the treatment of diurnal variations, which can have serious effects (Chapter 7).

5. *Development of 2- and 3-D models*--Improvement in the details of 1-D calculations will not necessarily compensate for the approximations inherent in a 1-D average of chemical and photochemical rates (Chapter 7). A 2-D (height and latitude) calculation does not suffer from the same limitation although in principle it is not so rigorous in determining chemical and photochemical rates as a full 3-D treatment. The primary distinction between the 2-D and 3-D models is that the transport of a 2-D model is derived largely from curve fitting to global chemical observations. Thus, its transport can never be more reliable than the observations used to develop it. The 3-D models, on the other hand, insofar as they generate their motions from first principles, are less dependent on observations of chemical species. They are the only suitable tools for examining feedback among the motions, temperature, and chemical processes.

In any case, the application of 2- and 3-D models to problems in stratospheric chemistry such as the CFMs should be expedited. In the development of such models, consideration should be given to using a chemically inert tracer such as [85]Kr to determine transport rates (Telegadas and Ferber, 1975) without the complications of chemical reaction. Another possibility is the controlled injection of specially designed tracers such as [13]CD_4 (Cowan *et al.*, 1976).

Concern about potential damage to the stratosphere will continue for the foreseeable future, and we should prepare ourselves to deal in greater confidence with other problems as they arise. Efforts to improve atmospheric models should encourage diversified approaches, because it is still too early to know which approaches are best for what purposes.

C. Atmospheric Chemistry and Photochemistry

The importance of accurate data for chemical and photochemical reactions of the CFMs and related species is
evident in Chapter 8 and restated in Section I. Here we
give more details and emphasize some of the more general
aspects.

   6. *Rate constants for reactions of ClX*--Accurate rate
constants for a large number of chemical reactions involving Cl, ClO, ClO_2, $ClONO_2$, $ClNO_2$, and atmospheric
species such as O_3, O, NO, NO_2, HO, HO_2 are required over
the temperature range 200 to 300 K. Although a number
of the key rate constants of the ClO_x catalytic chain
are now well established, wide gaps still exist in our
knowledge. This is particularly true for several reac-
tions of HO_2 such as $HO_2 + HO \rightarrow O_2 + H_2O$, $HO_2 + O \rightarrow HO +$
O_2, $HO_2 + Cl \rightarrow HCl + O_2$, whose rates control the overall
HO_x chemistry. Because of its interlocking with both the
NO_x and ClO_x cycles, the HO_x chemistry is crucial to our
understanding of the stratosphere, and the relevant rate
constants must be accurately measured. Stratospheric
NO_x and HO_x are produced in reactions of $O(^1D)$, which,
in turn, is formed in O_3 photolysis. The remaining uncertainties in the quenching and reaction rate constants
of $O(^1D)$ and in their temperature dependence must be
resolved.
   7. *Reactions of HO with halocarbons*--Much kinetic information on the reactions of HO with halocarbons of all
types has recently been obtained. Reaction-rate parameters for all important chlorocarbons and fluorocarbons
must be known with good accuracy over the tropospheric
and stratospheric temperature range in order to determine
removal times (cf. Section II.2).
   8. *Rate constants and photolysis of BrX*--Consideration
has recently been given to the equivalent catalytic destruction of ozone by bromine atoms, where the Br arises
from the photolysis and reactions of CH_3Br, now widely
used as a soil fumigant but probably also of natural origin (Appendix A, Section III.E). As the formation of
HBr from $Br + H_2$ or $Br + CH_4$ is endothermic, Br will tend
to remain in its more highly catalytic forms and may,
therefore, be more efficient in removing O_3. Although
the amount of CH_3Br released to the atmosphere is only
about 1 percent of that for the chlorocarbons, it might
increase and become a more immediate matter of concern.
If so, kinetic data will be needed for all the important

reactions of Br (i.e., Br + O$_3$; BrO + O, NO, NO$_2$, and O$_3$; Br + HO$_2$; CH$_3$Br + HO) over the temperature range of interest. Also, the photolytic properties of the various bromine-containing species will be needed.

9. *Measurements of solar flux and its variability*--The photolysis rates of the halocarbons needed for input to the models are determined by their absorption cross sections, their quantum yields, and the uv irradiance (184-225 nm) in the altitude region from 20 to 50 km, where photolysis is important. The irradiance appears to be known with insufficient accuracy, especially at wavelengths greater than 175 nm. Therefore, measurements of solar flux and scattering parameters are needed for the stratosphere. Also, in view of the fact that some of the important processes are nonlinear in the radiation intensity (Chapter 7), the variation in the solar flux, and effects of the variations on stratospheric chemistry should be investigated.

10. *Photochemistry of halocarbons and ClX*--The photolysis parameters for CFCl$_3$ (F-11) and CF$_2$Cl$_2$ (F-12) seem to be fairly well resolved, both as to absorption cross section and quantum yield, although there is some question as to whether all three Cl atoms will be released from F-11. Nonetheless, further study of them could help to reduce the uncertainties of the projected ozone reductions (Chapter 8) and of tropospheric removal (Appendix A, Section II.A). The temperature coefficients of the absorption spectra are important. This is particularly true when photolysis occurs predominantly in the tail of an absorption band (e.g., F-12 in the solar window). For polyatomic species, information on produce channels is needed in addition to absorption cross sections and quantum yields. Laboratory studies of the photolysis of OClO, ClOO, and ClNO$_2$ are needed, and further study of ClONO$_2$ could be helpful. Also, it seems desirable to have photolysis parameters for other halocarbons (such as CHFCl$_2$, CCl$_4$, CH$_3$CCl$_3$, CH$_2$Cl$_2$, and CH$_3$Cl) for the reasons mentioned in Section I.2.

D. *Atmospheric Measurements*

A serious deficiency that we see at present is in data on actual concentrations of CFMs and other critical trace components in the troposphere and stratosphere. Such measurements are difficult, costly, and often lie at the limits of present experimental capability. Some important

trace components such as $ClONO_2$ and HO_2 have yet to be
detected in the atmosphere, and even rough measurements
indicating whether they are present at approximately predicted levels would be valuable. On the other hand, for
CFMs and other halocarbons and for the main reactive species, it is becoming evident that further data, unless
their accuracy can be defined and improved, will be of
little value. What is needed most urgently is a thorough-
going standardization and calibration of analytical methods before further widespread data-gathering projects
are launched.

11. *Definition and establishment of a global network*--
In Section I.3, it was pointed out that the use of a
materials balance approach to identifying sinks for CFMs
requires a careful study to determine the minimum set of
observations needed for an *accurate* global average of
their atmospheric concentrations. Atmospheric measurements for other purposes (e.g., Sections I.2, I.6, II.1,
II.3, and II.4) have different requirements. In all cases,
there is little point in making measurements unless they
have the necessary sensitivity and the techniques are
standardized and calibrated so that the results are accurate enough for the purpose at hand and they are readily
comparable with those of other stations. Furthermore,
better coverage of the globe is needed, especially in the
southern hemisphere but also in the Soviet Union and some
parts of Asia.

12. *Measurements of particular species*--Besides the atmospheric measurements already mentioned in several other
contexts, it is important to measure the variation with
height of constituents of the upper atmosphere that are
related to the chemistry of chlorine, including HO, HCl,
HF, and NO. Some measurements exists for HCl below 30 km,
some for HO and for NO; none have yet been reported for
HF. These species are significant indicators of the
stratosphere processes involved.

13. *Global monitoring of ozone*--Because of its central
role, the need for long-term observations of stratospheric
ozone merits special comment. It is hoped that the measurements will become sufficiently sensitive, accurate,
and complete to detect small changes due to man's activities. A significant aspect of this is determination of the
causes for the long-term variations in total ozone (Chapter 9). Satellite observations appear attractive but apparently will require greater sensitivity and/or better
calibration. The possibility of detecting the effects of
pollutants by changes in the ozone profile (Chapter 8)

should be explored. Another approach is to look for localized disturbances, from natural events such as volcanic
eruptions or solar flares (Chapter 7) or perhaps even
from a controlled experiment.

E.   Other Concerns

14.   *Water vapor feedback*--A preliminary evaluation
(Chapter 9) of the direct and indirect effects of CFMs
on the temperature of the tropical tropopause indicates
that, *provided the "cold trap" hypothesis is valid,*
there could well result significant increases of stratospheric water vapor concentrations and corresponding increases in the ozone reduction. This problem needs
further attention using more rigorous modeling approaches.
15.   *Increased N_2O from fertilizer use*--More detailed
studies of the reduction in stratospheric ozone associated with the use of nitrogen fertilizers are essential,
especially of the biological production of N_2O and the
mechanisms for its removal from the troposphere. The
studies should include the effects of the interaction between the NO_x and ClO_x cycles (cf. Chapter 9).

REFERENCES

Cowan, G. A., D. G. Ott, A. Turkevich, L. Machta, G. J.
   Ferber, and N. R. Daly. 1976. Heavy methanes as atmospheric tracers, *Science* 191:1048-1050.

Telegadas, K., and G. J. Ferber. 1975. Atmospheric concentrations and inventory of krypton-85 in 1973, *Science*
   190:882-883.

GLOSSARY

absorption cross section -- a wavelength-dependent parameter that describes the absorption of radiation by a
given molecule

active removal -- halocarbon destruction processes that
lead to products that destroy ozone

aerosol -- in meterology, minute particles (solid or
liquid) suspended in the atmosphere

atmospheric circulation -- large-scale movements of the
air in the atmosphere

atmospheric residence (removal) time -- it is the time
taken for the atmospheric concentration of a given species to fall to $1/e$ of its initial value ($e = 2.72$),
from the time that all sources of the species are removed

carbon tetrachloride (CCl_4) -- a chlorinated methane found
in the atmosphere, which is used industrially in the
manufacture of chlorofluorocarbons and tetrachlorethyl-
ene and as a cleaning agent and solvent

catalysis -- the alteration of the rate at which a chemical reaction occurs, by the introduction of a substance
(catalyst) that remains unchanged at the end of the
reaction

chlorocarbon -- a carbon compound containing chlorine atoms

CFMs (chlorofluoromethanes) -- compounds containing one
carbon atom and at least one chlorine and one fluorine
atom; F-11 ($CFCl_3$) and F-12 (CF_2Cl_2) are the two CFMs
that are released in the largest amounts and are the
main subject of this report

ClO_x -- Cl and ClO

ClX -- Cl, ClO, HCl, $ClONO_2$

column density -- the number of molecules of a particular
species above a unit area of the earth's surface

diurnal -- over a period of 24 hours

350

eddy-mixing coefficient -- a parameter used to describe
transport in atmospheric models, empirically determined
from the distribution of trace species in the atmosphere
(also called eddy diffusion or vertical transport coefficient or rate)

fluorocarbon -- a carbon compound containing fluorine
atoms

halocarbon -- a carbon compound containing one or more of
the halogen elements (fluorine, chlorine, bromine, and
iodine)

HO_x -- HO, HO_2, H_2O_2

hydroxyl radical (HO) -- a molecular fragment containing
one oxygen and one hydrogen atom

inactive removal processes -- halocarbon destruction processes that lead to products that do *not* destroy ozone
(tropospheric removal) or other processes that remove
the halocarbon without affecting stratospheric ozone

infrared radiation -- electromagnetic radiation with wavelengths (0.8-1000 μm) longer than those of visible light
and shorter than those of microwaves

materials balance -- a comparison between the amount of
a compound released into the atmosphere and the amount
actually detected; for CFMs (after taking account of
the expected destruction in the stratosphere) this has
been proposed as a method for determining the magnitude
of any additional sinks

methyl chloride (CH_3Cl) -- a chlorocarbon found in the
atmosphere, produced both naturally and by man

NO_x -- NO and NO_2

NO_y -- NO, NO_2, and HNO_3

One-dimensional model -- a mathematical model of the atmosphere in which only the vertical direction is considered

$O(^1D)$ -- an electronically excited oxygen atom that is
highly reactive

$O(^3P)$ -- a ground-state oxygen atom

ozone -- a reactive oxygen compound (O_3) found in the atmosphere, which shields the earth's surface from uv
radiation

ozone layer (or shield) -- that part of the atmosphere in
which the concentration of ozone is greatest. The term
is used in two ways: (i) to signify the layer from
about 10-50 km in which the ozone concentration is appreciable; (ii) to signify the much narrower region
from about 20-25 km in which the concentration generally reaches a maximum

ozone column -- this term is used to describe the ozone
content of the atmosphere. It gives the equivalent
thickness of ozone in the vertical column at standard
temperature and pressure, between an observer on the
ground and the top of the atmosphere. It usually lies
in the range of 2-4 mm

partial residence (removal) time -- in the absence of
sources, it is the time taken for the atmosphere concentration of a given species to fall to 1/e of its
initial value (e = 2.72), assuming only a *particular*
removal process is operating

photochemical reactions -- chemical reactions that are
initiated, assisted, or accelerated by exposure to
light

photodissociation -- a chemical decomposition of a molecule resulting from the absorption of light

photolysis -- the chemical decomposition of a molecule
resulting from the absorption of light

rate constant -- a coefficient that relates the speed of
a chemical reaction to the concentrations of the reacting substances

residence time -- the time taken for the concentration of
an inert tracer in a given atmospheric region to fall
to 1/e of its initial value (e = 2.72) after cutoff

scale height -- the vertical distance over which the
atmospheric density changes by a factor e (= 8.4 km at
the earth's surface)

self-healing -- a process whereby reduction of ozone at
higher altitudes is partially compensated by an increase at lower altitudes

stratosphere -- that region of the atmosphere lying above
the troposphere in which, in contrast to adjoining regions, temperature does not increase with increasing
height. The stratosphere extends from the tropopause
to a height of about 50 km, where the temperature reaches
a maximum

three-dimensional model -- a mathematical description of
the atmosphere in which the vertical, latitudinal (north-
south), and longitudinal (east-west) directions are
considered

tracer -- a property or a substance that "labels" a particular mass of air and so makes it possible to infer
the movement of that mass over a period of time

tropopause -- the atmospheric boundary between the troposphere and stratosphere, defined as the level at which
the decrease in temperature with height becomes $2^{\circ}C/km$
or less, over at least 2 km

troposphere -- the lowest layer of the atmosphere below the tropopause extending from the earth's surface to about 8 km at the poles and 16 km at the equator

tropospheric sink -- a destruction or other removal process occurring in the troposphere. In particular, for chlorocarbons, a tropospheric sink limits the amount of these compounds entering the stratosphere, thus proportionately limiting their effect on stratospheric ozone

two-dimensional model -- an atmospheric model in which the vertical and latitudinal (north-south) directions are considered

ultraviolet (uv) radiation -- electromagnetic radiation lying beyond the violet end of the visible spectrum, with wavelength less than 400 nm

uv-B radiation -- electromagnetic radiation with wavelength in the range 280-320 nm, which affects plants and animals

volume mixing ratio -- the number of molecules of a particular species in a given volume divided by the total number of molecules of all species in that volume, commonly given in such units as parts per million (ppm)